Sustainable Forest Management

Sustainable Forest Management provides the necessary material to educate students about forestry and the contemporary role of forests in ecosystems and society. This comprehensive textbook on the concept and practice of sustainable forest management sets the standard for practice worldwide.

Early chapters concentrate on conceptual aspects, relating sustainable forestry management to international policy. In particular, they consider the concept of criteria and indicators and how this has determined the practice of forest management, taken here to be the management of forested lands and of all ecosystems present on such lands. Later chapters are more practical in focus, concentrating on the management of the many values associated with forests.

Overall the book provides a major new synthesis which will serve as a textbook for students of forestry as well as those from related disciplines such as ecology or geography who are taking a course in forests or natural resource management.

John L. Innes is Forest Renewal BC Chair in Forest Management and Dean of the Faculty of Forestry at the University of British Columbia, Canada.

Anna V. Tikina is an Adjunct Professor in the Department of Forest Resources Management and a consultant specializing in Sustainable Forest Management and International Environmental Governance at the University of British Columbia, Canada.

The Earthscan Forest Library

This series brings together a wide collection of volumes addressing diverse aspects of forests and forestry and draws on a range of disciplinary perspectives. Titles cover the full range of forest science and include biology, ecology, biodiversity, restoration, management (including silviculture and timber production), geography and environment (including climate change), socio-economics, anthropology, policy, law and governance. The series aims to demonstrate the important role of forests in nature, peoples' livelihoods and in contributing to broader sustainable development goals. It is aimed at undergraduate and postgraduate students, researchers, professionals, policymakers and concerned members of civil society.

Series Editorial Advisers:

John L. Innes, Professor and Dean, Faculty of Forestry, University of British Columbia, Canada.
Markku Kanninen, Professor of Tropical Silviculture and Director, Viikki Tropical Resources Institute (VITRI), University of Helsinki, Finland.
John Parrotta, Research Program Leader for International Science Issues, US Forest Service – Research & Development, Arlington, Virginia, USA.
Jeffrey Sayer, Professor and Director, Development Practice Programme, School of Earth and Environmental Sciences, James Cook University, Australia, and Member, Independent Science and Partnership Council, CGIAR (Consultative Group on International Agricultural Research).

Recent Titles:

Sustainable Forest Management
From Concept to Practice
Edited by John L. Innes and Anna V. Tikina

Gender and Forests
Climate Change, Tenure, Value Chains and Emerging Issues
Edited by Carol J. Pierce Colfer, Bimbika Sijapati Basnett and Marlène Elias

Forests, Business and Sustainability
Edited by Rajat Panwar, Robert Kozak and Eric Hansen

Climate Change Impacts on Tropical Forests in Central America
An Ecosystem Service Perspective
Edited by Aline Chiabai

Rainforest Tourism, Conservation and Management
Challenges for Sustainable Development
Edited by Bruce Prideaux

Large-scale Forest Restoration
David Lamb

Forests and Globalization
Challenges and Opportunities for Sustainable Development
Edited by William Nikolakis and John Innes

Additional information on these and further titles can be found at
http://www.routledge.com/books/series/ECTEFL

Sustainable Forest Management

From Concept to Practice

Edited by John L. Innes and Anna V. Tikina

Routledge
Taylor & Francis Group

LONDON AND NEW YORK

earthscan
from Routledge

First published 2017
by Routledge
2 Park Square, Milton Park, Abingdon, Oxon OX14 4RN

and by Routledge
711 Third Avenue, New York, NY 10017

Routledge is an imprint of the Taylor & Francis Group, an informa business

British Library Cataloguing in Publication Data
A catalogue record for this book is available from the British Library

Library of Congress Cataloging in Publication Data
Names: Innes, John L., editor. | Tikina, Anna V., editor.
Title: Sustainable forest management : from concept to practice / edited by
     John L. Innes and Anna V. Tikina.
Description: London ; New York : Routledge, 2017. | Includes
     bibliographical references and index.
Identifiers: LCCN 2016024920 | ISBN 9781844077236 (hbk) |
     ISBN 9781844077243 (pbk) | ISBN 9780203126547 (ebk)
Subjects: LCSH: Sustainable forestry. | Forest management. | Forest
     ecology. | Forest conservation.
Classification: LCC SD387.S87 S83835 2017 | DDC 634.9/2—dc23
LC record available at https://lccn.loc.gov/2016024920

ISBN: 978-1-84407-723-6 (hbk)
ISBN: 978-1-84407-724-3 (pbk)
ISBN: 978-0-203-12654-7 (ebk)

Typeset in Sabon
by Apex CoVantage, LLC

Printed and bound in Great Britain by Ashford Colour Press Ltd

'*Sustainable Forest Management* provides a unique and up-to-date synthesis of the state of knowledge on sustainable forest management from a variety of environmental, economic, social, cultural, and governance perspectives. Highly recommended as an interdisciplinary teaching text for university courses in environmental sciences, particularly in forestry, ecology, geography.'

John A. Parrotta, *Vice President, International Union of Forest Research Organizations*

'This new book provides a broad, well-researched and up-to-date introduction to many aspects of sustainable forestry. The book covers a wide range of topics that address both ecological and social perspectives on managing these critical resources.'

Marc McDill, *Associate Professor of Forest Management, Penn State University, USA*

'This book provides expert guidance for students on how the many dimensions of sustainable forest management, ranging from the maintenance of ecosystem health, biodiversity and soils to providing income and spiritual values, can be addressed in a comprehensive approach.'

Jürgen Bauhus, *Professor of Silviculture, Freiburg University, Germany*

'Today's students need to be selective and devote their precious reading time to the very best books. Therefore, my recommendation to all forest students is simple: This is a book you must read!'

Björn Hånell, *Professor of Silviculture, Swedish University of Agricultural Science*

Contents

Preface

Forests cover a third of the land surface of the Earth, and we are losing them at an alarming rate. According to the 2015 Global Forest Resources Assessment coordinated by the Food and Agriculture Organization of the United Nations (FAO), the rate of loss of the total forest area has declined in recent years, but the net figures for forest area conceal the fact that natural forests continue to be lost and are only partly being compensated by the development of planted forests. When we also consider the rate of forest degradation, which is many times greater than the rate of deforestation, but much less well-recognized, there are serious causes for concern.

In many instances, the blame for deforestation is placed on poor forest management. However, the situation is much more complex. Most deforestation is the result of a deliberate change of land use, generally from forest to range or agricultural land. Forestry may hasten the process by allowing access to previously inaccessible areas, and in some cases, the felling of forests for wood without any attempt to reforest has indeed resulted in deforestation. Generally, however, it is in the best interests of a forester to ensure that there is a future resource, and much forest management is based on the premise that if practiced correctly, forestry is sustainable.

This book is intended to provide guidance on what needs to be considered when managing a forest. It cannot be all-encompassing, and is intended to show the breadth of knowledge expected of a forester. The reader will, however, need to go into some of the further reading provided at the end of each chapter to obtain the depth of knowledge expected of a forest manager. Readers are expected to have a basic understanding of forestry, such as might have been gained in the initial years of an undergraduate forestry program and, as a result, important areas such as forest measurements and writing a forest management plan have not been included.

While a stand-alone volume, this book is also the text that can be used to accompany the online course 'Sustainable Forest Management in a Changing World', a project of the Asia-Pacific Forestry Education Coordination Mechanism, which is an organization set up under the auspices of the Asia-Pacific Network for Sustainable Forest Management and Rehabilitation (APFNet). Links to this course can be found on the APFNet website (http://www.apfnet.cn).

Acknowledgements

Figures 3.1, 3.7, 15.1, and 15.12 were prepared by Isobel Houde. The sources of all graphics have been acknowledged in the figure titles. All photographs were taken by John L. Innes unless otherwise indicated. Phil Grace and René Reyes made important contributions to Chapter 9, and Arvyas Lebedys at the FAO was particularly helpful with this chapter. Verena Griess provided many helpful comments on Chapter 13. The Asia-Pacific Network for Sustainable Forest Management and Rehabilitation provided generous funding that enabled all the figures to be reproduced in colour.

Notes on authors

Juan A. Blanco, Universidad Pública de Navarra, Dep. Ciencias del Medio Natural, Campus de Arrosadía, 31006, Pamplona, Navarra, Spain

Janette Bulkan, University of British Columbia, Faculty of Forestry, 2424 Main Mall, Vancouver, B.C., V6T 1Z4, Canada

Fred Bunnell, University of British Columbia, Faculty of Forestry, 2424 Main Mall, Vancouver, B.C., V6T 1Z4, Canada

Brett Eaton, University of British Columbia, Department of Geography, 1984 West Mall, Vancouver, B.C., V6T 1Z2, Canada

Hosny El-Lakany, University of British Columbia, Faculty of Forestry, 2424 Main Mall, Vancouver, B.C., V6T 1Z4, Canada

Takashi Gomi, Department of International Environmental and Agriculture Science, Tokyo University of Agriculture and Technology, Saiwai 3–5–8, Fuchu, Tokyo 1585809, Japan

Richard Hamelin, University of British Columbia, Faculty of Forestry, 2424 Main Mall, Vancouver, B.C., V6T 1Z4, Canada

Howard Harshaw, University of Alberta, Faculty of Physical Education and Recreation, 8840 114 Street, Edmonton, Alberta, T6G 2H9, Canada

Ngaio Hotte, University of British Columbia, Faculty of Forestry, 2424 Main Mall, Vancouver, B.C., V6T 1Z4, Canada

John L. Innes, University of British Columbia, Faculty of Forestry, 2424 Main Mall, Vancouver, B.C., V6T 1Z4, Canada

Robert Kozak, University of British Columbia, Faculty of Forestry, 2424 Main Mall, Vancouver, B.C., V6T 1Z4, Canada

Bruce Larson, University of British Columbia, Faculty of Forestry, 2424 Main Mall, Vancouver, B.C., V6T 1Z4, Canada

Yueh-Hsin Lo, Universidad Pública de Navarra, Dep. Ciencias del Medio Natural, Campus de Arrosadía, 31006, Pamplona, Navarra, Spain

Anne-Hélène Mathey, Natural Resources Canada, Policy, Economics and Industry Branch, Ottawa, Canada

R. Dan Moore, University of British Columbia, Faculty of Forestry, 2424 Main Mall, Vancouver, B.C., V6T 1Z4, Canada

Harry Nelson, University of British Columbia, Faculty of Forestry, 2424 Main Mall, Vancouver, B.C., V6T 1Z4, Canada

Craig R. Nitschke, University of Melbourne, Department Forest and Ecosystem Science, Melbourne School of Land and Environment, Australia

Roy C. Sidle, University of the Sunshine Coast, Sustainability Research Centre, Sippy Downs 4556, Queensland, Australia

Anna V. Tikina, University of British Columbia, Faculty of Forestry, 2424 Main Mall, Vancouver, B.C., V6T 1Z4, Canada

Patrick O. Waeber, Swiss Federal Institute of Technology Zurich, Department of Environmental Sciences, Switzerland

Clive Welham, University of British Columbia, Faculty of Forestry, 2424 Main Mall, Vancouver, B.C., V6T 1Z4, Canada

David Wilford, BC Ministry of Forests, Lands and Natural Resources Operations, Smithers, B.C., Canada

Figures, tables and boxes

Tables

Boxes

Chapter 1

Sustainable forest management

From concept to practice

John L. Innes

Introduction

The deliberate management of forests for the provision of goods and services is a phenomenon that dates back to the origins of the human species. The major products of forests have been food, fuel and medicines, and this is still the case for many indigenous societies today. In fact, almost half of all wood harvested is still used as firewood, for cooking and heating. Although humans evolved from forest-dwelling primates, early humans appear to have been predominantly associated with grasslands and other open areas from about 500,000 years ago. In this environment, early humans (*Homo sapiens*) and the related human species *Homo neanderthalensis* hunted meat and gathered plants and other foods such as shellfish. It is likely that, perhaps like one of their predecessors, *Homo erectus*, 1.5 to 1.7 million years ago, they burned grasslands as part of a hunting strategy, and this might have shifted the balance at the grassland-forest border in favour of grassland (Pyne, 1995). However, there is no evidence yet that they either made use of or managed forests during this period, and any impacts would have been small in relation to the large-scale changes in global vegetation patterns that occurred in response to glacial and interglacial periods over the past 2 million years.

The use of fire appears to have been one of the earliest forms of landscape management. Fire was used for a variety of reasons, including to increase the area of grasslands, to provide fresh regeneration that would attract game, to prevent animals from using cover and to drive them out of that cover, and to promote and harvest insects and edible and medicinal plants (Sands, 2013). Some of the most advanced techniques appear to have developed in Australia, where 'fire-stick farming' created by frequent, low-intensity fires encouraged the development of an open countryside, excluding forest from many areas (e.g. Gammage, 2011). The exact role of human-induced fire in creating or maintaining open landscapes is still debated, but it appears to have been important in many parts of the world, including Australia, New Zealand, South Africa, southern South America and elsewhere. However, distinguishing between forest and grassland mosaics created by deliberate fire as opposed to those occurring naturally through fires created by lightning is difficult, and more so when past landscapes are being considered.

Early interactions between humans and forest ecosystems

Between about 40,000 and 60,000 years ago, humans spread out across much of Africa, Europe and Asia, and even reached Australia. The Earth was in the middle of the last glacial period at this time: much of present-day Europe and northwest Russia was covered in ice, and sea levels were much lower. The early hunter-gatherers were nomadic and generally lived in small groups, moving on as resources were depleted locally (Bush, 1997). As humans spread, so did their impacts, and this period is one that was associated with the loss of many of the world's larger animals (the so-called megafauna). As with fire, there is considerable debate over the precise role that humans played in such losses, and some extinctions may have been a direct or indirect consequence of climate change. However, there are examples of direct correlations between the arrival of humans and the loss of local megafaunas. For example, in Australia, the loss of a wide range of species, including a giant flightless bird (*Genyornis newtoni*), giant kangaroos, a horned terrestrial tortoise and animals resembling giant sloths, rhinoceroses and lions, occurred very soon after the arrival of humans on the continent between about 53,000 and 60,000 years ago (Bowler *et al.*, 2003), and the megafauna had all apparently been lost by 40,000 years ago. High-resolution analysis of sediment cores from northeast Queensland has further explained what may have happened. According to Rule *et al.* (2012), the megafauna disappeared from the site 41,000 years ago, when no change of climate was recorded, but humans had arrived. An increase in charcoal in the sediments is evident, but only after the loss of the megafauna, suggesting that hunting alone was responsible. After the loss of the megafauna, there appears to have been a brief surge in rainforest vegetation at the site as a result of the cessation of herbivory. However, this was soon followed by fire, and a change in the vegetation away from rainforest towards more grasses and sclerophyllous trees. The findings are not unique, and Turney *et al.* (2008) have argued that the similar loss of the megafauna in Tasmania occurred later than on mainland Australia because of the later arrival of humans there.

The extinctions in North America occurred much later, about 13,000 years ago, with those in South America occurring about 500 years after that (Haynes, 2009). In Europe, a significant proportion of the megafauna was also eliminated at roughly the time of human arrival, with 7 out of 24 genera being eliminated (Williams, 2003). The European extinctions do not appear to have been as severe as in Australia and North America, possibly because humans had already been affecting the environment there for some considerable time.

In North and South America, there is debate over the relationship between the arrival of humans and the loss of the megafauna that dominated much of the continent. North America lost 33 out of 45 genera and South America lost 46 out of 58 genera (Williams, 2003). Humans seem to have been well established on the west coast of North America at or just prior to the end of the last glacial period, about 12,000 years ago, and some populations may have been present in ice-free areas such as the Haida Gwaii (at one time known as the Queen Charlotte Islands) of western Canada several thousand years earlier. With the melting of the continental ice sheets, humans spread rapidly over the continent and into South America, and this period of expansion coincides with the loss of the megafauna, which was most obvious between 11,500 and 11,000 years BCE. Whether or not the extinctions were caused by hunting or by climate change is hotly debated (see, for example, Grayson and Meltzer, 2002), and there are questions over the validity of some of the ecological modelling that has been done to support either hypothesis. Evidence derived from dating suggests that some megafauna in South America survived long after the climate warmed, and also long after humans arrived. Since humans could continue to exert hunting pressure, the available evidence in South America points more to a human role than it does to climate change.

Forestry and the emergence of civilizations

The Mediterranean region and the Levant

Sometime after about 20,000 years ago, agriculture emerged in western Asia and gradually spread east, west and north. The exact date is uncertain: agriculture was certainly present 10,000 years ago in western Asia, but there are records extending back as far as 18,000 years ago (Sands, 2013). The development of agriculture had a number of important consequences, not least of which was that humans were now able to live in permanent settlements in much bigger groups. This was because a number of founder crops (einkorn and emmer wheat, barley, peas, chickpeas, lentils and bitter vetch) had been domesticated, although exactly how and when remains uncertain. Essentially, this enabled the development of villages, towns and cities. The construction of buildings required the use of building materials, and wood appears to have featured prominently. It was also needed for furniture and most of the other uses that it is still put to today. As a result, forests around these early settlements were rapidly depleted for buildings, for charcoal and for the construction of ships and other wooden objects (Meiggs, 1982; Westoby, 1989). Charcoal in particular was needed for cooking and heating, and for smelting. References in the Bible to the primeval forests of Syria and in particular to the trans-Jordanian Forest of Ephraim suggest the presence of major forests in areas that are now desert. While, with a few notable exceptions, initial impacts seem to have been quite limited, large parts of southern Europe appear to have lost their forest cover in Roman times, with the cleared areas being used for agriculture and the wood being used for smelting, road construction (especially corduroyed roads), ship-building and other purposes (Perlin, 2005; Thirgood, 1981).

The Romans appear to have been particularly active in depleting the forests of the Mediterranean, although the relative importance of deforestation and climate change are still disputed. Our understanding today is that the loss of substantial areas of forest could actually have triggered some of the changes in climate that are believed to have occurred at this time (He *et al.*, 2014; Runyan *et al.*, 2012). However, the development of glass-making, smelting, ceramics and other industries, combined with the needs for wood in construction, heating and cooking, created huge demands for wood, and the forests of Italy and then the surrounding regions were rapidly depleted.

Severe deforestation also occurred in Roman times as a result of agricultural expansion. The development of cities created a demand for food that was supplied by lands cleared of their forests. Cultivation practices were generally quite poor, and although terracing was developed in some areas to try to conserve soil, the loss of forests was accompanied by a period of severe soil erosion (Vita-Finzi, 1969). This spread outwards from cities, initially in the heartlands around Rome, then throughout the Western Empire and eventually through the Eastern Empire. Areas that

Figure 1.1 Carved wooden chair from Tutankhamun's tomb, Egypt, dating from ca. 1325 BCE. The construction of Egypt's cities and monuments would have consumed large amounts of wood. Perlin (2005) indicates that Egypt was importing cedar (by wooden ship) from Phoenicia (modern Lebanon) as early as 2650 BCE.

3

Figure 1.2
The Parthenon in Athens, Greece. This building was completed in 438 BCE, at the height of Classical Greece. According to Perlin (2005), the use of stone rather than wood was because wood supplies in the region were already dwindling following large-scale depredations during the war with Persia 30 years earlier.

Figure 1.3
Fresco from the ruined Roman city of Volubilis, near Meknes, Morocco. The fresco shows trees and a variety of wild animals, and was made at a time (the 2nd century CE) when the area was an important agricultural reserve for Rome. Today it is semi-desert, as is the case with much of North Africa.

were once covered in forest became infertile desert, or reverted to scrub after the agriculture was abandoned.

While forests in areas such as the Taurus Mountains of Turkey have shown signs of recovery and appear relatively healthy, other forests in the Mediterranean have done less well. Figure 1.5 shows a Cedar of Lebanon (*Cedrus libani* subsp. *libani*) stand in the Tannourine Cedars Forest Nature Reserve of Lebanon. In the background, the extent of deforestation and desertification, typical of much of this area, is evident. Reestablishing forest in such a degraded environment will be very difficult, even though the current stands demonstrate that the area can still support forests.

Western Europe

After the decline of the Roman Empire, Europe experienced a long period of instability. Whereas the classical period saw major deforestation around the Mediterranean, it was during medieval times (ca. 500–1500 CE) that the forests of temperate western and central Europe were largely cleared (Williams, 2003). Much of the deforestation occurred in the middle period of the Medieval (1000–1300); during the late Medieval (1300–1500) a series of wars and plagues resulted in reduced rates of deforestation. Much of the deforestation was due to the expansion of agricultural land, but industrial use (e.g. smelting and glassmaking) were important, as was removal of wood for buildings and ships. Cattle, sheep and goats were grazed in the forests surrounding villages, and pigs were turned out to forage for acorns and beechmast. In many European countries, the opportunity to raise animals in forests was vigorously defended through a system of communal rights held by villagers. In some forests, game populations were maintained at artificially high levels, resulting in severe browsing damage and a lack of regeneration, a problem that persists in some European forests to this day.

Figure 1.4 Cedar of Lebanon (Cedrus libani subsp. libani) forest in the Taurus Mountains of southwest Turkey. These forests were an important source of wood in classical times, with wood being exported as far away as Egypt.

Figure 1.5 Cedar of Lebanon (Cedrus libani subsp. libani) stand in the Tannourine Cedars Forest Nature Reserve in Lebanon. The contrast with the stand in the Taurus Mountains (Figure 1.4) is striking. (Photo by Hosny El-Lakany.)

*Figure 1.6
Actively
managed coppice
stool in Garston
Wood, England.
This woodland
has been termed
an 'ancient
woodland', as
it has been in
continuous
existence since
before 1600.
An excellent
account of such
woodlands
is provided
by Rackham
(2006).*

Much of the landscape present today in Europe was formed in the middle Medieval period. For example, many of the woodlands described in the Domesday Book, produced in 1086 in England, are still present today, and many field boundaries remain unchanged. Various forms of forest management seem to have become widespread at this time, including coppicing and pollarding. These two forms of harvesting were used in Roman times, and gradually became more widespread, peaking in the Medieval period. This was also a period that saw the delineation of large areas of royal hunting forest (which often contained a variety of habitats rather than being purely forest); this is described in detail in the next section.

The medieval origins of land tenure legislation

Many features associated with governance of today's forests can be traced back over the last thousand years or longer. The history of forest tenure in England is particularly important, as it is the foundation for current forest governance in many parts of the world. The term 'land tenure' relates primarily to the legal regime under which an individual has access to land. The term dates back to the time when a monarch or equivalent (often referred to as the Crown) held all the land and private owners were tenants. The system emerged in Europe and elsewhere during the first millennium, and developed in particular with the emergence of the feudal system. When, for example, Duke William II of Normandy assumed the crown of England (as King William I) after defeating King Harold Godwinson at the Battle of Hastings in 1066, he took over all lands held by the Anglo-Saxon nobility. At this time, he acquired, by conquest, the sovereignty of England. This means that he had absolute and exclusive power. He introduced Forest Law to England that was superimposed on the Common Law, although pleas might be heard under one or the other system, depending on the nature of the offence (Nail, 2008). The king subsequently distributed some of the land that he acquired to his nobles in return for services (generally the raising of armed troops in times of war, particularly cavalry, but also including infantry; this form of service was known as knight-service or enfeoffment). He retained the remainder of the lands for his own use, especially for hunting, and as a source of income: these lands were termed the royal demesne. The land was granted to the nobles, but could just as readily be taken back. The nobles then allocated most of the land they had received to their subordinates in return for household or other services (known as serjeanty) or for fees, generally in kind (known as socage). Any land that was retained was termed the demesne, and generally included the main residence (often a castle) and the surrounding land. Some definitions of socage restrict it to a direct relationship between a tenant and a king; others include intermediaries. Another category was frankamolin: land held by an ecclesiastical body in return for a religious service such as saying prayers for the soul of the grantor. Under this last form of tenure, there was no other obligation to the landlord.

The area of land reserved by the king was often substantial. In England, in the 12th and 13th centuries, between a quarter and a third of the country was reserved in the form of Royal Forests for hunting. The land covered by Royal Forests included both land owned by the king (*boscus dominicus regis*) and lands that had been given to nobility and the clergy (*boscus baro, boscus priori, boscus mili*). Much of the land was farmed and had substantial populations. The word 'forest' seems to be derived from the Latin word *foris*, meaning 'outside' (the law of the land, i.e. Common Law), and royal hunting forests included not only land with tree cover, but also heathlands and agricultural land. Rackham (2006) suggests that only about half of the Royal Forests at this time were actually wooded. The game populations, particularly the venison (red, roe and fallow deer and wild boar) were protected by a category of people known as foresters, who would bring transgressors to one of the courts dealing with Forest Law. The foresters were assisted by under-foresters, who later became known as rangers. The foresters were also responsible for ensuring that the habitat of the game (the vert) was protected, and were supported by Verderers, who were responsible for investigating minor offences. The term 'forester' was actually used to describe a number of roles, ranging from the person in charge of a forest (also known as a warden) to very junior positions. Senior forester positions were considered valuable, and people would pay the king to be given the position because of the rights associated with it.

Figure 1.7 Fallow deer (Cervus dama) grazing in the New Forest in England. The New Forest was a Royal Forest established by King William I, and many of the laws and governance structures that were established then still exist. Fallow deer were introduced to England in about 1100.

Forest boundaries were maintained by people known as regarders, who ensured that there were no encroachments by local people. At the time, the action of encroachment was known as assarting, and was considered a serious crime. It was the most serious type of trespass, with the least serious, but still potentially punishable, being the presence of an individual in a Royal Forest carrying hunting equipment or having a dog (unless its front nails had been removed). In the royal demesne, it was not necessary to prove that a person was doing harm by their presence, a principle that has extended to today in some common law jurisdictions (such as England and the USA). In some civil law jurisdictions (such as Scotland and Switzerland), landowners did not have the right to exclude people from their land through laws of trespass unless harm could be demonstrated. Roads passing through Royal Forests could generally be used, except during the time of year when deer were fawning, when a charge could be levied (known as cheminage).

The land held in Royal Forests was far greater than could ever be used for hunting, even if this was the stated aim of the forests. While hunting was clearly a benefit, the Royal Forests represented a major source of income for the king. In return for payment, various rights could be bestowed on others, including the sale or leasing of assarts (areas of forest cleared for agriculture) and the development of forges and tanneries. Considerable income was also generated by forest amercements (fines for offences against Forest Law). Any removal of land from a forest was termed a disafforestation, and was done when the king was particularly short of funds, although considerable disafforestation occurred as a result of the charters mentioned next. In many cases, these disafforested areas might still contain wooded land, and might still be used for hunting. Larger areas were known as chases, and smaller areas as parks. A key point is that such lands were then governed under Common Law rather than Forest Law. Schama (1996) describes a system of enforcement and payment that was so extreme that he refers to it as 'sylvan gangsterism' (p. 148).

Much of the law related to Royal Forests was codified in the Assize of Woodstock in 1184 in a document called *Constitutiones de Foresta* (the original Charter of the Forest). Young (1979) summarizes the provisions of the law as follows:

1 Forest offenses will henceforth be punished not just by fines but by full justice as exacted by Henry I.
2 No person shall have a bow, arrows, or dogs within the royal forests.
3 No wood is to be given or sold from any woods within a royal forest, except wood may be taken for the owner's use.
4 Persons who have woods within a royal forest must name their own foresters and give security that they will commit no acts against the king.
5 Royal foresters shall have a care for the woods of knights and others within a forest.
6 All royal foresters must swear to uphold the assize of the forest.
7 Within each county with a royal forest there shall be chosen twelve knights to keep the venison and the vert and four knights for agisting the woods and collecting pannage.
8 A forester responsible for demesne woods of the king shall be arrested for any unexplained destruction.
9 No clerk shall transgress in hunting or by breaking other forest regulations.
10 Assarts, purprestures, and waste in the forest shall be inspected and recorded.
11 All men shall heed the summons of the chief forester to come and hear the pleas of the lord king concerning his forests.
12 For the first two transgressions safe pledges shall be taken, but for a third offense the person of the transgressor shall be taken.

Young (1979, pp. 28–29)

The absolute power held by the king was ultimately challenged by the nobles, resulting in an agreement in 1215 known at the time as the *Articles of the Barons* which, after several modifications, became the *Magna Carta*. This agreement between the king and nobles covered the protection of nobles from illegal imprisonment, protection of church rights and limitations to feudal payments to the Crown, as well as some conditions related to the governance of the Royal Forests. The terms were subsequently broken by both sides, leading to civil war. Following the death of King John in 1216, the civil war continued, and a new version of the agreement was drawn up, which became known as the Great Charter, but this version was also unacceptable to the nobles. In 1217, a peace agreement was reached between Henry III (John's son) and the nobles, and a new charter was drawn up. This dealt with many of the issues that the nobles had been concerned about, but failed to deal with some, especially disagreement over the extent and management of the Royal Forests. As a result, a second, smaller agreement was developed, known as the Charter of the Forest (*carta de foresta*). The larger agreement became known as the great charter of liberties (*magna carta libertatum*), or the *Magna Carta*. These two documents, and the agreements that they covered, represented the ceding of some of the sovereignty that had previously been held by the king.

The 1217 Charter of the Forest was designed to control some of the excesses of the Forest Law established by William's son, William Rufus, although the management of the vert and venison had always operated outside the Common Law system. William I had established a legal system specifically for the governance of the forests. Special courts, the court of attachment, the court of regard, swainmotes and the court of justice-seat had been established specifically to deal with infringements of the Forest Law, headed by two justices, initially termed 'justices of the forest' but later termed 'justices in eyre' ('eyre' meaning circuit, indicating the movement of the court around the Royal Forests). Punishment of those transgressing the Forest Laws had been severe, often involving death or mutilation (blinding and castration). Lesser offences were considered by a court known as a swainmote. The Charter established the rights of free men to access forests, and limited the punishments that they would be subject to if they transgressed the Forest Law. The Charter also established Commoners' rights:

> Every free man shall agist his wood in the forest as he wishes and have his pannage. We grant also that every free man can conduct his pigs through our demesne wood freely and without impediment to agist them in his own woods or anywhere else he wishes.
>
> Every free man may henceforth without being prosecuted make in his wood or in land he has in his forest, a mill, a preserve, a pond, a marl-pit or a ditch, or arable outside the covert in arable land, on condition that it does not harm any neighbor.

In these statements, the term 'agist' means to use the forest for the grazing of livestock. Some of the governance aspects established by the Charter, such as Verderer's Courts, still exist in England in the New Forest and the Forest of Dean, and the New Forest still has official positions of verderer and agister.

The rights held by Commoners are among the best documented. These rights are particularly well described in England, and date back a thousand years or more. They refer to the rights held by a person who occupies land that have attached Common Rights to take particular materials or products from somebody else's land. When such rights were formally established, the land to which the people were being given access to was generally owned by the Crown or nobility. Such rights might include pasturage (the right to graze domestic stock), estovers (the right to take wood for minor works on buildings, for making farm implements and hurdles [portable barriers made of wood] and for firewood), turbary (the right to cut peat or turf for fuel), pannage (the right to graze pigs on acorns, beechmast, chestnuts or other nuts in woodlands, usually for a specific period in autumn), common in the soil (the right to extract certain geological materials, such as marl, sand and walling stone) and piscary (the right to take fish). These rights were strictly protected, as was the right of the forest owner to protect the vert and the venison.

This concept has endured: in the legal systems that exist in most countries, the true ownership of the land rests with the Crown or State. Land rights, whether they be fee simple, leased or in any other form, involve claims to specific areas of land and are defined as the estate for that area. The claims that define the estate as private property can best considered as a bundle of rights (usually including the right to use the land, the right to earn income from the land, the right to transfer the land to

Figure 1.8 Beech (Fagus sylvatica) woodland in the New Forest, England. The presence of open woodland with grassland, termed 'wood-pasture', is believed to be typical of what many woodlands in Europe may have looked like in the Medieval period. The absence of low branches on the trees is due to browsing pressure.

others and the right to enforce property rights). This concept may seem obscure, but it becomes more understandable when some practical examples are considered. For example, in many jurisdictions, fee simple land ownership does not include rights to the minerals under that land. Even on fee simple land, there may not be exclusive rights: regulations, codes of practice and other instruments can restrict what a landowner can do with the land, and often restrict what can be done with trees. Even in cities, privately held land can still be subject to local by-laws, which may for example restrict the felling of trees. The recent privatization of forest land in China has involved the transfer of certain rights of use for a defined period (70 years), rather than transfer of fee simple ownership.

The Medieval system of land ownership gradually broke down over time, especially as the feudal system was dismantled. However, many of the laws remained in practice, and in particular the principle that the Crown or State owns the land is today reflected in the fact that more than 80% of the world's forests are considered to be in public (as opposed to private) ownership. This has allowed governments to allocate land to various interests, generally associated with a bundle of rights (although the nature of these rights varies dramatically between countries and even within countries). In England, the Royal Forests were transferred to the Forestry Commission, a government organization, by the *Forestry (Transfer of Woods) Act* in 1923. The *Wild Creatures and Forest Laws Act* of 1971 abolished the sovereign's rights to wild creatures and revoked the Forest Law over 900 years earlier.

Asia

Elsewhere in the world, expanding populations also led to deforestation – in Asia, Central America, the Pacific and elsewhere. Surprisingly little is known about the history of forest development in Asia, particularly in China. Population growth there is known to have occurred much earlier than in Europe, yet there is a paucity of records that would enable this development to be reconstructed (Williams, 2003). The early influences of humans are difficult to separate from the effects of climate: between 9000 BCE and 3000 BCE, China experienced a weakening of the East Asian monsoon (Liu, 2007). The changes in the forests that this climatic change caused, combined with steadily increasing conversion of forests for agriculture, led to the southward retreat of animals such as the Asian elephant (*Elephas maximus*) from a maximum range at 5000 BCE (when it occurred throughout most of China) to today's minimal range in the southern part of Yunnan Province in southwest China (Edmonds, 1994; Elvin, 2004).

It seems likely that the main phase of deforestation occurred earlier in China than in Europe, reflecting the earlier development of significant iron production during the Song dynasty (910–1126 CE), a large ship-building industry and the presence of major cities early in its history. For example, Shandong is described in 845 as being heavily wooded, but was denuded within a few hundred years, with any remaining forest being managed through coppicing and pollarding (Ennin, cited in Williams, 2003). Firewood was brought in from across China, particularly from Sichuan, Hunan and Fujian. By 1100, there were at least five cities with populations in excess of 1 million, including Kaifeng (formerly known as Bianjing), the capital of the Northern Song dynasty, and Hangzhou (formerly known as Lin'an), capital of the Southern Song dynasty (Kracke, 1954).

Much more information is available about the forest history of Japan. Japan experienced a major phase of forest destruction about 1,000 years ago, during the construction of cities and wooden temples. The complexes at Nara and Heian (Kyoto), which

Figure 1.9
China has a history of major engineering projects, all of which would have required significant amounts of wood. At Dujiangyu in Sichuan, a flood control project started in 256 BCE involved diverting a river using Zhulong – *stones wrapped in bamboo (left) – anchored by* Macha – *wooden tripods (right). The project involved tens of thousands of workers, took about 14 years to complete and is still in use today (after several major modifications).*

began in the 8th century, consumed large amounts of wood. In particular, both temples and castles required very large logs of *hinoki* or Japanese cypress (*Chamaecyparis obtusa*). This resulted in severe deforestation in the mountains surrounding the plains where these developments occurred, causing soil erosion and subsequent flooding. Records of forest management extend back to the 17th century, when both the central government and individual landowners instigated controls on forest use, including seedling protection, selective cutting and punishment for illegal wood cutting. Iwamoto (2002) describes such activities having being instigated in the Kiso area by a relative of the Tokugawa Shogun in 1665. The first reform was quickly abandoned because of the significant loss in income from the forest, but a second reform was instigated in 1724, which resulted in a 60% reduction in timber production. The reform succeeded, enabling the forests to recover much of their standing volume. It is notable that forest inventory and production planning began in Kiso in 1779, at roughly the same time that such techniques were being developed in central Europe.

The situation in Japan in the 17th century is complicated by the feudal nature of Japanese society at that time. Central Japan was governed by the shogun's military regime, whereas southern Japan was governed by a large number of barons (*daimyo*), each of whom had considerable autonomy (Totman, 1989). Early regulations were related to the reservation of forest areas for the exclusive use of the government. However, there were also detailed regulations: peasants, for example, were not permitted to use in the construction of their buildings two of the most highly valued woods, *sugi* and *hinoki*. Later regulations protected forests on steep slopes and in riparian areas in an effort to prevent damage to low-lying agricultural areas. Wood demands were considerable, driven by the isolationist policy of the Tokugawa Shogunate (1603–1867),

Figure 1.10 Left: Hinoki (Chamaecyparis obtusa) *and right:* Sugi (Cryptomeria japonica), *two commercially important tree species that were being planted for timber production in Japan as early as the 17th century.*

which prevented any timber from being imported. Much of the demand was driven by the construction that occurred in this period. Each *daimyo* constructed a castle, around which towns were constructed (Iwamoto, 2002). All of these required wood, and the towns required wood fuel. Edo (which later became Tokyo) and Osaka were built at this time, with Edo having over 1 million inhabitants, making it the world's largest city in the late 17th century.

The late 17th century also saw the development of intensive plantation forestry in Japan. This was practiced by farmers, who planted *sugi* (Japanese cedar, *Cryptomeria japonica*) and *hinoki* on common lands. The way that this was done is interesting, and Iwamoto (2002) describes an example from the Yoshino area. Farmers planted the seedlings on common land and tended them. Although they were planted on common land, the trees were owned by the farmers. Often, the most valuable trees were sold to merchants or rich landowners, who would then hold the trees until they were large enough for their timber to command a high price. In the meantime, the farmers were paid to maintain the trees. Farmers also planted broadleaved trees, which were harvested and turned into charcoal. Over time, local merchants gradually cut out the farmers, paying local labourers to establish tree plantations which the merchants owned throughout the growth cycle.

Central and South America

In Central America, the rise of Mayan civilization has often been associated with deforestation of the areas surrounding their cities. The Classic Maya civilization emerged about 600 BCE and peaked at about 800 CE before abruptly collapsing. While the reasons behind the collapse of the Mayan civilization remain disputed (Diamond, 2005), it appears that deforestation and subsequent soil erosion played an important part. However, Demarest (2004) has argued that the agricultural practices utilized by the

Maya were highly sophisticated and adapted to rainforest environments. While shifting agriculture was practiced, it was not as destructive as modern swidden systems, as useful trees (e.g. fruit-bearing trees) were left in the clearings. This enabled more rapid regeneration during fallow periods. The careful selection of species and varieties being used in agriculture and the generally sensitive approaches to land management enabled large populations to be maintained for two millennia. Deforestation and erosion did occur, but were restricted to specific regions at specific times. As in the Mediterranean, deforestation may have changed the climate, helping precipitate the collapse (Cook *et al.*, 2012). However, unlike the Mediterranean, the forest recovered after the collapse of the population. The factors governing such recoveries (or not) are complex (Runyan *et al.*, 2012), and are as relevant today as they were during the historic period.

A particular feature of many Central and South American forests is the abundance of tree and other plant species of value to Indigenous peoples. These include fruiting species, trees and other plants used for specific flavours such as vanilla (*Vanilla planifolia*) and allspice (*Pimenta dioica*); trees used for construction materials, including thatching; and trees used as sources of wood with particular properties, such as sapodilla (*Manilkara zapota*), mahogany and cedar. The available evidence strongly suggests that such species have been favoured by Indigenous peoples, and as a result, they are now common in the forest (Levis *et al.*, 2012).

European expansion into other parts of the world

In Europe and in many other parts of the world, successive waves of people moved across the land, bringing with them their customs and governance systems. In most cases each new group acquired the land rights by conquest or treaty, as in the case of the Norman conquest of England. This was not the first time that land had been acquired in this way: previous invasions by the Romans, Danes, Anglo-Saxons and others had occurred, each with their own systems of land governance. Similar changes occurred over the rest of Europe and in Asia, although in areas where there is no history of written records, details are obscure or even lost.

Much as Europe saw successive waves of conquest and colonization, the Europeans themselves expanded into other parts of the world. This expansion had major implications not only for the world's forests, but also for Indigenous peoples. The expansion occurred from the 14th century onwards, initially by Spain and Portugal, and subsequently by France, Holland and Great Britain. Germany, Belgium and Italy were also important colonial powers in some parts of the world.

Many modern-day issues in forests between forest authorities and Indigenous peoples, both claiming jurisdiction over the forest, relate to the assumption of ownership rights by colonial powers. This occurred over much of Central and South America, Africa, South and Southeast Asia, and Australia and New Zealand. In some cases, this was achieved through conquest. Elsewhere, treaties were signed with Indigenous peoples, such as the Treaty of Waitangi in New Zealand in 1840, signed by over 500 Maori chiefs (Orange, 1987). The most important means seems to have been through a system known as the 'discovery doctrine', whereby the rights to the land lay with the government whose citizens had occupied the land. If the people occupying the land were not the subjects of a European Christian monarch, their rights were not recognized. This doctrine was formalized in 1494 by the Treaty of Tordesillas, which divided newly discovered lands outside Europe between Portugal and the Crown of Castile. Although other countries ignored the treaty, and the ability of Portugal and Castile to defend the lands claimed under it gradually declined, it has been cited in some modern-day cases, such as when Indonesia claimed Western New Guinea in 1960. Historically, the Doctrine of Discovery was widely used by all colonial powers, and in a US Supreme Court case in 1823 (*Johnson v. M'Intosh*) it was used to justify the government's alienation of Indigenous lands in the USA.

The Doctrine of Discovery originated from the belief that Christians had the moral duty and imperative to occupy lands and convert the Indigenous peoples to Christianity. While this belief has been maintained in some quarters, it was largely replaced after the Reformation and during the Age of Enlightenment by the belief that Europeans were superior to Indigenous peoples, and therefore should colonize their lands in order to help those peoples develop. This view continues, despite the independence of most colonies, and has governed relationships between central governments and Indigenous peoples in countries such as Australia, India, Indonesia, Brazil and Peru.

The Doctrine of Discovery was closely related to the concept of *terra nullius*, used to describe land that was never subject to the sovereignty of a state. In many cases, Indigenous sovereignty was simply not recognized. The doctrine derived from a papal bull issued by Pope Urban II in 1095, which permitted European Christian states to claim land occupied by non-Christian peoples. Subsequent rulings confirmed the lack of rights; the 1455 papal bull *Romanus Ponifex* denied that Indigenous peoples had rights to their land, and the Spanish papal bull of 1493, known as the *Inter Caetera divinai*, declared that non-Christians could not own land when it was claimed by European Christian sovereigns (Venne, 1998). While most European countries accepted these principles, there were challenges. In the 16th century, two members of the Dominican religious order, Francisco de Vitoria and Bartolomé de Las Casas, argued that the Indigenous peoples of the Americas had legal rights and that the pope's grant to Spain of title to the Americas had no legal foundation (Williams, 1990). In 1550, the Council of the Indies was established by the Spanish king to resolve the issue: it decided that Indigenous peoples were to be converted to Christianity and that any unbaptized Indigenous person could be killed by a Christian. The Indigenous people were recognized as humans (previously they had been considered as animals), but they were not recognized as legitimate peoples.

When the British defeated other colonial powers in North America, a distinct change occurred, in particular through the Royal Proclamation of 1763. This referred to Indigenous peoples as 'nations', recognizing them as distinct societies with whom treaties had to be established. It gave the British Crown the duty to protect Indigenous lands, established that settlement of Indigenous lands could only occur through treaty, and recognized that Indigenous nations had inalienable rights to their lands. The Proclamation created a clear boundary between the British colonies and the American Indian lands west of the Appalachian Mountains. The boundary was not intended to be permanent, and a series of treaties (such as the Treaty of Fort Stanwix and the Treaty of Hard Labour in 1768 and the Treaty of Lochaber in 1770) extended the boundary westwards.

At the same time, the British were adopting other policies elsewhere. In the 'Ceded Islands' of St. Vincent and Tobago in the Caribbean, the 1764 Grenada Governorate Ordinance, the 1765 Barbados Land Ordinance and the 1791 King's Hill Forest Act all aimed to conserve forest, regardless of the interests of the local Carib population in the land. These legal instruments were based on the premise that colonial annexation and acquisition of 'sovereignty' were justified by the exercise of forest clearance and cultivation. The reservation of forests was based on a Royal Proclamation issued in 1764 that specifically set aside forests, partly as a source of timber for fortifications and public buildings and partly to prevent the droughts that were believed to occur following the removal of forests (Grove, 1995). The steps taken by the British in the Caribbean to protect forests on colonial lands were partly based on the experience of the French in Mauritius, where forest removal had been accompanied by severe erosion, and were to be copied in a number of other areas colonized by the British, including India and Burma.

The Doctrine of Discovery was used in particular in Britain's colonization of Australia, since the Aboriginal people had entirely oral traditions and no clear sovereign(s) that the British could negotiate with. The implications were considerable: an 1835 proclamation by the governor of the New South Wales colony indicated that Indigenous Australians could not sell or assign land, nor could an individual or group acquire it, except from the Crown. This rule was only really overturned in Australia with the Mabo Case, when the Australian High Court found that the Mer people (from Murray Island) had owned their land prior to annexation by Queensland (Russell, 2005).

Today, native title is used in some countries, such as Australia, to recognize that some Indigenous people have certain land rights deriving from traditional customs and laws and that these persisted after the assumption of sovereignty by the colonial governments. In some cases, different Indigenous groups can exercise native title over the same area of land. In Australia and elsewhere, Indigenous people have strong religious ties to the land and sense of responsibility for their 'country', based on ancestry (giving rise to the term 'ancestral domain'). Ancestral domain includes spiritual and cultural ties that can be completely independent of legally defined land titles. In practice, this may manifest itself in a sense of guardianship of the land, rather than ownership, and Aboriginal people frequently refer to themselves as custodians, rather than owners, of their country (e.g. Sveiby and Skuthorpe, 2006).

Rose (1996, p. 7) provides an interesting description of this relationship between Aboriginal people in Australia and the concept of country:

> Country in Aboriginal English is not only a common noun but also a proper noun. People talk about country in the same way that they would talk about a person: they speak to country, sing to country, visit country, worry about country, feel sorry for country, and long for country. People say that country knows, hears, smells, takes notice, takes care, is sorry or happy. Country is not a generalised or undifferentiated type of place, such as one might indicate with terms like 'spending a day in the country' or 'going up the country'. Rather, country is a living entity with a yesterday, today and tomorrow, with a consciousness, and a will toward life. Because of this richness, country is home, and peace; nourishment for body, mind, and spirit; heart's ease.

She goes on to add (p. 49):

> The relationships between people and their country are intense, intimate, full of responsibilities and, when all is well, friendly. It is a kinship relationship, and like relations among kin, there are obligations of nurturance. People and country take care of each other.

Elsewhere, there is generally an emphasis on the connections between Indigenous peoples and the land they have occupied. Many hold the view that they are intimately connected with elements of the forest ecosystem and a part of it. This contrasts markedly with some Western views that do not see humans as integral parts of the ecosystem, and has given rise to some of the conflicts between environmentalists and Indigenous peoples. Many Indigenous peoples also have sacred sites, with strict protocols associated with them. For example, in Arnhem Land, Australia, there are many sites linked with the Dreaming, a timeless concept associated with the presence of supernatural Ancestor beings such as Namarragon (Figure 1.11). These beings are associated with particular sites (Figure 1.12), and Namarragon is believed to live in the Arnhemland Escarpment.

Figure 1.11
Aboriginal rock art portraying Kangaroo with Lightning Man, Nourlangie Rock, Kakadu
National Park, Northern Territory, Australia. In the Dreamtime, Lightning Man, called
Namarrkon or Namarragon, lived in the sky. He carried a lightning spear. He had stone clubs
tied to his knees and elbows so that he was always prepared to hurl thunder and lightning.
He normally lived at the far ends of the sky, but during the wet season came down into the
atmosphere to monitor the humans. He settled in some cliffs in the Arnhemland Escarpment
(Figure 1.12).

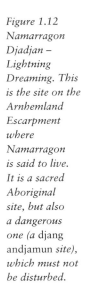

Figure 1.12
Namarragon
Djadjan –
Lightning
Dreaming. This
is the site on the
Arnhemland
Escarpment
where
Namarragon
is said to live.
It is a sacred
Aboriginal
site, but also
a dangerous
one (a djang
andjamun *site),*
which must not
be disturbed.

The links between Australian Aboriginals and the land that they occupy are so tight that instead of being described as owning the land, they have sometimes been described as being owned by the land. The different groups and clans held in sacred trust the land and the Dreaming sites that they believed their Ancestors had visited, following distinct tracks.

Moves towards more formal types of forestry

As can be seen from the discussion in the preceding sections, forestry in one form or another dates back thousands of years. However, it has taken two forms. On the one hand is the destruction of forests for timber, with little if any thought given to regeneration. This is more properly termed logging. On the other hand, there is the deliberate management of forests and their products, which is what the term forestry should be restricted to. Foresters practice forestry, and loggers practice logging. While often confused, even today, the two conduct very separate activities. Logging is a part of forestry, but is best restricted to the felling of trees.

Timber for construction was an important product derived from the forests, but it was not the main forest product at the dawn of humankind. Wood was involved in the construction of Egypt's pyramids, in many buildings of early China, in the stone circle at Stonehenge (and in the recently discovered circle of wooden poles that balanced the stone circle) and in many other structures. Early trade by the Phoenicians was dependent on the construction of seaworthy vessels, and among other products these ships were subsequently used to transport wood (primarily cedar of Lebanon, *Cedrus libani* subsp. *libani*) to Egypt, the Greek Empire and the Roman Empire. In many instances, these early examples of logging involved the destruction of forests, often with little awareness of the potential for adverse effects. On the other hand, the extent of logging was quite small, reflective of the size of the human population. Besides some radical examples of overharvesting (such as on Easter Island in the Pacific), the forests were not cleared beyond what was needed for swidden agriculture. Sands (2013) provides a particularly good account of some of the early relations between humans and forests.

Just as the timing of the period of forest destruction has varied around the world, the timing of the introduction of forest management practices and forest conservation has varied. What is evident, however, is that the success of forest renewal has varied substantially. Some countries, such as the Republic of Korea, have been very successful at reforesting after a period of extensive forest loss. Others, such as the eastern Mediterranean countries, have had much greater difficulty, most likely due to near-irreversible changes in the ecosystems (and local climate) following the loss of forest cover.

As the population of pre-industrial Europe decreased (due to famine, bubonic plague, the Hundred Years' War and the Thirty Years' War) and increased, the demand for forest products diminished and accelerated accordingly. For example, coppicing appears to have been undertaken on a large scale in parts of France as early as the beginning on the 13th century. In most cases, this involved controlling the way that animals used the forest, including not only sheep and cattle but also pigs. At this time, there were also considerable concerns about the regulation of coppicing, and a royal decree in 1346 in France required that masters of forests were to ensure that they were 'perpetually sustained in good condition' (Keyser, 2009). Eventually, with the improvement of agriculture in Europe, the demand for timber exceeded the demand for small wood (from coppice) and non-timber forest products. Timber was needed for mine shafts and ship-building, and most of the followers of economic liberalism theory viewed it purely in economic terms, neglecting the wider social importance of the forests.

Timber management did not transform instantaneously from selective logging (taking the trees that best reflect the demand for certain products; often simply the best trees) done predominantly before the 20th century to the extensive industrial-style clearcutting of the 20th century, and there are a number of examples of attempts at more careful silviculture. British forestry produced one of the first silvicultural books: John Evelyn's *Sylva: or a Discourse of Forest Trees and the Propagation of Timber in His Majesty's Dominions*, written in 1664; it described major British tree species and their uses. A later French publication by Henri-Louis Duhamel du Monceau, *Traité complet de bois et forêts* [*A Comprehensive Treatise on Woods and Forests*] (1755–1764), contained a discussion of coppicing silviculture and forest valuation.

At the same time as John Evelyn's work was being published in England, France was passing a major new forest law, the French Forest Ordinance of 1669. They are both considered to represent the turning point between a period when forests were mined for wood and the realization that there was need to take greater care of the forest, and to ensure that trees were planted to replace those being cut (Glacken, 1967; Williams, 2003). The French ordinance was aimed at sorting out previous forest laws and making forestry a separate part of the economy. It included a range of actions, mainly aimed at ensuring the future supply of wood for the French navy.

The early 18th century saw the emergence of two very different approaches to forestry. One was the so-called scientific approach that emerged in a number of German-speaking countries in central Europe, and which is examined in the next section. The other was the deliberate creation of wooded landscapes, as particularly developed in Britain. Proponents of this approach included Charles Bridgeman (1690–1738), William Kent (ca. 1685–1748) and Lancelot Brown (1716–1783), better known as Capability Brown. These individuals would probably be better considered landscape architects than foresters, although the landscapes they designed included numerous woodlands.

The British approach aimed to create landscapes around country houses that were aesthetically pleasing. The approach started in England and was quickly adopted in Scotland. However, there was also a strong economic incentive, since the large-limbed trees that characterized parklands could be used for a number of specialized cuts needed for wooden ships. The desire for aesthetically pleasing landscapes spread to cities, and many formal parks and avenues were laid out at this time – this was essentially the birth of urban forestry.

The introduction of 'scientific forestry'

Even though similar ideas were being developed in Japan at the same time, it is widely considered that German forestry crystallized forest science and influenced Western forestry to the greatest extent. One of the earliest works was by Hans von Carlowitz, a mining engineer from Saxony, who in 1713 published *Sylvicultura Oeconomica*. He argued that the large amounts of timber that were needed in mines were in jeopardy, and advocated that fast-growing plantations of conifers, primarily Norway spruce (*Picea abies*), be established. Another early forest scientist was W. Gottfried von Moser, who in 1757 published one of the first books on forest economics, *Grundsätze der Forstökonomie*. Among the rapidly growing number of German-speaking forestry scholars (who were mostly also practitioners), Heinrich Cotta (1763–1844) and Georg Hartig (1764–1837) stand apart. Their works provided a basis for the further development of silviculture and contained considerations of ecology, silvics and opportunities to meet the increasing demands for forest products. Hartig and Cotta drove the establishment of the shelterwood system that removed timber in several cuts, whereas before foresters relied primarily on clearcuts and natural regeneration

(Fernow, 1907). Heinrich Cotta is particularly remembered for the preface of his 1817 book *Anweisung zum Waldbau* (*Instruction on Forestry*). (The book is sometimes mistakenly dated to 1816. The preface is dated 21 December 1816, but the actual book was published the following year, and was then republished in a number of editions. It is available in English at the Forest History Society's website: http://www.foresthistory. org/publications/FHT/FHTFall2000/ cotta.pdf.) In this, he described many of the principles of modern forestry and warned of the dangers of over-exploiting forests. Hartig studied forestry at the University of Giessen from 1781–1783, and went on to found a forestry school at Hungen in the Principality of Solms-Braunfels (now part of Hesse in modern Germany). One of the first professors of forestry (at Schemnitz, then located in the Austro-Hungarian empire and now Banská Štiavnica, in Slovakia) was Heinrich David Wilckens (1763–1832). Another early forestry academic was Carl Ludwig Obbarius (1780–1860), a follower of the forest management principles espoused by Hartig and Cotta. He developed a forestry school for mining companies in Sweden (Brynte, 2002). An interesting aspect of these early approaches to scientific forestry is that they generally required full control of the land, including the exclusion of grazing animals and control of agricultural practices and hunting within the forests (Watkins, 2014). As such, they occurred alongside the extinguishment of the rights of commoners, and were often codified in new sets of laws that replaced those drafted in the early Medieval period.

The growing European population of the 17th and 18th centuries demanded ever-higher volumes of forest products. A significant volume of timber was traded from Scandinavia and Russia, which also adopted German approaches to forest management. Puettmann *et al.* (2009) argue that the German ideology of *Waldreinertragslehre* (forest landowners have social responsibilities to the greater community) has been used in Central and Northern

Figure 1.13

Frontispiece of an early (1796) German textbook on forestry: Über die örtliche progressive Wachsthumszunahme der Waldbäume: in Anwendung auf den möglichsten Ertrag eines Waldbodens *by Johann Leonhard Spath. This particular book is in the collection of rare books held in the library basement of the Food and Agriculture Organization in Rome.*

Europe, while an alternative approach, *Bodenreinertragslehre* (economic interest as a sole objective of forest management), later became more prevalent in North America.

Notwithstanding the early actions of William Penn in protecting the forests of Pennsylvania, some of the first warnings about forest destruction in the USA were sounded in 1847 by George Perkins Marsh, a US congressman. He warned of the problems associated with the logging industry, which was pushing westward as it progressively exhausted timber stocks in the east and in the Great Lakes states. However, at this time, there were no forestry schools in North America, and work in the woods consisted of logging, not forestry. The spread of German forest science in North America is often attributed to John Warder, who argued for the application of German forestry methods in his 1873 work *Forests and Forestry in Germany* (Fernow, 1907). However, it was Carl Schenck who in 1898 founded the Biltmore Forest School in Asheville, North Carolina, the first forestry school in the USA, predating the New York State College of Forestry at Cornell University (under the leadership of Bernhard Fernow) by a few weeks. Schenck trained in the University of Giessen, Hesse, and Fernow at the University of Königsberg and the Royal Prussian Academy of Forestry at Münden.

Carl Schenck, Gifford Pinchot, the first chief forester of the US Forest Service, and Bernhard Fernow, who went on to found the Faculty of Forestry at the University of Toronto, all popularized European approaches to forestry in the USA and Canada. However, the methods were only applied in some cases; more often there was no real attempt to use the knowledge of forestry that was imparted in the new forestry schools. The assumption of an infinite resource to be exploited for the benefit of the growing population of North America remained firmly entrenched in many peoples' minds.

The Germanic notions of 'the normal forest' and sustained yield remained prevalent in North America and many other parts of the world until the second half of the 20th century, and persists in some places to this day. Attributing all such development to German foresters is perhaps unfair: foresters in Switzerland and France also played a major role in promoting the methods espoused in 'German' forestry. Note that while numerous references are made to 'German' forestry and German foresters, Germany was only established as a country in 1871, having been a loose confederation of independent sovereign states since 1814. Hence the practice of forestry in central Europe actually predates both modern Germany and the earlier confederation. Hermann Knuchel, a forestry professor from Switzerland, described the strong feeling about these techniques: 'Normality is that practically attainable degree of perfection in a forest which we strive to secure in all parts of the forest and to maintain in perpetuity' (Knuchel, 1953, p. 28). Knuchel's book was originally published in 1950 in German, but was considered sufficiently significant to be translated into English only three years later.

Despite being frequently associated with some sort of perfection in forestry, sustained yield only pertained to producing timber volume in perpetuity, thus omitting or minimizing attention to other forest values and services, and even failing to take into account timber quality. Because of these omissions, the concept of sustained yield has caused considerable concern about the negative impacts of forestry on the environment. As a result, today's ideas about sustainable forest management have moved well away from the original concepts of sustained yield promoted in German forestry practices.

Different approaches to forestry

As discussed in the preceding section, much early scientific forestry was based on the concept of sustained yield, defined as the yield of a product (usually timber) that a forest can provide indefinitely. This approach has been adopted in central Europe for centuries. For example, the city of Erfurt (now the capital city of the State of Thuringia in modern Germany) passed legislation in 1359 requiring that harvests be planned specifically so that timber could be supplied indefinitely (Heske, 1938). Such approaches were based on the premise that the practice of forestry could be likened to farming, with crops being established, tended and harvested after set time intervals. This is still the basis of the various forms of plantation forestry, although there have been attempts in some countries to move plantation forestry away from such a strict focus on timber production. The model evolved because timber was in increasingly short supply in central Europe, and major concerns existed over future supplies.

Different approaches were adopted in other parts of Europe. In 17th century Britain, as described earlier, there was an increasing interest in the planting of trees on private estates. Trees were planted for a number of different reasons, including the demonstration of wealth, for aesthetics and to supply special timbers for the navy.

As Europeans spread to other parts of the world, they discovered vast forest reserves. Usually, these were seen as a resource to be exploited, a view that persists in many parts of the world today. Forestry focused on harvesting these resources, with much less attention being given to managing the forests sustainably. This attitude can be related to the apparently limitless nature of the resources. Mather (1990) describes a conceptual model for this phase of forest development, in which early colonists view a resource as limitless, resulting in a diminution of the value given to forested areas. As the forest resource is depleted, people eventually notice the loss, and steps are taken to conserve and sustainably manage the remaining forests. The forests also stood in the way of agricultural expansion, with many forests located on fertile soils ideal for agriculture, especially in the temperate zone. Clearing the forests became an important element of regional development, and in some cases, the timber that was harvested was simply burned as the supply so exceeded demand.

However, as this phase of forest destruction continued, concerns began to be expressed that many parts of the world would face the same issues as Europe had done, namely a significant loss of forest cover with associated environmental problems such as soil erosion and increased frequency of natural hazards. This resulted in interest in taking a more natural, or at least aesthetically pleasing, approach to forest harvesting and forest management. Silvicultural systems such as single-tree selection had been pioneered in France and Switzerland as a means of maintaining yield from a forest without losing the forest cover (Schütz, 1997), and even though these techniques created a forest that was far from 'natural', they were embraced by more environmentally minded forest managers.

More recently, international agreements such as the United Nations Convention on Biological Diversity (CBD) have resulted in new approaches to forestry. In these, forest management is no longer central but is critical to the success of the management of a landscape. This is evident in the 'Ecosystem Approach', which was endorsed at the fifth Conference of the Parties to the CBD in May 2000. This approach is best described as a strategy for the management of land, water and living resources that promotes conservation and sustainable use in an equitable way. As Sayer and Maginnis (2005b) point out, this approach is similar to sustainable forest management in many ways, but there are marked differences in emphasis. The emphasis on conservation and the requirement to use adaptive management are important components, and these are reflected in Chapters 3 and 15 of this book. There are 12 principles to the Ecosystem Approach, summarised in Table 1.1.

Since the Ecosystem Approach was formulated, there have been a number of initiatives aimed at integrating the conservation of biodiversity and forest management, in particular in relation to industrial development projects (including the harvesting of

Table 1.1 The 12 principles of Ecosystem Management elaborated for the Convention on Biological Diversity. A full discussion of these is provided by Smith and Maltby (2003).

Principle

1 The objectives of management of land, water and living resources are a matter of societal choice

2 Management should be decentralized to the lowest appropriate level

3 Ecosystem managers should consider the effects (actual or potential) of their activities on adjacent and other ecosystems

4 Recognizing potential gains from management, there is usually a need to understand and manage the ecosystem in an economic context. Any such ecosystem-management programme should:
    a) Reduce those market distortions that adversely affect biological diversity
    b) Align incentives to promote biodiversity conservation and sustainable use
    c) Internalize costs and benefits in the given ecosystem to the extent feasible

5 Conservation of ecosystem structure and functioning, in order to maintain ecosystem services, should be a priority target of the ecosystem approach

6 Ecosystems must be managed within the limits of their functioning

7 The ecosystem approach should be undertaken at the appropriate spatial and temporal scales

8 Recognizing the varying temporal scales and lag-effects that characterize ecosystem processes, objectives for ecosystem management should be set for the long term

9 Management must recognize that change is inevitable

10 The ecosystem approach should seek the appropriate balance between, and integration of, conservation and use of biological diversity

11 The ecosystem approach should consider all forms of relevant information, including scientific and Indigenous and local knowledge, innovations and practices

12 The ecosystem approach should involve all relevant sectors of society and scientific disciplines

natural forests). 'No Net Loss' (NNL) and 'Net Positive Impact' (NPI) are two such approaches; these focus on ensuring that there is no net reduction in the diversity within and among species and vegetation types, in the long-term viability of species and vegetation types or the functioning of species assemblages and ecosystems (Aiama *et al.*, 2015).

As discussed in the next chapter, there are many regional differences in the manner in which forestry is practiced. What is considered good practice in one area may be denigrated in another. This is not because the practice is inherently bad, but because forests differ in their nature and context. Many forestry textbooks devote a great deal of space to describing different types of forest based on their floristics. There are many different forest classifications, and these will not be described here. A central tenet of this book is that many principles of forest management are universally applicable. Silvicultural techniques can be adapted to meet local conditions, but those techniques are applied to meet higher goals, and it is these goals that we focus on here.

If there is to be any division of forest types, the most important from the point of view of sustainable forest management is based not on floristics or climate, but on purpose. A natural forest is usually very different from a planted forest, and it is the management objectives for the forest that most importantly define the practices that are used. For example, a forest that is managed for the conservation of a rare mammal species will have different management practices than a forest that is managed for timber production. In contrast, a tropical pine plantation managed for timber will share many of the same management characteristics of a temperate pine plantation, even though the climate, species, soils and growth rates may be very different.

Indigenous communities, forest-dependent communities and traditional ecological knowledge

As forestry became more firmly based on scientific principles, the need for skilled practitioners emerged. In many parts of Europe, foresters were and still are highly regarded, often as key persons in the community. Elsewhere, this has evolved into the concept of the professional forester, and in the most extreme cases, rights to practice legislation exists, meaning that only registered professional foresters are permitted to practice forestry. It is possible to trace some interesting connections between the development of professional foresters in different parts of the world. The first Inspector General of the Indian Forest Service (then called the Imperial Forest Service) was Sir Dietrich Brandis; Gifford Pinchot, who established the US Forest Service, trained under him.

While restricting the right to practice forestry to professional foresters has some advantages, it can also result in the disenfranchisement of local people. This was particularly evident during colonial times, when foresters were often trained in the colonial countries and then sent to manage forests in the colonies. Little account was taken of local knowledge and customs, and local people could be prevented from decision-making or even accessing forests in their localities.

Today, it is increasingly recognized that local people may have knowledge useful in the management of forests in their area. This is particularly true of Indigenous people, who have forest knowledge that has been accumulated over long periods of time, widely termed 'traditional ecological knowledge'. This knowledge is being increasingly recognized, and has been protected under the United Nations Convention on Biological Diversity. Article 8(j) states:

> Each contracting Party shall, as far as possible and as appropriate: Subject to national legislation, respect, preserve and maintain knowledge, innovations and practices of indigenous and local communities embodying traditional lifestyles relevant for the conservation and sustainable use of biological diversity and promote their wider application with the approval and involvement of the holders of such knowledge, innovations and practices and encourage the equitable sharing of the benefits arising from the utilization of such knowledge innovations and practices.

The recognition of the rights of Indigenous peoples extends to others living or working in or near forests, and is discussed in detail in Chapter 10. As a result, considerable emphasis is placed on this in criteria and indicator schemes (see Chapter 2), although as will be shown, it is often difficult to apply these ideas in practice.

Forests and gender

The traditions of forestry, as documented by (generally male) authors, make little reference to the role of women in forestry. Similarly, treatises and textbooks about forest management rarely mention the role that women have played, and continue to play, in the management of the world's forests. The stereotypical image of a forester is male, and the first foresters, described earlier, were without exception male. This trend has continued through time, with the vast majority of people in positions of forest governance throughout the world being male (Hoskins, 2016). However, because about half the world's population is female, and because in many societies women play a significant role in forests, it is important to recognise this bias. It is likely that change will be slow, but there is much evidence of change, including progressively higher proportions of women taking forestry degrees. Women are also increasingly taking on senior management roles. For example, in 2016, the province of British Columbia, Canada, appointed its first female chief forester. Simultaneously, the CEOs of the Council of Forest Industries and the Association of BC Forest Professionals and the president of the Truck Loggers Association were all women. Whether this really means that the province has moved from dealing with a lack of women in the sector to addressing the key issue of male dominance (Holmgren and Arora-Jonsson, 2016) remains to be seen.

It is important to differentiate between gender and sex. The latter refers to the physical differences between men and women and is based on the sexual and reproductive functions. Gender is a term that is generally used to 'conceptualize the socially and culturally constructed roles and relationships, personality traits, attitudes, behaviours, values, relative power and influence that society attributes to men and women' (Alber, 2015, p. 9). The bias against women means that unequal power relations are predominant, and women's identities, attitudes and behaviour are neglected. Bringing gender issues to the fore ('gender mainstreaming') means that all decisions need to be examined from the perspective of both men and women, and in theory should have the result that inequalities are not perpetuated.

The bias against women extends beyond governance: as a result of socio-cultural norms, women and men have differentiated needs, uses and knowledge in relation to the ecosystems within which they work (Colfer *et al.*, 2016). In some countries, women and men have unequal rights to the land, including the right to own and use land. This issue is particularly apparent in many African countries, resulting in the establishment of advocacy groups such as the African Women's Network for Community Management of Forests (*Réseau des Femmes Africaines pour la Gestion Communautaire des Forêts*, REFACOF). This group has successfully pressured some governments to reform their land tenure systems and to develop systems that increase the involvement of women in decision-making, especially at the community level. At the same time, there has been growing recognition through, for example, the UN Women's World Survey, that women's access to, use of and control of natural resources will play a critical role in meeting targets such as the Sustainable Development Goals described in Chapter 12 (Aguilar, 2016). Elsewhere, although women may have equal rights, the forest governance system remains heavily dominated by men, even in supposedly progressive countries such as Sweden (Holmgren and Arora-Jonsson, 2016).

To materialize the benefits that will be brought by the greater involvement of women in all aspects of forestry, policies will have to be much more proactive. Gender-responsive policies are needed that will directly promote

Figure 1.14 An example of a fairly frequent gender interaction in forestry, from Ramnagar in Uttarakhand, India. A man, who has just brought a bundle of firewood to a collection point with the help of a bicycle, allocates bundles to women who will then carry them to the point of sale. The reverse is rarely, if ever, seen.

the role of women in all aspects of sustainable forest management. There is ample evidence that this will be highly beneficial to all parties, yet forest management and the forest sector as a whole has been resistant to change. The issues associated with gender inequality need to be addressed within all aspects of sustainable forest management.

Forests as a part of the landscape

One of the most recent trends to emerge in forestry is the desire to view forests as a part of the landscape, rather than as unique features that should be managed independently (Sayer and Maginnis, 2005a). There are numerous flows to and from forests within a landscape, and their management requires the management of the landscape as a whole. This presents many challenges, and may require the establishment of multilevel governance (Watts and Colfer, 2011). In practice, this rarely happens, as many different managers are usually involved, and these individuals often have poor lines of communication between each other. As an example, deforestation is primarily an agricultural problem, not a forest management problem, whereas forest degradation is much more associated with forest management, even though the individuals responsible are usually from outside of a forest. Adapting to climate change will require a landscape approach (Minang *et al.*, 2015), which in turn will require breaking down many of the political, social and cultural constraints that prevent a truly holistic approach to the management of forested landscapes.

Related to this is the concept of cultural landscapes. Over time, humans have modified the landscape in many parts of the world, and the resulting mosaic of forests and other forms of land use has become a valued part of a country's heritage. This is particularly apparent in Europe, where attempts to change the mosaic of forest and agricultural land to one that is wholly forest (the natural situation) or agriculture would be vigorously resisted.

The need for stability of tenure

Throughout history, there have been various forms of ownership of forest property, and various approaches to the practice of forestry. The two are closely related: the more rights that an individual or organization has to the land, and the longer these are for, the greater the investment in the land. Many would argue that sustainable forest management is not possible without clear and long-term rights to the forest; FAO (2007, p. 2) specifically states that 'without tenure security, there cannot be sustainable forest management and the contribution of forestry to poverty alleviation or sustainable livelihoods will be limited'. It has also been argued convincingly that unclear property rights and weak local land-use governance can be directly linked to problems such as deforestation and forest degradation (e.g. Chomitz, 2007). Consequently, before examining the practice of sustainable forest management, it is important to have understanding of the different forms of tenure that exist. In cases where tenure is uncertain, for example when Indigenous peoples also have claim to the land, tenure holders may be reluctant to invest in forest operations that they may not benefit from. The increasing recognition of Indigenous rights, described in Chapter 10, is also changing the ways that some forest tenures are viewed, with the requirement for consultation, and more specifically for free, prior and informed consent, affecting the way that management decisions are taken. This is not an argument for complete privatization of land, but there are numerous examples in which tenure security has encouraged a longer-term approach to forest management.

Modern forest tenures

Modern forest tenures generally relate to the rights given by the Crown or State to private individuals, companies, communities and other groups in return for payment and, sometimes, other services in the public and collective interest. In many cases, a distinction is drawn between forests under the direct control of central governments and land controlled by regional, provincial and district governments (e.g. FAO, 2007). However, more commonly, a distinction is made between 'privately owned' land (land owned by individuals and families, either individually or under some form of cooperative agreement; by forest industries or by private organizations, including private corporations, cooperatives or institutions [religious, educational, pension or investment funds, nature conservation societies]), and 'publicly owned' land (land owned by the State, either at the central or provincial level, and communal forest land owned by communes, cities and municipalities) (Schmithüsen and Hirsch, 2010). Land held by Indigenous groups under common property arrangements is sometimes considered separately, but could also be considered to be 'publicly owned' under the preceding definition. The simplest form is a logging concession involving the rights to harvest timber from a specified area in return for a payment. The payment (termed 'stumpage') may be based on the volume of timber harvested or on the value of the timber. Progressively more complex forms of tenure involve requirements to replant after harvesting, requirements to construct infrastructure such as roads, and requirements to construct wood processing facilities such as sawmills and pulpmills. In many cases, these requirements were part of government policies aimed at using forestry to encourage regional economic development in areas that might otherwise remain undeveloped.

Leasehold and rented land refers to the case where an owner provides access to land and certain use rights in return for payment. The duration of the arrangement may be for periods of a few years, or much longer (in the United Kingdom, and its former colonies, leases lasting 999 years have been used for forest land and other properties). Leases have often been arranged between the Crown and private individuals or companies wishing to undertake activities on Crown lands. In recent years, the reverse has been occurring: in Scotland, Forestry Commission Scotland will lease land from a landowner for 10 years and will establish a forest on that land. At the end of the 10-year period, the land (and forest) will be handed back to the original owner. This is being done because the establishment of forest is seen as being in the public interest. Experiments have also been undertaken with the leasing of land to communities. For example, in Nepal, leasehold forestry was developed with the specific objective of reducing poverty and improving ecological conditions. The leases are rent-free, last 40 years, and are for degraded land. Groups of 5 to 10 households use plantation or natural regeneration methods to develop forests on these lands, and can harvest any production (Poudyal *et al.*, 2015).

The non-governmental organization Rights and Resources Initiative has paid particular attention to land rights associated with forests. In a 2014 report, they identify four principal categories of forest ownership:

* Forest land administered by governments;
* Forest land designated by governments for Indigenous peoples and local communities;
* Forest land owned by Indigenous peoples and local communities;
* Forest land owned by individuals and firms.

The fourth category constitutes allodial rights, which are relatively rare given the tendency for the Crown/State to retain ultimate ownership of the land.

The Rights and Resources system is based on the concept of the bundle of rights, mentioned earlier. They distinguish a number of different rights, based on the work of Schlager and Ostrom (1992) and Larson *et al.* (2010). They can broadly be divided into use, management and ownership rights. Use rights are primarily related to withdrawal rights, the ability to benefit from the resources on the land. Management rights cover the right of access, enabling a person to enter or pass through a particular area of land, the regulation of forest resources and territories (the right of management), and the ability to exclude others to be excluded from the land or excluded from utilizing its resources (the right of exclusion). Ownership rights include the right to due process and compensation ('extinguishability'), which enables a government's efforts to extinguish, alienate or revoke one, several or all of the rights held by an actor to be challenged; duration, which relates to the time that rights are held for, which may be for a specific period or in perpetuity; and the right of alienation, which enables the holder of rights to transfer them to another person or institution.

The National Aboriginal Forestry Association in Canada has developed a useful classification system for forest tenures on Crown Lands (Table 1.2). This illustrates some of the complexity of tenures, and emphasizes that many different forms can be present, even within the same jurisdiction.

The greatest problems occur when rights associated with tenures are not explicit or not enforced. Other problems occur when there are multiple tenure holders for the same land because the rights have been divided (for example, mineral rights separated from forest harvesting rights and/or grazing rights), or because the tenures are insecure

Table 1.2 The tenure classification system developed by the National Aboriginal Forestry Association in Canada.

Tenure Type	Description
Group 'A': Alternative, conservation-based tenures	– 'alternative' tenures where the tenure holder has flexibility to manage the forest for other values beyond sustainable supply of fibre and timber
Group I: Major long-term tenures with management responsibility	– usually 20 years (less in British Columbia) and renewable (evergreen) – licensee responsible for inventories, long-term management plans, operational planning, protection, roads, silviculture/reforestation – area-based – large scale in terms of area and volume
Group II: Significant timber volume supply	– long-term and renewable – licensee responsible for operational planning, protection, reforestation but not long-term management or inventories
Group III: Small enterprise–oriented tenures	– typically shorter term, or long-term but smaller volume – moderate to small-scale timber volume access (under 50,000 m³) – licensee responsible for operational planning, protection, reforestation but not long-term management or inventories (although small-scale Forest Management and Timber Supply Agreements in Quebec would be an exception) – term is variable, though generally short-term
Group IV: Minor tenures and special permits	– cutting permits for personal or small business/specialty use – includes firewood; Christmas trees; non-timber forest products – permit holder is not involved in management or reforestation

NAFA (2007).

or too short. There have numerous recommendations on how some of these problems might be resolved (e.g. FAO, 2013), but in recent years a lot of emphasis has been placed on the decentralization of rights and tenures (Colfer *et al.*, 2008; Moeliono *et al.*, 2009; Sikor and Stahl, 2011).

Decentralization of rights and tenures

In the 1960s, an influential paper was published by Garrett Hardin titled 'The Tragedy of the Commons'. In it, he argued that common property would be inexorably degraded as each user sought to maximize its use. The paper was probably misnamed, since as discussed earlier in this chapter many common rights in the past were highly regulated, and often prevented the sort of deterioration that Hardin proposed. Hardin's argument has frequently been assumed to indicate that commonly held land would inevitably be degraded without the intervention of an organizing force, such as a government, or the privatization of the land so that all decisions rest with a single interest. The work of Elinor Ostrom (e.g. Ostrom, 1990, 1999) demonstrated that there are numerous examples where resource users have cooperated to ensure that such degradation did not occur. She also argued that many of these communal management systems were better suited to managing the resource than systems imposed by higher-level governments designed to standardize management across larger areas.

Ostrom's body of work defined eight principles for managing common resources:

1   Clearly define the boundaries of the area being managed.
2   Match the rules for governing the use of common goods with local needs and conditions.
3   Ensure that those affected by the rules can participate in modifying the rules.
4   Make sure that the rule-making rights of community members are respected by outside authorities.
5   Develop a system, undertaken by community members, for monitoring the behaviour of members.
6   Use graduated sanctions for rule violators.
7   Provide accessible, low-cost means for dispute resolution.
8   Build responsibility for governing the common resource in nested tiers from the lowest level up to the entire interconnected system.

The arguments put forward by Ostrom and Hardin have been used to justify community-based management of land and privatization of land, respectively. In reality, forests remain primarily owned by governments: about 75% of global forest area is owned by governments (White and Martin, 2002). This varies regionally: 99% in Africa, 98% in Western and Central Asia, 90% in South and Southeast Asia, 82% in South America and 70% in North and Central America (FAO, 2015). Governments in this sense can mean a number of things: the Crown, the State, a provincial government, a city or even a community (although in the last two cases, there is some confusion over whether these constitute public or private properties).

Unfortunately, recent forest decentralization has often been set up in such a way that failure is inevitable. The most valuable forest lands are rarely placed under the control of communities. Instead, land that is handed over is often degraded or marginal. This problem has been documented in China, Vietnam, Malaysia and Nepal (FAO, 2007) but occurs in many other countries. In addition, the area of land transferred may be quite limited, precluding the establishment of a commercially viable operation by the community.

Figure 1.15 Public forests in France. A forêt domaniale *is a forest that is owned by the French state and managed under the national forestry law, derived ultimately from the Forest Law of Charlemagne (which King William I introduced to England). A* forêt communale *is a forest owned by a community.*

Community-based forestry

Community-based management or co-management is advocated by many, yet in practice, it has been slow to develop. According to Gilmour (2016), community-based forestry includes 'initiatives, sciences, policies, institutions and processes that are intended to increase the role of local people in governing and managing forest resources' (Gilmour, 2016, p. 2). It includes both formalized customary and Indigenous initiatives and government initiatives. Key characteristics include decentralized and devolved forest management, smallholder forestry schemes, community-company partnerships, small-scale forest-based enterprises and the Indigenous management of sites of cultural importance.

Community-based forestry arose out of a desire among many to see more rights assigned to local people: centralized management has been assumed to be 'bad', whereas local management has been assumed to be 'good'. Such attitudes have frequently been challenged, and many still consider that centralized management by government authorities is the only way that the multiple values associated with forests can be managed effectively, especially given that there are now multiple stakeholders with an interest in any particular forest. Decentralization has led to the creation of a number of different forms of participatory forest management (PFM), including community forest management (CFM), adaptive co-management (ACM) and joint forest management (JFM). There is a bewildering number of names for these attempts to vest more decision-making powers in local people, all related to the precise bundle of rights being allocated to the community. In the majority of cases, the aim has been to replace government responsibilities for managing a forest with community responsibility (Bluffstone *et al.*, 2015). This has been done with varying levels of success, as central and local governments are often unwilling to give up complete control, conferring what amounts to allodial rights to the land. There are also suspicions over the motivations of those assuming control of the forest, whether they be forest user groups or community forest associations.

In theory, transferring the rights to communities should result in better management that is more responsive to the needs of communities. There are many examples where this has indeed been the case. Such cases invariably involve the transfer of a large bundle of rights to the communities. However, community-based forestry seems to be failing to deliver its full potential (Gilmour, 2016), a problem that seems to be related to the rights that are actually transferred. For example, forest management rights might be transferred to a community, but the government retains the right to determine the annual allowable cut. In such cases, a community may be forced to harvest timber against its will.

Gilmour (2016) places this dilemma in terms of locks and keys. The locks are the factors that inhibit the effective implementation of community-based forestry. He argues that the keys are secure tenure, an enabling regulatory framework, strong governance, viable technology, adequate market knowledge and access, and a supportive bureaucratic culture. It is rare to find all of these conditions together. In contrast, FAO (2007) argues that the most positive results occur when the management plan for a community forest builds on an existing traditional structure, when there are clear provisions about rights, responsibilities and decision-making, and when local people have training. Funding also plays an important role in the success of community forests.

In many cases, the limited success of community forests can be directly related to the withholding of some of the rights to the forest by the public body controlling the forests. This has been documented in Orissa, India, in the case of joint forest management (FAO, 2007), and has been a major factor in the success of community forests in British Columbia, Canada. There are numerous other examples, all of which point to the retention of certain rights (such as determining the annual allowable cut) by the agency controlling the forest.

When discussing tenure transfers, Indigenous peoples and local communities are often considered together, although there are often distinct social, cultural and even legal differences. The Rights and Resources Initiative has been monitoring the area of forest owned by or designated for Indigenous peoples and local communities in 33 low- and middle-income countries. Within these countries, indigenous and community ownership has been recognized to 388 million ha, with a further 109 million ha designated with a more limited set of rights (RRI, 2016). The area of land under indigenous and local community control is increasing, and this important trend is one that will affect the nature and process of sustainable forest management.

Transfer of rights to private smallholders

Decentralization can also involve the transfer of rights to individuals. This has occurred recently on a very large scale in China. While it is often referred to as privatization, this is a misinterpretation. The land remains the property of the State, and it is the right to use the land, for a fixed period, that has been transferred to individuals. The rights can be sold or passed on to heirs, and individuals can obtain a mortgage based on their land rights, but there are strict limits on what can be done with the land. State-run centres have been established where the rights can be documented and then sold or exchanged – an attempt to prevent elite capture, although this is still occurring.

In Vietnam, 23% of the land is directly managed by individuals (FAO, 2007). Several different types of tenure exist, although all are restricted to access and land-use rights. The rights are detailed in Red Book Certificates and can be exchanged, leased, inherited or mortgaged.

Another area where large-scale transfer of property rights to smallholders has occurred is Brazil. Here, there has been both recognition of Indigenous rights to land, with over 100 million ha of land being transferred to Indigenous control, including to groups such as the Yanomami and Kayapo, and a transfer of property rights to millions of households that have settled in the Amazon. In the case of Indigenous control, the transfer appears to have resulted in a marked reduction in deforestation, primarily because of the exclusion of agricultural, mining and illegal logging interests (Nepstad *et al.*, 2006; Nolte *et al.*, 2013). Brazil has also granted timber concessions in specific areas, and Azevedo-Ramos *et al.* (2015) suggest that there would be 1.3 million ha of forest under concessions by the end of 2015. The design of these concessions took into account many of the problems that have plagued the design of such concessions globally, including inefficient mechanisms to collect rent, the exclusion of small- and medium-sized enterprises, corruption and a lack of institutional capacity.

Blended cases

As indicated earlier in this chapter, in many parts of the world Indigenous people have significant claims to forest land. Tenure systems that fail to recognize this will inevitably have problems, as witnessed by continuing court cases in Canada. Attempts have been made to blend modern tenure arrangements with traditional arrangements, with varying levels of success. This is because modern tenure systems tend to focus on individual ownership, whereas customary systems are often linked to management in the form of common property (FAO, 2007). Such cases require careful consideration, and often there is a failure on the part of the forest authorities to fully appreciate the nature of the customary ownership of the land and the procedures by which rights are allocated. Such issues, combined with many of the other elements described earlier, have given rise to a series of principles that should be applied during any attempt to reform tenure systems (Table 1.3).

Table 1.3 Principles to be incorporated when designing or reforming tenures.

Principle 1. Adaptive and multi-stakeholder approach	Effective tenure reform requires an adaptive, deliberative, reflective and multi-stakeholder approach
Principle 2. Tenure as part of a wider reform agenda	Forest tenure reform should be implemented as part of a holistic and integrated freeform agenda
Principle 3. Social equity	All aspects of tenure reform should give attention to the empowerment of marginalized groups, particularly women and the poor
Principle 4. Customary rights and systems	Relevant customary tenure systems should be identified, recognized and incorporated into regulatory frameworks
Principle 5. Regulatory framework	The regulatory framework to support policy changes associated with tenure reform should be enabling as well as enforcing
Principle 6. Tenure security	The regulatory framework should include mechanisms for making forest tenure as secure as possible
Principle 7. Compliance procedures	Compliance procedures should be as simple as possible to minimize transaction costs and maximize the regulatory framework's enabling efforts
Principle 8. Minimum standards for forest management	A minimum standards approach should be applied when developing management plans for smallholder or community use
Principle 9. Good governance	Forest governance systems should be transparent, accountable and participatory, including multi-stakeholder decision-making processes
Principle 10. Capacity building	Supportive measures should be in place to ensure that all stakeholders know their rights and responsibilities and have the capacity to exercise them effectively

From FAO (2011).

Structure of the book

As will be described in the following chapter, the many different pressures on the management of forests resulted in some profound changes in forest management in the late 20th century. In particular, the need to codify the different goals of forestry was recognized, and this resulted in the emergence of a series of criteria that helped to define the concept of sustainable forest management. The development of these criteria, and their associated indicators, is described in the following chapter. Subsequent chapters deal with individual criteria, such as the maintenance of biodiversity and the protection of forest soils. Finally, a series of chapters cover how these different criteria can be brought together into a comprehensive approach to forest management, termed sustainable forest management.

Further reading

Colfer, C.J.P., Basnett, B. S., Elias, M. (eds.) 2016. *Gender and Forests: Climate Change, Tenure, Value Chains and Emerging Issues*. London: Earthscan from Routledge.

Perlin, J. 2005. *A Forest Journey: The Story of Wood and Civilization*. Woodstock, VT: Countryman Press.

Sikor, T., Stahl, J. (eds.) 2011. *Forests and People: Property, Governance and Human Rights*. London: Earthscan.

Westoby, J. 1989. *Introduction to World Forestry*. Oxford: Basil Blackwell.

Williams, M. 2003. *Deforesting the Earth: From Prehistory to Global Crisis*. Chicago: University of Chicago Press.

Chapter 2

Criteria and indicators of sustainable forest management

John L. Innes

In the previous chapter, we examined some of the early interactions between humans and forests and how concerns about the supply of forest products led to the development of ideas about forest management. We also looked at how the evolving ideas about forest management were focused very much on the volumetric yield of timber (reflecting their origins in concerns about timber supply), whereas at the same time concerns were growing about environmental degradation in general, often triggered by concerns over deforestation and forest degradation. By 2007, and the *Non-legally Binding Instrument on All Types of Forests* (UN, 2007), these two streams of thought had reunited, with recognition that good forestry practices could play a major role in maintaining the environment and economic and social welfare. How did this happen?

The early concepts developed in central Europe about maintaining a continuous supply of timber formed some of the earliest ideas about sustainability. In such cases, it was strictly related to the development of a sustainable supply of timber. Forests were managed on an economic basis, with no concern about other values, such as biodiversity. This reflects the views of Heinrich Cotta that forests should be managed as crops: in agriculture, the crops were also seen purely from the point of view of the products they produced.

Such ideas were inconsistent with the views that forests could serve multiple purposes. In the second half of the 20th century, there was increasing recognition that forests could provide many different goods and services if managed well. This view has evolved over time, and today it is widely recognized that forests can supply these multiple goods and services, provided that a landscape view is adopted. There are however some differences of opinion here. Some believe that an individual stand cannot provided all goods and services, and instead a landscape viewpoint should be adopted. Others argue the opposite, insisting that every forest area can meet all needs. As in many forest-related questions the answer probably lies somewhere in between, and might best be couched as 'It depends . . . .'

Whichever spatial scale is being considered, forest managers still need to have an idea of why they are managing an area of forest. The growing environmental concerns about forest practices in the 1970s and 1980s led to attempts to define what the goals of forest management should be, with such definitions invariably being linked to sustainability, or sustainable forest management. One of the first international organizations to work in this area was the International Tropical Timber Organization.

The principles of sustainable forest management developed by the International Tropical Timber Organization (ITTO)

As a result of growing concerns about maintaining supplies of high quality timber, and potential threats of a boycott among consumer countries of wood derived from tropical forests, the International Tropical Timber Organization (ITTO) began to look at ways in which forests might be managed more sustainably. Some of this early work recognized that the whole idea of sustainable management depended on what value was being sustained (Poore, 1989). While the emphasis remained on production of timber, the importance of other goods and services provided by forests was recognized. An unpublished report commissioned by ITTO in 1988 concluded that 'the extent of tropical moist forest which is being deliberately managed at an operation scale for the sustainable production of timber is, on a world scale, negligible'. It went on to state that 'progress in establishing stable sustainable systems is still so slow that it is having very little impact on the general decline in quantity and quality of the forest' and 'the future existence of large areas of tropical forest, perhaps even the majority, and of highly significant ancillary goods and service of the forest, depends equally on the establishment of sustainable systems of management, many of which must have timber production as their basis' (cited by Poore, 2003).

Poore (1989) recognized many of the issues that subsequently found their way into the definition of sustainable forest management. It is worth quoting from his introductory chapter as his concerns bring together the issues that have exercised forest managers all over the world.

Figure 2.1 Deforestation and forest degradation near Bogor, Indonesia. The lack of sustainable management of tropical forests remains a particular concern of ITTO.

> The widespread concern about tropical forests is based on a number of issues: that these forests are disappearing at an alarming rate; that the loss of so much forest has potentially disastrous environmental effects – on soil, water, climate, the genetic richness of the globe and the supply of future possible forest products; that the uses to which the land is being converted are often not sustainable – that the forest in fact is being destroyed for no ultimate benefit, and that forest-dwelling peoples are being arbitrarily displaced.
>
> (Poore, 1989, p. 1)

As a result of the work done by Duncan Poore, ITTO was one of the first organizations to recognize the need for guidelines on the sustainable management of forests. In the first of its publications, *ITTO Guidelines for the Sustainable Management of Natural Tropical Forests*, 41 principles for the sustainable management of tropical forests for timber production were laid out, as well as some possible actions under these principles (ITTO, 1992a). At this time, the emphasis was still very much on timber production, but the principles clearly recognize the need to consider a broad range of factors if the timber production is to be sustainable. The 41 principles were very general, and there was recognition that these would need to be modified for local application. There was also a recognition by Duncan Poore that these principles would have little effect if there was no system of reporting, and that this reporting would need to have a sound baseline and regular evaluation. This however, proved to be too big a step, and even today no country has provided a comprehensive report covering all the areas proposed by Poore.

At the same time as these principles were being formulated, ITTO was also developing ways in which the sustainable management of forests could be defined and measured. ITTO commissioned a discussion paper from Duncan Poore and Mok Sian Tuan and then convened a panel of experts in The Hague in September 1991 to work on both a definition of sustainable forest management and a set of criteria and indicators by which it might be measured. Following more discussions and recommendations from a further panel of experts, ITTO published its *Criteria for the Measurement of Sustainable Tropical Forest Management* (ITTO, 1992b). This publication provided a revised definition of sustainable forest management and a provisional list of criteria and indicators for use at both the national scale and the scale of the 'forest management unit'. It was aimed at helping producer countries meet ITTO's Target 2000 – namely to ensure that all trade in tropical timber was sourced from sustainably managed forests by the year 2000. At this point, there was no clear definition of a criterion or indicator – it was simply taken for granted that people would understand these terms. The first national-level criterion was the 'forest resource base', characterized by several indicators such as the area of protection and production forests within the permanent forest estate. The remaining criteria were the continuity of flow, the level of environmental control, socio-economic benefits and institutional frameworks. At the scale of the forest management unit, the criteria were less descriptive, with one being 'An acceptable level of environmental impact', and an indicator for this criterion was the extent of soil disturbance. The remaining criteria were resource security, the continuity of timber production, the conservation of flora and fauna, socio-economic benefits, and planning and adjustment to experience.

The criteria and indicators presented in 1992 were subsequently revised in 1998. By the time of this revision, a number of other criteria and indicator schemes had been developed to help define and measure sustainable forest management, and ITTO was able to draw upon these. The resulting criteria (Table 2.1) thus show a number of similarities to the other schemes developed elsewhere (ITTO, 1998). The revision acknowledged that the first version of the criteria and indicators had focused on the production of timber, and stressed that the 1998 revision was intended to cover the full range of forest goods and services, including non-timber values such as biological diversity. As in the first document, indicators were proposed for both the national level and the level of the forest management unit.

ITTO (1998) also clearly specified the nature and purpose of criteria and indicators. It argued that a criterion 'describes a state or situation which should be met to comply with sustainable forest management', whereas indicators are designed such that 'a change in any one of them would give information that is both necessary and significant in assessing progress towards sustainable forest management' (ITTO, 1998, p. 3).

The 1998 criteria and indicators presented by ITTO were accompanied by two manuals on their implementation (ITTO, 1999a, b), dealing with implementation at the national level and implementation at the scale of the forest management unit. The manuals were intended to give countries the tools that they needed to measure progress towards sustainable forest management, and through this, progress towards meeting Target 2000.

ITTO continued to work on criteria and indicators of sustainable forest management, and in 2003 worked with the African Timber Organization to combine two parallel sets of criteria and indicators applicable to tropical forests in Africa (ITTO, 2003). This represents one of the few successful attempts to merge independently

Table 2.1 Criteria of sustainable forest management as published by ITTO in 1998.

1.	Enabling Conditions for Sustainable Forest Management	This consists of the general institutional requirements for sustainable forest management to succeed. It addresses the policy, legislation, economic conditions, incentives, research, education, training and mechanisms for consultation and participation. Many of the indicators are descriptive. Taken together, the information gathered indicates the extent of a country's political commitment to sustainable forest management.
2.	Forest Resource Security	The extent to which a country has a secure and stable forest estate, which could include plantations, to meet the production, protection, biodiversity conservation and other social, cultural, economic and environmental needs of present and future generations.
3.	Forest Ecosystem Health and Condition	The condition of a country's forests and the healthy biological functioning of forest ecosystems. Forest condition and health can be affected by a variety of anthropogenic actions and natural occurrences, from air pollution, fire, flooding and storms to insects and disease.
4.	Flow of Forest Produce	Forest management for the production of wood and non-wood forest products. This can only be sustained in the long term if it is economically and financially viable, environmentally sound and socially acceptable. Forests earmarked for production should be capable of fulfilling a number of other important forest functions, such as environmental protection and the conservation of biological diversity. These multiple roles of forest should be safeguarded by the application of sound management practices that maintain the potential of the forest resource to yield the full range of benefits to society.
5.	Biological Diversity	The conservation and maintenance of biological diversity, including ecosystem, species and genetic diversity. At the species level, special attention should be given to the protection of endangered, rare and threatened species. The establishment and management of a geographic system of protected areas of representative forest ecosystems can contribute to maintaining biodiversity. Biological diversity can also be conserved in forests managed for other purposes, such as for production, through the application of appropriate management practices.
6.	Soil and Water	The importance of the protection of soil and water in the forest is twofold. It has a bearing on maintaining the productivity and quality of forest and related aquatic ecosystems (and therefore on the health and condition of the forest, Criterion 3), and it also plays a crucial role outside the forest in maintaining downstream water quality and flow and in reducing flooding and sedimentation. The environmental and social effects of mismanagement (landslides, flooding, water pollution) can be very significant.
7.	Economic, Social and Cultural Aspects	The economic, social and cultural aspects of the forest, besides those mentioned under Criteria 4, 5 and 6, are included here. The forest should have the potential, if sustainably managed, to make an important contribution to the sustainable development of the country.

Adapted from ITTO (1998).

developed sets of criteria and indicators. However, comprising 1 principle, 5 criteria, 33 indicators and 45 sub-indicators at the national level, and 3 principles, 15 criteria, 57 indicators and 140 sub-indicators at the forest management unit (FMU) level, the ITTO system typifies one of the problems associated with such indicator sets: a large number of indicators and sub-indicators are specified with no indication of how these will actually be measured in countries where access to finances and technologies is severely limited.

The difficulties associated with the proliferation of indicators led to some serious rethinking of ITTO's proposed criteria and indicators, and in 2005 it made further revisions to its lists (ITTO, 2005). The revisions took into account the many developments in the use of criteria and indicators around the world, and the experience gained by ITTO in the training of numerous parties in the use of criteria and indicators (C&I). While there are always problems associated with the revision of indicators that are designed to measure change over time, if carefully done revisions can increase the value of a monitoring system. This was recognized by ITTO, and ITTO (2005) stated:

> The ITTO C&I should continue to be reviewed and refined to benefit from experience and to reflect new concepts of sustainable forest management. Revision should take into account evolving knowledge about the functioning of forest ecosystems, human impacts on forests, whether planned or unplanned, and the changing needs of society for forest goods and services. Moreover, the capacity to measure indicators will increase and knowledge will improve about the nature of the 'best' indicators with which to assess, monitor and report on forest management.
>
> (ITTO, 2005, p. 8)

While similar to the criteria used in ITTO (1998) (see Table 2.1), the revised criteria more closely fitted 'common thematic areas' of sustainable forest management that had

Figure 2.2 The merging of the ITTO criteria with those developed in Africa means that the full range of tropical forests is included, from humid to semi-arid. These heavily grazed Acacia woodlands in northern Kenya would now be included.

been identified during conferences of experts in 2002 and 2004, organized jointly by ITTO and the Food and Agriculture Organization of the United Nations (FAO). These common thematic areas are:

- Extent of forest resources;
- Biological diversity;
- Forest health and vitality;
- Production functions of forest resources;
- Protective functions of forest resources;
- Socio-economic functions;
- Legal, policy and institutional framework.

Most recently, there have been several changes in ITTO's approach (ITTO, 2015), and four objectives and seven principles are recognized in the voluntary guidelines for the sustainable management of natural tropical forests (Table 2.2). The seven principles are accompanied by 60 guidelines, each of which has up to eight suggested actions. Some of these are quite specific; others such as 'Ensure that forest management plans have provisions for biodiversity monitoring and that managers understand and are responsive to the outcomes of such monitoring' will encounter the types of problems described in Chapters 3 and 15.

A full account of the history of ITTO and its role in promoting the sustainable management of tropical forests can be found in Poore (2003).

Table 2.2 Objectives and principles for the sustainable management of natural tropical forests, and their relationship to ITTO's criteria and indicators.

Objective	Principle	C&I for SFM
Providing the enabling conditions of SFM	Principle 1: Forest governance and security of tenure	Criterion 1: Enabling conditions for SFM
	Principle 2: Land-use planning, permanent forest estate and forest management planning	Criterion 1: Enabling conditions for SFM
		Criterion 2: Extent and condition of forests
Ensuring forest ecosystem health and vitality	Principle 3: Ecological resilience, ecosystem health and climate-change adaptation	Criterion 3: Forest ecosystem health
Maintaining the multiple functions of forests to deliver products and environmental services	Principle 4: Multipurpose forest management	Criterion 4: Forest production
	Principle 5: Silvicultural management	Criterion 5: Biodiversity
		Criterion 6: Soil and water protection
Integrating social, cultural and economic aspects to implement SFM	Principle 6: Social values, community involvement and forest-worker safety and health	Criterion 7: Economic, social and cultural aspects
	Principle 7: Investment in natural forest management and economic instruments	

The Montreal Process

At the 1992 United Nations Conference on Environment and Development (UNCED), held in Rio de Janeiro, Brazil, discussions about forests were particularly protracted. Conference discussions eventually led to the production of a document called the *Non-Legally Binding Authoritative Statement of Principles for a Global Consensus on the Management, Conservation and Sustainable Development of All Types of Forests*, otherwise known as the Forest Principles. This document described in broad terms the principles by which forests should be managed, but had no legal force.

The Montreal Process originated in a meeting in Montreal in September/October 1993 organized by the CSCE Seminar of Experts on Sustainable Development of Temperate and Boreal Forests. At this meeting, the concept of developing criteria and indicators of sustainable forest management was discussed, using ITTO's documents as an example and with strong encouragement from Duncan Poore, who had played such a large part in developing the criteria and indicators for the ITTO. Another influential document at the meeting was a US Forest Service report titled *Ecosystem Management: Principles and Applications*. This report was one of the outcomes of the *Eastside Forest Ecosystem Health Assessment*, published by the US Forest Service in 1994. A draft of the chapter ('An Overview of Ecological Principles for Ecosystem Management') by P. Bourgeron and M. E. Jensen, who were both at the meeting, was available and was used heavily in the formulation of the meeting documents.

The meeting divided into two main parts, one looking at the physical and biological aspects of forest management, and the other dealing with the socio-economic and cultural aspects. By the end of the meeting, criteria and indicators for the physical and biological aspects had been identified and agreed upon, but the socio-economic group failed to reach any level of significant agreement. This problem has persisted throughout the history of the Montreal Process, and to this day the criteria and indicators for the socio-economic and cultural aspects of forest management remain quite unsatisfactory.

In December 1993, in a meeting attended by representatives from Canada, the USA, Japan, Russia, the UK, Germany and Finland, an agreement was reached to continue the development of criteria and indicators of sustainable forest management in at least two ways: in Europe, through the European Ministerial Conference on Forests (termed at the time the Helsinki Process), and informally with the non-European countries (Brand, 1997). The Helsinki Process drew up a list of criteria and indicators which was agreed upon in June 1994, with the list largely based on the environmental criteria developed at the previous year's Montreal meeting, with the addition of a sociological criterion. The criteria and indicators were accompanied by Pan-European Operational Guidelines for Sustainable Forest Management aimed at guiding operations at the management unit scale. At the Third Ministerial Conference on the Protection of Forests in Europe in Lisbon in 1998, the final Pan-European Criteria, Indicators and Operational Guidelines for Sustainable Forest Management were approved (Resolution L2). After the Montreal meeting, the non-European temperate and boreal countries agreed to continue in an informal manner, and a group of negotiators began the task of drawing up a list of criteria and indicators of sustainable forest management. This list, which drew heavily on the report of the 1993 meeting, was finalized quite quickly and was published in February 1995 in the form of the Santiago Declaration (see http://www.montrealprocess.org/documents/publications/techreports/1995santiago_e.pdf). It contained five environmental criteria, a socio-economic criterion (as in the European criteria) and a seventh criterion dealing with the legal, institutional and economic

*Figure 2.3
Badly damaged
forest in the
Boonoo Boonoo
area of New
South Wales,
Australia.
Deforestation
and forest
degradation is
not restricted
to tropical
areas; it is also
widespread in
temperate and
boreal forests,
and C&I
schemes such
as the Montreal
Process are
intended to
measure progress
towards the
sustainable
management
of these forests.*

framework for forest conservation and sustainable management. Since then the Working Group has continued to meet, and the criteria and indicators have continued to evolve. The Montreal Process agreement concerns the sustainable management of forests in a range of temperate boreal countries, currently Argentina, Australia, Canada, Chile, China, Japan, Republic of Korea, Mexico, New Zealand, Russian Federation, the USA and Uruguay. The forests in these 12 countries represent 60% of the world's forests and 90% of the world's temperate and boreal forests.

The most recent (2015) edition of the Montreal Process criteria and indicators identifies seven criteria:

- Conservation of biological diversity;
- Maintenance of productive capacity of forest ecosystems;
- Maintenance of forest ecosystem health and vitality;
- Conservation and maintenance of soil and water resources;
- Maintenance of forest contribution to global carbon cycles;
- Maintenance and enhancement of long-term multiple socio-economic benefits;
- Legal, institutional and economic framework for forest conservation and sustainable management.

Around the world, similar processes were occurring at the same time as ITTO and the Montreal Process were developing their criteria and indicators. A vast number of indicators have been proposed, and many of these have been listed online (see http://www. sfmindicators.org; a Chinese version is available at http://www.sfmindicators.cn). This website was initially set up as an interactive site that would enable anyone to add new indicators, but following the abuse of this policy, it is now a read-only site.

Other criteria and indicator initiatives

In 2000, Castañeda reported that there were nine major processes dealing with criteria and indicators. These were:

- Pan-European Forest Process on Criteria and Indicators for Sustainable Forest Management;
- Montreal Process on Criteria and Indicators for the Conservation and Sustainable Management of Temperate and Boreal Forests;
- Tarapoto Proposal for Criteria and Indicators for Sustainability of the Amazon Forest;
- Dry Zone Africa Process;
- Near East Process;
- Lepaterique Process of Central America;
- ITTO Initiative on Criteria and Indicators;
- African Timber Organization;
- Regional Initiative for Dry Forests in Asia.

(Castañeda, 2000)

As indicated earlier, the African Timber Organization and ITTO merged their schemes in 2005. Of the remainder, enthusiasm for their use has varied. However, even if they have not been fully implemented in national reporting schemes, they have resulted in the development of a general consensus over the definition and characteristics of sustainable forest management (McDonald and Lane, 2005).

Political realities

Criteria and indicators as a concept has long been viewed as both important and useful. It is in the forest sector's best interest to measure progress towards achieving sustainable forest management, a goal that applies to all managed forests. In 1997, the Intergovernmental Panel on Forests endorsed the concept, urging countries 'to integrate suitable criteria and indicators for sustainable forest management, as appropriate, into the overall process of the formulation, implementation, monitoring and evaluation of national forest programmes' (Commission on Sustainable Development, 1997, p. 7). As a result, the concept was pursued, with regional politics resulting in the development of parallel processes around the world. When assessed in 2000, 150 countries appeared to have been involved in the development of criteria and indicators in some form.

In 2011, only six of the nine international sets of criteria and indicators were still in use (Grainger, 2012), and by 2014, it was evident that a lot less effort was being expended on criteria and indicators of sustainable forest management than earlier. In particular, the level of reporting was extremely variable, and many indicators that had been suggested had been dropped. Even within the Montreal Process countries, there was considerable variation in the level of reporting, and the quality and content of the few reports that had been published was extremely variable (Chandran and Innes, 2014). Countries were inconsistent in the indicators they chose to report, and with those indicators they frequently used differing methods to derive the data, making comparison almost impossible.

Criteria and indicators were designed specifically as monitoring tools to measure progress towards the achievement of sustainable forest management. This goal has only been partially fulfilled, but this does not mean that the concept has been a failure. In fact, it has achieved remarkable success in bringing into focus what is meant by sustainable forest management, and what components are a part of the management of forests. As a result, the criteria, which are broadly similar across the schemes, have been incorporated into a number of national and international forest policies and regulations (Rametsteiner and Mayer, 2004). Caswell (2014) argues that criteria and indicators are widely used for a variety of purposes, including monitoring specific projects, regional reporting, certification and other purposes. She found that important uses included developing a global understanding of sustainable forest management, increasing awareness and appreciation of non-timber forest benefits and values, improving and expanding forest monitoring and assessment, developing management plans and standards, communicating trends in forest conditions to policy-makers and the public, communicating and engaging with stakeholders, and improving forest databases and inventories. The concept has also been adapted to meet local needs (e.g. Hickey and Innes, 2008), with varying levels of effectiveness. An important adoption has been by certification standards: all the important standards are based on one or more of the criteria and indicator systems. Since these certification standards determine whether a management unit can be independently certified as being managed sustainably, criteria and indicators are playing a major role in the day-to-day management of forests around the world.

It is worth noting that while there has been relatively good agreement internationally on the meaning and content of the criteria, one criterion has proven to be stubbornly resistant to treatment, namely the maintenance and enhancement of long-term multiple

*Figure 2.4
Social and
cultural values
of forests have
not been handled
well in criteria
and indicator
schemes. The
photo shows
a canopy walk
in the research
forest of the
Experimental
Forest, College
of Bio-Resources
and Agriculture,
National Taiwan
University. Such
raised walks
have become
increasingly
popular around
the world.*

socio-economic benefits (Gough *et al.*, 2008). Twenty-one years after the initial Montreal Process meeting, there are still major concerns over this criterion, and it contains many noticeable gaps, such as the absence of any indicators dealing with the aesthetic values of forests. There are a number of reasons for this, but one of the most important is that the numerous expert panels that have been convened to define criteria and indicators worldwide have been remarkably short on socio-economic expertise, reflecting the traditional training of foresters in the natural sciences.

Despite the problems associated with the reporting of indicators, the criteria that have been developed still represent an attractive way to examine the various facets of sustainable forest management, and they form the basis of the different chapters in this book. The indicators also provide some interesting guides as to the types of information that need to be considered when assessing progress towards achievement of the criteria. As will be shown, it may be very difficult to gather some of this information, but good forest managers should be aware of them when making management decisions. For example, while it is almost impossible to measure genetic diversity, it *is* possible to avoid taking actions that are likely to reduce genetic diversity. In fact, achieving the conservation of biological diversity is one of the biggest challenges facing forest managers, and this is examined in the following chapter.

While criteria and indicators are useful in defining the scope of sustainable forest management, for framing forest policy at a national scale and for preparing forest management plans at the scale of a forest management unit, they are generally missing a critical component. The indicators are designed to measure progress towards sustainable forest management, but no scheme so far has gone so far as to specify target values for a particular indicator. When planning the management of a forest area, having such targets is critically important. However, they are extremely difficult to set, and there is a general recognition that setting a target for one indicator will have an impact on what can be achieved with other indicators. Consequently, defining such targets has become a complex process, involving not only the incorporation of state-of-the-art scientific knowledge but also stakeholder consultation and complex modelling to optimize the potential solutions (Mendoza and Prabhu, 2003). Some of the techniques that can be used to do this are addressed in Chapter 16.

Conclusions

There is general agreement in the forest sector and beyond that we must manage the world's forests sustainably. This agreement was formalized in 1992 at the World Conference on Environment and Development in the form of a set of non-legally binding principles of forest management (see Chapter 12). Since then, great effort has gone into codifying the nature of forest management. Criteria and indicators have become a widely accepted means of doing this. The criteria and indicator sets that have been developed are applicable at a range of scales (national, regional, forest management unit), although not every criterion or indicator is applicable at every scale. In practice, these different sets are actually quite similar, and convergence continues to occur as the sets are refined over time. While any of the sets of criteria and indicators could have been used as a framework for this book, the Montreal Process set has been chosen because its geographic coverage is so broad (its member countries represent 80% of the temperate and boreal forests of the world), and because the criteria are equally applicable to tropical forests. While the indicators are mentioned, little emphasis is given to them in this book. They provide some guidance for the type of phenomena that managers should consider, but the appropriateness of indicators will vary from forest to forest and situation to situation. There is a substantial literature on the selection of indicators, and considerable care is needed when developing them.

Further reading

ITTO 1999a. *Manual for the Application of Criteria and Indicators for Sustainable Management of Natural Tropical Forests. Part A: National Indicators.* ITTO Policy Development Series No. 9. Yokohama: International Tropical Timber Organization. Available at: http://www. itto.int/policypapers_guidelines/

ITTO 1999b. *Manual for the Application of Criteria and Indicators for Sustainable Management of Natural Tropical Forests. Part B: Forest Management Unit Indicators.* ITTO Policy Development Series No. 10. Yokohama: International Tropical Timber Organization. Available at: http://www.itto.int/policypapers_guidelines/

ITTO 2005. *Revised ITTO Criteria and Indicators for the Sustainable Management of Tropical Forests Including Reporting Format.* ITTO Policy Development Series No. 15. Yokohama: International Tropical Timber Organization. Available at: http://www.itto.int/policy papers_guidelines/

Poore, D. 2003. *Changing Landscapes: The Development of the International Tropical Timber Organization and Its Influence on Tropical Forest Management.* London: Earthscan Publications.

Chapter 3
Forest biodiversity

Fred Bunnell

The first criterion described by the Montreal Process is the conservation of biological diversity. Virtually all criteria and indicator schemes emphasize the importance of conserving biodiversity, and provide a number of indicators that might be used. The indicators listed in the Montreal Process are provided in Table 3.1, but even a cursory look at these indicators reveals that their monitoring presents major problems for a forest manager. Within the context of ecosystem diversity, area and percent of forest by ecological and socio-economic classifications is relatively easy to assess, but as discussed later in the chapter, fragmentation is a much more difficult concept. Both species diversity and genetic diversity are required, yet there is no forest in the world where all species present have been listed. In fact, most invertebrates remain undescribed, so listing them is impossible. When many of the species have yet to be documented, it makes little sense to ask that genetic diversity within them should be documented. However, considerable advances have been made in genomics in recent years (see, for example, Allendorf *et al.*, 2011), and future indicators may take this new knowledge into account.

The concept of biodiversity itself is vague: is biological diversity synonymous with biodiversity, and how does it relate to terms such as 'wildlife'? This chapter starts off with an examination of these questions.

Defining biodiversity

The term 'biological diversity' was first used in a popular book advocating conservation by Dasmann (1968). 'Natural diversity' was more commonly used among the scientific community until Thomas Lovejoy used biological diversity in his introduction to Soulé and Wilcox's book, *Conservation Biology* (1980). Wilson (1988) credits Dr W. G. Rosen with coining the term 'biodiversity', as a synonym for biological diversity, during planning for the 1986 National Forum on Biodiversity. The United Nations Conference on Environment and Development (UNCED '92 or Earth Summit) gave the term so much visibility that within three years there were at least 85 published definitions for biological diversity and biodiversity. Many of these were crafted by governments and international agencies and intended to provide guidance to management actions.

Table 3.1 Indicators associated with the criterion 'Conservation of biological diversity'.

1.1	Ecosystem Diversity	1.1a	Area and percent of forest by forest ecosystem type, successional stage, age class, and forest ownership or tenure
		1.1b	Area and percent of forest in protected areas by forest ecosystem type, and by age class or successional stage
		1.1c	Fragmentation of forests
1.2	Species Diversity	1.2a	Number of native forest-associated species
		1.2b	Number and status of native forest-associated species at risk, as determined by legislation or scientific assessment
		1.2c	Status of onsite and offsite efforts focused on conservation of species diversity
1.3	Genetic Diversity	1.3a	Number and geographic distribution of forest-associated species at risk of losing genetic variation and locally adapted genotypes
		1.3b	Population levels of selected representative forest-associated species to describe genetic diversity
		1.3c	Status of onsite and offsite efforts focused on conservation of genetic diversity

Because international agreements emerging from UNCED '92 committed land managers to sustaining biodiversity, managers and researchers sought guidance from the definitions, initially with little success. Some authors observed that science had played only a minor role in the maintenance of biological diversity (Weston, 1992) or that there was no agreed-upon approach for evaluating management strategies and approaches to maintain biodiversity (e.g. Murphy, 1990). Workers noted that the term biodiversity was 'little more than a brilliant piece of wordsmithing' (Bowman, 1993) or 'an environmental fashion statement' (Merriam, 1998). Others wrote that the term is 'clearly defined' (Lyons and Scott, 1994), or that they 'find it shocking that we are still trying to define biological diversity after all the efforts of the Office of Technological Assessment and E. O. Wilson's book, *Biodiversity*' (Soulé, see Anonymous in Hudson, 1991).

The literature of the 1990s reveals that much of the difference in views of the utility of the 85 or so definitions was associated with having, or not having, direct responsibility for taking action. Those writers most familiar with management actions frequently noted that despite the many definitions of biodiversity, the term offered few measurable goals or objectives (e.g. Salwasser, 1988). Those less directly involved with management activities often believed the term sufficiently well-defined to guide actions (e.g. Soulé, 1985).

The Convention on Biological Diversity and associated documents obligated foresters to sustain biodiversity, but appeared to provide little help in defining biodiversity. The Convention states: 'Biological diversity means the variability among living organisms from all sources including, inter alia, terrestrial, marine and other aquatic ecosystems and the ecological complexes of which they are part; this includes diversity within species, between species and of ecosystems' (United Nations, 1992b). It distinguishes biodiversity from biological resources by noting that the latter 'includes genetic resources, organisms or parts thereof, populations, or any other biotic component of ecosystems with actual or potential use or value for humanity'.

Defining biodiversity operationally

There were two major challenges to creating an operational definition of biodiversity that could guide management. First, the pace with which biodiversity moved from a new concept to incorporation into international agreements was uncommonly rapid. Second, emphasis on variability eliminated simple targets and suffers from the problem of reification: the treatment of an abstract idea as if it were a thing. Both challenges hindered creation of operational goals and created a form of 'paralysis by complexity' (Bunnell, 1998a).

Biodiversity is not a thing but a 'concept cluster' (West, 1994). The welter of definitions following UNCED in 1992 revealed five features of the concept of biodiversity that encouraged paralysis by complexity among practitioners. These merit summary because they have not gone away (see Bunnell and Kremsater, 1994, and Bunnell, 1998a, from which the following is summarized).

1   *Components of the target are numerous and ill-defined or 'invisible', even as separate entities, let alone when combined.* In tropical, temperate, and boreal regions, forests are the richest terrestrial ecosystems (Wilson, 1988). Some processes within forests are defined as much by methodology as by our sensory apparatus. Nitrification, for example, proceeds through at least four known modes, none readily visible. Most genetic variation also is not readily visible, and thus susceptible to 'secret extinction' (Greig, 1979). Moreover, at any given time a complete genetic inventory would provide only a static 'snapshot' of processes know to be dynamic.

2   *The spatial boundaries of an appropriate target are ill-defined.* The term biological diversity emphasizes living organisms, rather than fixed units of space. Although they have an asymptote, species area curves are continuous, with larger areas containing more species. That observation creates two scale problems: (1) there rarely are tidy boundaries to guide planning efforts, and (2) legally defined planning units such as forest tenures, even when large, are unlikely to represent discrete units of natural variation.

3   *The temporal boundaries of an appropriate target are ill-defined.* Both the specific species present and their relative abundance in an area change naturally with time. That is especially obvious in forested systems where successional trends create habitat for a changing array of species. Different species are naturally present at different times.

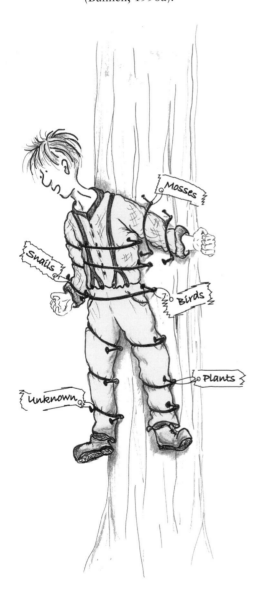

Figure 3.1
Paralysis came readily. (Figure prepared by I. Houde.)

4 *Entities and processes that constitute biodiversity span an enormous range of scale in both time and space and interact.* Forest practices alter forests from the scale of a treatment unit (10^0–10^2 ha) to a planning unit (10^3–10^6 ha). Species contributing to biodiversity, from cryptogams and canopy insects to migratory ungulates and wide-ranging carnivores, exploit space over the same range. Time scales are equally variable: demographic events, natural succession, natural disturbance regimes, and genetic change all modify biodiversity, but each follows its own schedule. Moreover, rare events such as long-distance dispersal appear to be very important in the structure of communities. Interaction occurs across all scales as forest practices and natural succession alter the spatial structure of the forest to which species are responding by habitat selection, survival, and dispersal.

5 *The necessary equation of diversity with variation or variability creates potential barriers to definition.* First, except for cases of parthenogenesis and identical twinning, virtually no two members of the same species are genetically identical due to the high levels of genetic polymorphism across many of the gene loci (Selander, 1976). The degree of difference creating variability is necessarily arbitrary. Second, specification of variability represents a departure from traditional approaches to classification and inventory. Third, explicit emphasis on variation implies both an acceptable range and a moving target.

In short, focus on the scope and complexity of elements in definitions of biodiversity implies targets that are not readily visible, with ill-defined boundaries in time and space that result from sets of constantly shifting processes. It was this condition that led Bunnell (1998b, p. 822) to write: 'We are proceeding at great speed over difficult terrain, towards an unknown, unquantifiable goal'. It also led to initial paralysis or hesitancy in management efforts because no clearly defined target was apparent. Bunnell (1998c) reviewed existing definitions and barriers to effective definition and concluded that (1) effective management cannot ignore the challenges to definition but profits from separating the processes from species richness, and (2) the simplest and most meaningful operational definition of biodiversity is the number of species and subspecies, or species richness. That approach has become widely adopted. Reasons why species richness is a useful measure are clarified in the following section.

Why biodiversity is important

UNCED '92 was a pivotal point in our concerns about biodiversity. Going into the conference at Rio de Janeiro, there was concern about climate change, but most attention was focused on biodiversity and loss of species. Governments rarely sign international agreements without compelling public pressure, yet five significant international agreements emerged from UNCED '92. Species loss has been increasing (Secretariat of the Convention on Biological Diversity, 2010), but an obvious question is why were so many people concerned about species loss? There are two broadly different answers; both exemplify why biodiversity eludes simple definition.

The first answer is a diverse collection of subjective reasons, including moral and aesthetic, that devolve to many of us thinking that it is wrong to preside over the loss of individual species. These reasons may not have any scientific underpinning, but are strongly held. Scientific underpinnings include the genetic foundation on which species diversity depends and the fact that we often cannot distinguish keystone species from other species until they are gone. There are two broad metaphors for species loss: airplane passengers that are simply along for the ride (Walker, 1992), and rivets

*Figure 3.2
American
beaver (*Castor
canadensis*), an
example of a
keystone species
because of its
impact on treed
landscapes.*

holding the plane together so that the loss of a rivet is critical to the flight and the integrity of the airplane (Ehrlich and Ehrlich, 1981). We call the latter 'keystone species' because they largely determine the existence, function and contributions of ecosystems. Some keystone species are obvious – without beavers (*Castor canadensis*) there would be far fewer wetlands and species reliant on wetlands. Some are less obvious – as cougars (*Puma concolor*) and wolves (*Canis lupus* ssp. *lycaon*) disappeared from forests of eastern North America, so have shrubs and shrub-nesting birds, while white-tailed deer (*Odocoileus virginianus*) have burgeoned (McShea *et al.*, 1997). We almost certainly do not need all species; our problem is that species do not come neatly labelled as passengers or rivets. That is exposed only when they are gone.

The second answer is more profound and now captured in the mundane term 'ecosystem goods and services' from the Millennium Ecosystem Assessment Reports of the United Nations Environment Programme. These goods and services include fish, pollination, flood regulation, water purification provided by wetlands and timber, air quality improvement, and sustenance of biodiversity by forests. They are actually gifts. Species and ecosystems charge us nothing for providing them. We appear to be the only species on Earth that is not fully employed at doing something useful for other species. Other species and ecosystems, however, provide our most basic needs. Consider our most basic needs: the oxygen we breathe (about 50% comes from land plants and 50% from algae in the ocean); the water we drink (potable water is dependent on filtration by organisms and soil) and rainfall in many areas is directly tied to the presence of forests; the soil that filters water and grows crops (soil is not just eroded rock but is built by organisms); the food we eat (all was once a living organism); the clothes we wear (we recognize cotton, linen, rayon, and wool as plant or animal products, but even those derived from petroleum were also once alive); the fuels we use to do so much (wood, peat, coal, oil, gas – all were once alive). We have become efficient at digging up and modifying the remains of the dead, but they were once living organisms; only nuclear fuel derives from something not once living. We are wholly dependent on nature for our most basic needs. Forests are a renewable, sustainable source of those needs.

What is less clear is that these gifts are dependent on genetic variation. That dependence is recognized explicitly in the international agreements to emerge from UNCED '92. Five significant international agreements emerged from UNCED '92: the *Convention on Biological Diversity*, *Agenda 21*, *Guiding Principles on Forests*, *UN Convention to Combat Desertification*, and the *UN Framework Convention on Climate Change*. The

Figure 3.3
*The common muskrat (*Ondatra zibethicus*) is similar to the beaver in its habitat requirements, and will share the beaver's lodge, but is not considered to be a keystone species as it is much smaller, does not build dams, and has less impact on the landscape and other species.*

term 'guiding principles' reflects that document's complete title: *Non-legally Binding Authoritative Statement of Principles for Global Consensus on the Management, Conservation, and Sustainable Development of All Types of Forests.* The *Convention on Biological Diversity* addresses biodiversity directly; both *Agenda 21* and *Guiding Principles on Forests* include the conservation of biological diversity as goals. Climate change is increasingly being recognized as a threat to biodiversity. Careful reading of the documents reveals that the public concerns or fears propelling creation of the agreements are repeatedly addressed in different documents. They were fears about loss and include:

- Loss of species
- Loss of productivity (e.g. soil degradation)
- Loss of present and future options
- Loss of economic opportunities
- Lack of local participation and influence in decision-making.

Remarkably, all five major fears or concerns raised at UNCED are closely related in two significant ways. First, each stresses the enormous significance that nature has in our lives, from ameliorating potential site degradation to providing future options and economic opportunities. The request to participate in decision-making affirms our wish to have a say about the future of nature and us. Second, loss of productivity, present and future options, and economic opportunities are each related to loss of species (Figure 3.4).

When we sustain the variety of species and their populations, we also sustain the only renewable, self-replicating parts of nature and genetic variation. Retaining a variety of individuals and species permits the genetic adaptability necessary to respond to changing environments, such as those created by global warming. The capability to respond to changing environments helps sustain future productivity, which in turn facilitates future economic opportunities. Too few of us have learned that the economy is a subsystem of the biosphere, rather than the other way around. Only variety can beget new combinations of variety that can respond to changing futures and thus help meet present and future options. A species does not have to disappear to reduce diversity; as populations of a species become smaller, genetic diversity within the population declines.

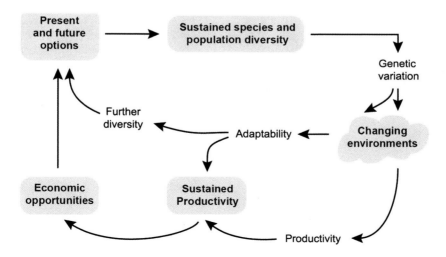

Figure 3.4
Relationships
among sustained
species and
population
diversity and
desired outcomes
of sustained
productivity,
economic
opportunities,
and present
and future
opportunities.
(From Bunnell
et al. (1998).)

The *Convention on Biological Diversity* was correct to emphasize variability. That variability is essential to adapt, respond, and survive. Species richness is a meaningful and measurable index of the concept cluster of biodiversity simply because species and subspecies are the only self-replicating units of genetic variation.

The scale problem

Species richness can serve as an index of biodiversity, but management efforts to sustain biodiversity must include all processes and activities influencing persistence of those species (Bunnell and Huggard, 1999). These processes and activities span a broad range of spatial and temporal scales. The space-time diagrams of Figure 3.5 illustrate the way we often think about scales. They illustrate the range of scales; the concepts associated with them are loose. For example, it is not clear what is intended by time. 'Time' can refer to duration or lifespan, or to frequency of occurrence, or to a loose impression of rate and duration of change. Some forest elements (leaves and crowns in Figure 3.5) are highly discrete units with definable lifespans; for example, 5–15 years for conifer needles. Natural disturbances represent the 'frequency of occurrence' measure of time. Their duration is less important than their return interval. The representation of disturbances, such as budworm infestation or fire, in Figure 3.5 is an amalgam of duration of events (short), natural return intervals (variable), and duration of consequences (often long-lasting).

It is apparent that influential processes occur at different scales depending on whether we are considering genetics, species, or ecosystems. It is for this reason that Levin (1992, p. 1943) argued that 'the problem of pattern and scale is the central problem in ecology, unifying population biology and ecosystems science and marrying basic and applied ecology'. Levin's argument is now more than 20 years old, but it is still valid. Genetics, species, or ecosystems are all features of biodiversity, so a large range of scales must be considered if forest management is to sustain biodiversity.

The challenge is not simple, but the practice of forestry has an advantage

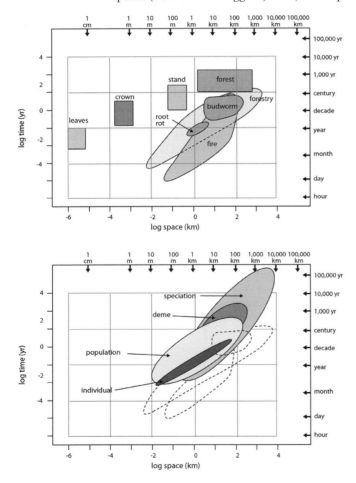

Figure 3.5
The way in which space-time domains of entities and concepts relevant to maintaining biodiversity in forests are commonly perceived. From top down: components of forests and effects of modifying forests, forest dwelling organisms and processes affecting them, and concepts related to conserving biodiversity. The dotted lines indicate overlapping domains of concepts and processes. (Adapted from Bunnell and Huggard (1999), with permission from Elsevier Ltd.)

other practitioners of land management do not have: forestry has a long history of planning over both short and long time frames.

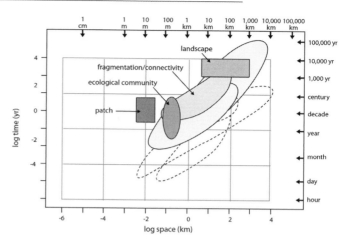

Figure 3.5
(continued)

Forests – stands, landscapes, and the forest matrix

The terms 'stands', 'landscapes', and 'matrix' reflect our descriptions of various scales and our efforts to communicate elements of the scale problem. Forest management changes amounts and distribution of different kinds of habitat. There is ample evidence that both the amount and distribution of habitat influence the species present. To discuss the processes underlying those influences, researchers use other terms: 'fragmentation', 'patch', 'connectivity', and 'edge effects'. Communication is not always clear, in part because our thinking is sometimes muddled, and in part because nature's variability defies tidy categorization. The simplest way to illustrate both the problems and partial solutions is to begin with conventional definitions.

Forest stand: A stand is an area of forest sufficiently uniform in species composition and age that it is easily recognized and can be managed as a single unit. Because stands are planning units, they are intended to be a management convenience. Maps of forest cover rarely depict 'stands' less than 1 ha in size or greater than 100 ha (we combine small units and find reasons to divide large ones).

Forest landscape: There is less consistency among definitions of forest landscapes than for forest stands. The definition of boreal forest states 'a portion of the land that the eye can see in one glance and in which the forest is the most dominant element' (http://www.borealforest.org). That definition is helpful for managing for visual quality objectives. For most forest planning, forest practitioners tend to view the forest landscape as the areal extent of the tenure for which they have responsibility, and often exclude areas within the tenure for which they have no responsibility or are non-forested.

Patch: A patch of habitat is typically smaller than most stands and may be non-forested. They sometimes are designated as separate treatment units within a stand. Because stands are a unit of convenience, when researchers describe edge effects, they typically use the term 'patch size' rather than 'stand size'.

Stands and landscapes are entities that can be seen and are directly relevant to forest practice – they describe the forest as we perceive it. Patches are seen and their size can be measured, but like other terms used to reflect concepts or processes important in sustaining forest-dwelling organisms, they describe the forest as we believe other organisms perceive it.

Forest matrix: The forest matrix is most commonly used to describe areas perceived as 'non-habitat' or the area of forest between or around that suitable forest practices. That, of course, cannot be consistent for all organisms.

Connectivity: Connectivity is typically cast in terms of a hostile matrix and suitable patches. It is intended to describe the degree to which a species can pass through the matrix and so exploit patches of habitat.

*Figure 3.6
A complex
mosaic of
different stands
making up
a section of
forest landscape
in Ibaraki
Prefecture,
Japan.*

Figure 3.6 A complex mosaic of different stands making up a section of forest landscape in Ibaraki Prefecture, Japan.

Edge effects: Edge effects can be both positive and negative, but negative edge effects have received the most attention (e.g. Laurance *et al.*, 2007). Negative edge effects shrink the effective size of the patch by making an area around its perimeter hostile.

Fragmentation: Fragmentation has proven most difficult to define, particularly in forests. If you drop a fragile plate on the floor, three things happen simultaneously: it is broken apart (the dictionary definition of fragmentation), each piece is smaller than the original plate, and each piece is farther from other pieces than when the plate was entire. If we consider the scattered pieces of plate in terms of habitat, there will be loss of habitat provided that smaller pieces of habitat host fewer of the species (edge effects), the scattered pieces invoke a distance effect (matrix effects and connectivity), but there will be no loss of total habitat unless a piece slides under the stove. For many species, the practice of forestry always generates habitat loss because forest stands are converted into forest matrix. Our problem is that much of the early literature gathered area effects, distance effects, and loss of habitat into the single term – fragmentation. Fahrig (1997) offered an early and clear distinction between habitat loss and habitat fragmentation and noted that habitat loss and habitat fragmentation have different management implications.

There are two major difficulties with terms that rely on our perception of how forest-dwelling biodiversity uses habitat patches. One is that the richness of forest-dwelling biodiversity ensures that there is a wide range of responses among organisms, even if we could define the terms well. The second is that there are major difficulties in defining the terms well. In the introduction to a book on forest wildlife and fragmentation, Bunnell (1999a) noted four barriers to clear guidelines and implementation.

1   *We impose distinctions of startling clarity where they do not exist.* Even if we talk about fragmentation as a gradient, we do not analyze it that way. This is primarily a consequence of our desire to use tractable models. Because we seek them, we find discrete habitats, seral stages, or structural stages that are amenable to simple models of patch dynamics or meta-populations. The problem arises when we apply ideas generated from crisply defined patches to the untidy continua of the real world.

Figure 3.7
Patches and
scales are
variable. (Figure
prepared by I.
Houde.)

2 *We measure or scale things relative to us.* Just as we scale the environment according to human perceptions, other organisms rely on their own perceptions that are different from ours. Just what a patch or fragment is becomes remarkably slippery and is certainly not consistent across species. In Figure 3.7, the whole area is habitat for a black bear – there is no patch. There is a patch of snags for woodpecker feeding and nesting, a log for beetles and fungi, older forest for some songbirds, recent clearcut for those species preferring younger seral stages, and riparian for still more species. Each of those patches of habitat are different sizes. Group selection or variable retention silvicultural systems may create a more heterogeneous environment for a wide-ranging carnivore, but discrete patches for a salamander.

3 *We act as if nature were far too well-behaved and well-intentioned to break something into pieces without our help.* Some habitats are naturally scattered – talus slopes, young areas in fire-dominated forests, wetlands, emergent trees in a tropical forest. Apparently fragmentation is not wholly unnatural.

4 *We encompass all the effects that happen when something is reduced to pieces into a single term, thereby reducing our ability to distinguish processes.* For a group of species that prefers older forest, several things happen simultaneously when a forest without much natural edge is disturbed: (1) total area of older forest decreases, (2) amounts of edge initially increase, (3) amounts of interior

habitat decrease, (4) isolation of patches of older forest increases, (5) number of patches of older forests increases, and (6) average patch size of older forest decreases.

Bunnell (1999a) offered suggestions for reducing confusion that were directed primarily to researchers designing studies and interpreting data. Practitioners must continue to make decisions during planning and while selecting practices. Several important points can be distilled from the wide range of work that has been done on this.

There is little evidence that vertebrates perceive old stands as discrete patches. If habitat is not perceived as discrete separate patches, then the total amount of habitat is more important than the distribution of habitat.

Many vertebrates appear to perceive habitat in terms of habitat elements. If organisms are responding primarily to habitat elements, such as large snags, large down wood, and fruiting trees, that implies practices should be designed to maintain those habitat elements. Maintaining those elements in the 'matrix' reduces the effects of both habitat loss and fragmentation.

Many elements creating associations with late successional stages can be sustained within harvested areas by management practices. Harvesting and silvicultural systems, such as variable retention or group selection, can be designed to maintain particular habitat elements.

Stand age remains an issue. Habitat elements important to some organisms are a product of stand age – large cavity sites, depth of litter or organic matter, slowly shedding and deeply fissured bark, and time for sufficient long-distance dispersal events (e.g. arboreal lichens).

Connectivity is important, but the degree of necessary exchange cannot be known. For some threats to small populations (e.g. inbreeding depression) the amount of necessary interchange is surprisingly low. However, the major threats to small populations are chance events, which cannot be known. Many efforts to maintain connectivity devolve to risk management.

The jury is out on the relative advantages of corridors or matrix management to maintain connectivity. Dangers have been documented for corridors, but the small species that are most likely to benefit are generally poorly studied. However, it is clear that for some small organisms, matrix management, such as variable retention, does distribute the species more widely through the forest (for mycorrhizal fungi see Outerbridge and Trofymo, 2004, and Kranabetter *et al.*, 2013).

No single approach is sufficient. For some species, clearcutting creates a high degree of isolation among patches. In many natural forests, sustaining the entire array of biodiversity is most likely to be a blend of conventional forestry (clearcutting), long-rotations and reserves, and creative efforts to establish and maintain late-successional attributes in managed stands (e.g. variable retention; Figure 3.8).

While providing some guidance, the aforementioned points fail to answer two important and frequently asked questions: 'How much is enough?' and 'How important is landscape analysis?' 'How much is enough?' is usually asked of old-growth and other relatively undisturbed forests, but could as easily be applied to how many headwater streams or wetlands should be buffered from forest harvest. Climate change will encourage more buffers simply because we also rely on water. Fortunately, most organisms for which strict dependence on late-successional stages has been documented are small – insects, bryophytes, lichens, and fungi. That implies that individual patches of old forest can usually be relatively small, though edge effects have been documented for some species in all these groups. It also reveals that even small patches of older forest reserved from harvest can contribute significantly to maintaining biodiversity. It does not, however, answer 'How much is enough?'

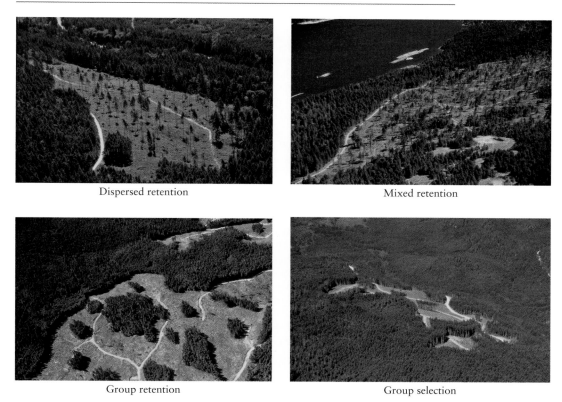

Dispersed retention

Mixed retention

Group retention

Group selection

Figure 3.8
Variable retention blurs the distinction between habitat and non-habitat, yielding more suitable habitat.

We are most likely to answer the question by the undesirable consequence of having species disappear. Not only is the question an example of a wicked problem with social and economic as well as ecological dimensions (see Chapter 15), but it is extremely difficult to quantify. There are only three things we know with reasonable certainty: (1) some species disappear locally as old growth (or other specific habitats) disappears; (2) retaining habitat elements in the forest matrix between older stands helps maintain some species; and (3) well documented estimates of 'How much is enough?' vary widely in different regions. A more productive approach than considering the more academically attractive questions of size and shape of reserves is to attempt to reserve some of each distinct ecosystem type from loss (see the review by Huggard and Kremsater, 2009).

Ideally, we would use landscape analysis to guide the amounts, sizes, and distances between areas reserved from harvest. Landscape analysis can be helpful, but only when we are aware of its limitations. Those limitations are largely conceptual, primarily a result of our persistently stark and incorrect distinction between habitat and non-habitat, but include the other three barriers noted earlier. However, even if our concepts were sound and included gradients, the costs of building and running such models plus analyzing and interpreting the data are beyond what many forest managers could afford. Nor in most instances do we have reliable data on which to base the models. The compromise described later, in the section 'Towards solutions', may be the best that can be done in many instances.

Coarse and fine filters

Approaches to sustaining biodiversity commonly invoke coarse and fine filters in planning and actions. Such approaches explicitly recognize that organization in the natural world occurs at different scales, and therefore planning and implementing management efforts occur at different scales (see, for example, Lindenmayer and Franklin, 2002). Coarse and fine filters often are described as ecosystem-based and species-specific approaches, respectively.

The ecosystem or coarse-filter approach plans, manages, and asks questions at the scale of the forest or landscape by considering broad concepts, such as the representation of ecosystems, natural disturbance processes, and natural succession within stands. It is integrative and relies on broad assumptions; for example, managing the extent and distribution of a particular age class of forest in a watershed will maintain numerous smaller habitats, communities and life forms associated within those habitats, and the processes ongoing within the habitats. It is intended to protect ecosystem function and provide a broad range of habitats for a broad range of species.

Fine-scale or species-specific approaches acknowledge that a forest-level or coarse-filter approach is unlikely to maintain specific needs for all species or all vegetation communities. Examples of management strategies taking into account fine-filter concerns include avoiding leks of sharp-tailed grouse (*Tympanuchus phasianellus*), avoiding mountain goat (*Oreamnos americanus*) winter range, maintaining large mountain ash (*Eucalyptus regnans*) to sustain Leadbeater's possums (*Gymnobelideus leadbeateri*), and protecting natal ponds for tiger salamanders (*Ambystoma tigrinum*).

Appropriate planning can accommodate the broad goals intended by the coarse-filter approach. Specific management actions are usually necessary to implement the fine-filter approach. Planning is often cheaper than specific actions and, when well implemented, can reduce the need for specific actions. Using coarse and fine filters is not, however, a simple exercise of reducing costs. They are attempts to address a pervasive characteristic of the real world – processes maintaining biodiversity occur at different scales.

Figure 3.9 Maintenance of large eucalypts in a mosaic of young eucalypt plantations, Otway Ranges, Victoria, Australia.

To learn from specific planning and management activities, and to evaluate their success, indicators of success need to be established for both coarse-filter and fine-filter approaches. The indicators associated with criteria and indicator schemes are generally too broad to achieve this, and locally specific targets need to be designed. Chapter 15 provides examples in more detail.

Towards solutions

There are ample challenges to sustaining biodiversity in managed forests. Among the largest are the number of species involved and the challenges of dealing with the scale problem. We do know some things with near certainty: species richness is an informative and useful measure or indicator of success (see the section 'Why biodiversity is important'); processes maintaining species are likewise important; those processes occur at a variety of scales; and some common assumptions are well documented (e.g. species often are restricted to particular communities or by specific habitat elements). Here an approach is summarized that exploits what we know to guide actions and facilitate learning through monitoring (see also Chapter 15). It has been effectively employed in temperate rainforest, boreal forest, and temperate dry forests of the province of British Columbia in western Canada, and portions of the approach applied to the very different forests of Australia, Argentina, and Iran. Some elements likely apply to most forests.

The focus is on species simply because they are the most common and encompassing indicator of success and the one most clearly identified by the public. Other measures critical to success are described more completely in Chapter 15 on adaptive management. For brevity, the examples are limited to vertebrates. One of the most common reasons we fail to learn from our management is the expense of monitoring (Chapter 15). Among the advantages of the approach is that it focuses limited resources for monitoring success (or learning) on those species that most require it and helps expose potential problem areas that can be addressed by coarse-filter approaches. It is most useful when coupled with elements of adaptive management. It is known as the Species Accounting System (SAS).

The system recognizes three broad indicators of success in sustaining biodiversity (discussed more fully in Chapter 15), and is intended to:

1   Estimate approximate amounts and location within the forest area of suitable habitat for most forest-dwelling vertebrates;
2   Permit 'scaling up' of monitoring findings over the entire area, providing estimates of the amount of suitable habitat, including where and when, over the entire area;
3   Provide credence to indicators assessing ecosystem representation (Indicator 1) and habitat (Indicator 2) by evaluating species associations with those measures;
4   Provide trend estimates for species as data are accumulated (Indicator 3);
5   Focus more expensive effectiveness monitoring on areas of greatest uncertainty;
6   Be self-correcting, thus increase the credibility of the system as data are acquired.

Species identified, but whose habitat is not projected by the system, are those whose habitat is too finely discriminated to be included in GIS layers. These species are gathered in Group 4 of the Species Accounting System (Table 3.2). The system is not self-correcting in the sense of a self-cleaning oven; correction requires monitoring and interpretation of the acquired data. The initial step is to assign species to 'monitoring' classes that reflect both species' attributes and the costs of monitoring.

Group 1 species are those which inhabit many habitat types or respond positively to forest practices. The designation 'generalist' refers to forest cover type. For example,

Table 3.2 *Species Accounting System (SAS) as applied to vertebrates of Tree Farm License (TFL) 48 in northeastern British Columbia.*

SAS Group	Attributes	Monitoring
1 Generalists	inhabit many habitat types	none necessary
2 Forest types	can be statistically assigned to broad forest types	GIS; tabular look-up
3 Habitat elements	strong dependencies on specific habitat elements	guided practices; implementation monitoring
4 Specialized or localized habitats	restricted to specialized and highly localized habitats	develop guidelines
5 Distribution important	patch size or connectivity important	GIS analysis
6 Non-forested	do not require forest	none necessary

the barred owl (*Strix varia*) is designated 3c/1. That indicates that during monitoring, the species must first be recognized as a cavity nester (3c; see Group 3), and success in maintaining it depends foremost on planning and practices designed to sustain cavity sites. However, with respect to cover forest type, the species is a generalist (Group 1) and will select cavity sites in virtually any broad forest type. Few of these species need to be monitored directly because they either benefit from forest practices or occur across a wide range of forest cover types.

Group 2 species are those that can be accounted for by habitat type and for which no other feature of habitat use dominates the appropriate monitoring approach. For most jurisdictions at least two broad classifications are available: forest cover type for inventory purposes and ecosystem type as specified by an ecological classification system. In British Columbia, both approaches were exploited because both were available province wide. Forest cover types included:

- NV: non-vegetated upland, less than 5% tree cover (other vegetation may be present) and includes roadsides and oil and gas developments but excludes lakes, rivers, and ponds;
- NT: non-treed or <10% tree cover;
- RD: recent disturbance or 0–30 years post-disturbance in the boreal;
- H1: young hardwoods, 31–90 years old and at least 75% hardwoods;
- MW2: old mixed woods, >90 years old, and neither hardwoods nor conifers attain 75%;
- C2: old conifers, >140 years old and at least 75% conifer cover;
- R: riparian forest around streams, lakes, and rivers, but not wetlands.

Ecosystem type used an ecosystem classification system developed for British Columbia, Canada, namely the variant level of the Biogeoclimatic Ecosystem Classification (BEC) (Meidinger and Pojar, 1991). Most Group 2 species can be usefully monitored by GIS 'table look-up' of the amounts and distribution of the appropriate habitat type. Both forest cover type and ecosystem type are informative (see the section 'Using the Species Accounting System').

Group 3 consists of species with strong dependencies on specific habitat elements (cavity sites (3c), downed wood (3dw), and understory (3u) including shrubs, riparian areas (3r), and wetlands (3w)). Species assigned to Group 3 sometimes show strong affinities for particular forest cover types. The association of Group 3 species with particular habitat elements is sufficiently critical that practices affecting these elements must be assessed. Provided these practices sustain the element, many of these species can then be monitored as for Group 2 by tracking appropriate habitat types. That is, practices must first be evaluated before associations with habitat type are employed. It is for this reason that these species are assigned to Group 3 rather than Group 2. It is also for this reason that readily monitored species in this group can be particularly useful in effectiveness monitoring of specific practices (e.g. appropriate size of retained wildlife trees). Periodic follow-up implementation monitoring is important to ensure that appropriate practices are being maintained. Types of monitoring are discussed in Chapter 15.

Group 4 species are those whose habitat is highly specialized and often localized. We do not expect any of these species to show statistical relations with the habitat types tested for two reasons: (1) their habitat is typically too specialized to be accessible to coarse-filter approaches, and (2) most occur too uncommonly to allow statistical evaluation. Because of the specialized nature of the habitat, it is not readily captured by coarse-filter measures of forest cover or ecosystem types. For this reason the species are generally not monitored, but managed by specific regulatory measures (e.g. provincial Wildlife Habitat Areas or Ungulate Winter Ranges) or company guidelines that prescribe appropriate actions to be taken when the habitat is encountered. Once it is clear that the practices can be implemented to ensure provision of habitat, monitoring of these species is primarily through implementation monitoring of the practices.

Group 5 consists of a small group of species for which at least some literature suggests the distribution of habitat is as or more important than the amount of habitat. Distribution of habitat includes features such as patch size and associated negative edge effects and connectivity. These features often are lumped under the term fragmentation. For forest-dwelling species, fragmentation has been documented to be a much more serious concern where the practices replace forest cover with different habitat rather than just alter forest cover – agriculture, human settlement, and wide, well-travelled roads, for example (Bunnell, 1999b; Freemark and Merriam, 1986; Kremsater and Bunnell, 1999; Trombulak and Frissell, 2000). The simplest form of monitoring for Group 5 is to track patch size distribution of the appropriate habitat type (GIS) and visual evaluation of the degree to which patches of habitat are becoming disjunct.

Group 6 species are those that rely on non-forested habitats. Early versions of the Species Accounting System did not include this group. It was added because members of the public frequently requested knowledge of their status. These species are not monitored directly, but their affiliation with broad habitat types is usually known and both amounts and distribution of preferred habitat can be quickly summarized in GIS. Group 6 species are not monitored by the forest tenure holder because they are unaffected by forest practice.

A useful adjunct to help focus effort is to note conservation priority designations assigned to species provincially (or by individual states) and federally. The province of British Columbia employs a Conservation Framework to rank species for conservation efforts. There are six potential ranks, from 1 = highest to 6 = lowest. The province ranks each species within three broad goals for conservation. These goals are:

1   To contribute to global efforts for species conservation. This goal is intended to ensure that some provincial resources are assigned to conserving species globally at risk, even when these are widely distributed.

2   To prevent species from becoming at risk. Goal 2 is intended to be proactive and provide early detection of threats, thereby reducing the need for costly recovery actions. It is facilitated by including all native species in assessments of priority, rather than focusing solely on those already 'at risk'.

3   To maintain the richness of native species. Goal 3 represents efforts to sustain all native species, even when only jurisdictionally rare and abundant elsewhere. It is intended to ensure that challenging, jurisdictionally rare species will not be ignored in pursuit of Goal 2.

National designations are specified by the Committee on the Status of Endangered Wildlife in Canada (COSEWIC) and are broadly equivalent to those of the US Endangered Species Act,

---

Box 3.1 The IUCN Red List classification system used to describe the conservation status of a species.

The conservation goals in the text should not be confused with international rankings for individual species. Two systems are commonly encountered. The IUCN Red List provides an estimate of the conservation status of every species for which such an assessment has been done.

Extinct (EX)

No reasonable doubt that the last individual has died. A taxon is presumed Extinct when exhaustive surveys in known and/or expected habitat, at appropriate times (diurnal, seasonal, annual), throughout its historic range have failed to record an individual.

Extinct in the Wild (EW)

Known only to survive in cultivation, in captivity, or as a naturalized population (or populations) well outside the past range.

Critically Endangered (CR)

The best available evidence indicates that it meets any of the criteria for Critically Endangered (see http://jr.iucnredlist.org/documents/redlist_cats_crit_en.pdf), and it is therefore considered to be facing an extremely high risk of extinction in the wild.

Endangered (EN)

The best available evidence indicates that it meets any of the criteria for Endangered (see http://jr.iucnredlist.org/documents/redlist_cats_crit_en.pdf), and it is therefore considered to be facing a very high risk of extinction in the wild.

Vulnerable (VU)

The best available evidence indicates that it meets any of the criteria for Vulnerable (see http://jr.iucnredlist.org/documents/redlist_cats_crit_en.pdf), and it is therefore considered to be facing a high risk of extinction in the wild.

Near Threatened (NT)

A taxon has been evaluated against the criteria but does not qualify for Critically Endangered, Endangered, or Vulnerable now, but is close to qualifying for or is likely to qualify for a threatened category in the near future.

Least Concern (LC)

A taxon is Least Concern when it has been evaluated against the criteria and does not qualify for Critically Endangered, Endangered, Vulnerable, or Near Threatened. Widespread and abundant taxa are included in this category.

Data Deficient (DD)

A taxon is Data Deficient when there is inadequate information to make a direct, or indirect, assessment of its risk of extinction based on its distribution and/or population status. A taxon in this category may be well studied, and its biology well known, but appropriate data on abundance and/or distribution are lacking. Data Deficient is therefore not a category of threat. Listing of taxa in this category indicates that more information is required and acknowledges the possibility that future research will show that the threatened classification is appropriate. It is important to make positive use of whatever data are available. In many cases great care should be exercised in choosing between DD and a threatened status. If the range of a taxon is suspected to be relatively circumscribed, and a considerable period of time has elapsed since the last record of the taxon, threatened status may well be justified.

Not Evaluated (NE)

A taxon is Not Evaluated when it has not yet been evaluated against the criteria.

Box 3.2 The classification system used by NatureServe to describe the conservation status of a species.

The NatureServe system primarily deals with the USA, but is expanding. It is important to ensure that major costs are not incurred protecting a particular species that is rare in a specific area, but common elsewhere. At the same time, efforts should be taken to ensure that rare habitat types within an area are maintained.

GX **Presumed Extinct** (species)/**Eliminated** (ecological communities and systems) – Species not located despite intensive searches and virtually no likelihood of rediscovery. Ecological community or system eliminated throughout its range, with no restoration potential.

GH **Possibly Extinct** (species)/ **Eliminated** (ecological communities and systems) – Known from only historical occurrences but still some hope of rediscovery. There is evidence that the species may be extinct or the ecosystem may be eliminated throughout its range, but not enough to state this with certainty.

G1 **Critically Imperiled** – At very high risk of extinction due to extreme rarity (often five or fewer populations), very steep declines, or other factors.

G2 **Imperiled** – At high risk of extinction or elimination due to very restricted range, very few populations, steep declines, or other factors.

G3 **Vulnerable** – At moderate risk of extinction or elimination due to a restricted range, relatively few populations, recent and widespread declines, or other factors.

G4 **Apparently Secure** – Uncommon but not rare; some cause for long-term concern due to declines or other factors.

G5 **Secure** – Common; widespread and abundant.

Using the Species Accounting System

The Species Accounting System has been applied to more than 10 forest tenures distributed throughout British Columbia. Specifics are discussed for three adjacent forest tenures in northeastern British Columbia: Tree Farm License (TFL) 48 covering 643,239 ha of sub-boreal forest; Fort St John Timber Supply Area (TSA) covering 4.67 million ha of sub-boreal and boreal forest; and the Fort Nelson TSA (9.8 million ha of boreal forest). The primary focus is on TFL 48. The example discussed is for vertebrates, but it has been extended to non-vertebrates where data permit. A major value of the approach is to guide coarse-filter analyses. Broad findings of those analyses are summarized; specifics are found elsewhere (Bunnell *et al.*, 2009d, 2009e, 2010). Sources cited in this section are for specific forest tenures, but can be found at the BC Ministry of Forests, Lands, and Natural Resource Operations electronic library. Examples of conservation priority designations are provided.

Generalists (Group 1)

Each of the three tenures hosts the same 47 native generalist species; examples include pine siskin (*Spinus pinus*), red-tailed hawk (*Buteo jamaicensis*), red fox (*Vulpes vulpes*), and southern red-backed vole (*Myodes gapperi*). When habitat generalists occur they typically occur commonly. For example, on TFL 48, only six of the habitat generalists are casual; 20 are considered common, and 22 uncommon. Only six rank 1 or 2 within the provincial Conservation Framework for at least one Goal. The highest ranking is 2 and consistently occurs within Goal 2, which is intended to be proactive (e.g. dusky grouse [*Dendragapus obscurus*], rufous hummingbird [*Selasphorus rufus*], porcupine [*Erethizon dorsatum*]). The grizzly bear (*Ursus arctos*) ranks highly in Goal 3 – to maintain native species richness within the province. It has been designated of 'Special Concern' by COSEWIC, as has the wolverine (*Gulo gulo*). Other habitat generalists have not been evaluated by COSEWIC or have been evaluated and found 'not at risk': golden eagle (*Aquila chrysaetos*), red-tailed hawk, black bear (*Ursus americanus*), Canada lynx (*Lynx canadensis*), and grey wolf (*Canis lupus* ssp. *occidentalis*).

No generalist species was monitored directly, but they were recorded when encountered during monitoring. That was useful primarily to evaluate the initial classification as generalist. The fact that monitoring of 'listed' generalist species (grizzly bear and wolverine) is considered a government responsibility rather than the forest industry's may appear contrary to a commitment to sustain biodiversity. It actually exposes a social issue. With one exception, forest practices tend to be favourable conditions for those species. The exception is access, and its impact can be large. As with oil or gas exploration and extraction, forestry increases access, thus both legal and illegal hunting and trapping. About 95% of land in British Columbia is Crown or public land and it has proven very nearly impossible to control access to Crown land once even crude access has been provided by industry.

Habitat types (Group 2)

Across all three areas that were used to test the system, 42 species could be accounted for solely by forest cover type. For example, one of the areas contains 41. Although red squirrel (*Tamiasciurus hudsonicus*) and northern flying squirrel (*Glaucomys sabrinus*) frequently use cavities, use is not consistent and they were assigned to Group 2. The northern hawk owl (*Surnia ulula*) sometimes uses cavities, but that use is opportunistic and only about one-third of nests are reported from cavities. It also was assigned to Group 2.

More bird species showed strong associations with hardwood or mixed wood types (e.g. American redstart [*Setophaga ruticilla*], least flycatcher [*Empidonax minimus*], ovenbird [*Seiurus aurocapilla*]) than with conifer-leading types (e.g. ruby-crowned kinglet [*Regulus calendula*], Townsend's warbler [*Dendroica townsendi*]). A few showed significant associations with sparsely treed (NT) cover types (e.g. blackpoll warbler [*Dendroica striata*]). For others hardwood cover is sought primarily in riparian areas (Baltimore oriole [*Icterus galbula*], black-and-white warbler [*Mniotilta varia*]). Some species are strongly dependent on additional habitat elements: understory under canopy, cavity sites, and wetland or riparian habitat. For these latter species, coarse-filter evaluation of effects of forest practice must first address the key elements – understory, dying or dead wood, and riparian. Only three mammals were designated Group 2: hoary bat (*Lasiurus cinereus*) (conifer), northern flying squirrel (older mixed wood and older conifer), and red squirrel (conifer and older mixed wood). Other mammal species show some affiliation with hardwood or mixed wood cover, but the association is not pronounced and is a product of affinity for riparian areas (e.g. western jumping mouse [*Zapus princeps*]) or because they forage on young hardwood trees, shrubs, and herbaceous vegetation that often is more abundant in hardwood than in conifer stands (e.g. long-tailed vole [*Microtus longicaudus*], elk [*Cervus canadensis*] during winter).

No species assigned to Group 2 showed a statistical preference for an ecosystem variant, although the American redstart consistently was relatively more abundant on the same variant across all three tenures. That reveals that forest cover type dominates habitat use by these group members. Some members of all other groups tested did show apparent selection for specific ecosystem variants; Group 4 could not be tested.

Most Group 2 species are uncommon or casual. For example, of the 40 species associated with forest type that occur within TFL 48, 20 are uncommon and six are casual. The ability to track habitat of less common species by GIS is cost-effective. Both amounts and distribution of habitat are tracked with relative little cost (Table 3.3). Eight of these species rank 2 within Goal 2 (proactive); two rank 2 within Goal 3 (maintain species richness). The latter two species are uncommon or casual within TFL 48.

Table 3.3 Assignment of species to kinds of monitoring for TFL 48 in northeastern British Columbia.

SAS Group	Number of species	Monitoring and guided practices	Relative cost
1 Generalists	47	none necessary	low
2 Habitat types	41	GIS; tabular look-up	low
3 Habitat elements		guided practices; implementation monitoring	
3c Cavity sites	39	VR[1]; stubbing; implementation monitoring	moderate
3dw Downed wood	5	normal inventory interpreted by diameter	low
3u Understory	26	normal inventory adequate for early seral	moderate
3r&w Riparian and wetlands	60	implementation monitoring	moderate
4 Specialized or localized habitats	10	guidelines for local practices	moderate
5 Distribution important	6	GIS analysis	low
6 Non-forested	28[2]	none necessary	low
	262		

Note: [1] variable retention [2] peregrine falcon (Group 4) also seeks non-forested habitat.

This group is a clear example of how monitoring can correct errors in the initial assignment that is based on literature and expert opinion. As data are acquired, selection for or against particular forest cover types and BEC variants is tested using observed versus expected (available) locations. Across British Columbia, some species show regional shifts in apparent selection for both habitat type and BEC variant. The shift in variant use is expected because different variants are available for occupancy. The shift in apparent preference for forest cover is largely a product of different vertebrate community composition.

Key findings from this work are:

- In each of the three northeastern tenures there are species demonstrating statistical preference for each of the three broad forest types – hardwood-leading, mixed wood, and conifer-leading.
- Analysis suggests no limitation due to the amounts of older conifer-leading stands, or to the manner in which conifer-leading forest is distributed (e.g. patch size), nor do projections of planned harvest schedules, when they were possible, suggest any shortfall over the next 20 years.
- The most commonly expressed statistical preference was for hardwood-leading stands. In one area, there is no apparent negative trend in amounts of hardwoods, but the area in larger patches decreases over the 20-year projection.
- Among the three broad forest types, amounts of older mixed wood appear most likely to decline. Under the definition employed here (any stand with <75% conifer or hardwood), mixed wood declines sharply in localized areas within one of the three study areas. Harvest scheduling there should be evaluated to determine its likely impact on old mixed wood and opportunities for modification.

Habitat elements (Group 3)

A total of 39 bird and mammal species likely to occur within any of the areas use cavities as nest or den sites more than 50% of the time, often 100% (3c in Table 3.3). Other species use cavities more opportunistically (e.g. common grackle [*Quiscalus quiscula*], harlequin duck [*Histrionicus histrionicus*], Pacific-slope flycatcher [*Empidonax difficilis*], northern hawk owl). Cavity-using bats are poorly known and their presence is most often inferred rather than documented. Not all species designated 3c use conventional cavities. The bats and brown creeper [*Certhia americana*] use cracks in or behind deeply furrowed bark as well as conventional cavities. The trees they seek, however, are similar to those providing cavity sites. The fisher (*Martes pennant*) and American marten (*Martes americana*) are designated 5/3c. Cavities as den sites are important for these species and need to be considered. Some literature, however, argues that habitat distribution is important for these two species, and that even if den sites were available the species would not inhabit the landscape where habitat was inappropriately distributed. For that reason they are first treated as a Group 5 species with habitat distribution as potentially the most limiting habitat feature.

Breeding Bird Surveys (the simplest approach to monitoring birds) record most cavity users poorly, so few statistically significant associations with either habitat type or BEC variant were established. Where records were adequate (e.g. black-capped chickadee [*Poecile atricapillus*], Pacific wren [*Troglodytes pacificus*], yellow-bellied sapsucker [*Sphyrapicus varius*]), associations are evident. The majority of cavity using species are uncommon. In one area, only 9 of 39 species occur commonly; 14 are uncommon and 16 are casual – the bats, however, have been inadequately surveyed. A high proportion (15/39 = 38%) rank highly (ranks 1 or 2) within the Conservation Framework. Most harvest on the northeastern tenures is by feller-bunchers, which makes 'high-topping' or 'stubbing' a convenient way to create potential future cavity sites during harvest.

Although many species exploit down wood, only five vertebrates in northeastern British Columbia appear to have strong relations with down wood (3dw, Table 3.3). None of the amphibians present breeds in down wood. The pileated woodpecker (*Dryocopus pileatus*) forages intensively in rotted down wood for carpenter ants (*Camponotus* sp.); the Pacific wren frequently nests in upturned root wads; the ruffed grouse (*Bonasa umbellatus*) requires down wood as 'booming' or 'hooting' sites, and both the American marten and fisher use down wood as resting sites (and as den sites in more southerly forest types). The wren and the woodpecker are treated first as cavity users; the two mammals are considered first as species for which habitat distribution must be evaluated (Group 5). The dusky grouse does hoot from down wood, but is not nearly as restricted as the ruffed grouse, making do with stumps and rocks. American marten, fisher, and ruffed grouse rank highly within the provincial Conservation Framework.

There have been hundreds of studies documenting sizes of trees or logs used by both vertebrates and invertebrates (see review by Bunnell and Houde, 2010). Median diameter is more informative than mean diameter, because species also use large but relatively sparse trees or snags that skew the distribution. This information can be coupled with that of the Species Accounting System to guide coarse-filter analyses. Median diameters can be related to growth curves for the tenure to define which age of cover type classes provide useful habitat. That permits a cheap and biologically credible way of assessing the amount and distribution of suitable habitat. Likewise, companies typically assess down wood following logging. Modest refinement of that assessment can provide useful information with little additional cost. Key findings for species using cavity sites and down wood were:

- Minimal diameters specified in the Sustainable Forest Management Plan for retained trees were less than they should be if all cavity-using species are to be retained. In retention patches <10 ha, guidelines for retained wildlife trees should be tree sizes >23 cm dbh and patch-wise densities of >3 per ha. Greater densities of small snags also should be retained. The utility of stubbing and diameters of trees retained in Wildlife Tree Patches had not been documented.
- Anchor points for retention should acknowledge potential limitations (e.g. older mixed wood, hardwoods >23 cm dbh; conifer trees >30 cm dbh).
- Guidelines for retention of down wood were inadequate or lacked sufficient specificity. Volume of down wood is an inadequate guideline for efforts attempting to sustain biodiversity. Waste management guidelines should ensure that a few larger conifer pieces (>17.5 cm diameter; random not top-end diameter) are retained where stand conditions permit; pieces >2 m long. Pieces of down wood >17.5 cm diameter (random) should not all be piled; some should be left scattered on site if poorly dispersing species of lichens, bryophytes, fungi, and invertebrates are to be sustained.
- Some cavity nesters are strongly associated with riparian and, to a lesser degree, wetland habitats. Most of these species seek out hardwoods because of the decay patterns in hardwoods. The degree to which hardwoods are maintained in Riparian Management Zones is unclear and should be documented.

A total of 21 forest-dwelling bird species are strongly associated with understory (designated 3u, Table 3.3). The emphasis is on shrubs. Other bird species showing strong associations with understory are assigned to Group 6 because they occur primarily in grasslands or shrublands, including alpine and subalpine areas. An additional five mammal species, mostly large ungulates, show strong associations with understory as forage. Several small mammal species, classified as generalists in response to forest type, show a preference for early seral stages where they profit from understory species.

Most species designated 3u show statistically significant associations with forest cover type, generally non-vegetated or non-treed, both non-commercial types. Although these species respond strongly and often are dependent on understory, there were few significant selections for recently disturbed areas. This is because many of these species seek shrubs under canopy. When older forest cover types are preferred, they are more often hardwood-leading or mixed wood types than conifer-leading, and more often older than younger types. That reflects the importance of radiation regimes and their influence on shrubs under canopy. The selection for particular habitat types is also revealing. No understory associate showed selection for a habitat type where understory was not abundant; selection for habitat type appears to reflect greater abundance of hardwood-leading and mixed wood types. Seven of these 26 species rank 2 within Goal 2 (proactive) of the Conservation Framework; the Connecticut warbler (*Oporornis agilis*) ranks 2 within Goal 3 (maintain native species richness). Of the 25 understory associates occurring in one of the areas, most are common (17), the veery (*Catharus fuscescens*) is at the edge of is range and casual, and seven are uncommon. Key findings were:

- Amounts of early seral stages are an inadequate index of habitat suitability of understory associates in this region. Of 52 statistical preferences documented for birds across the three northeastern tenures, only 15% were for recently disturbed areas. A more commonly expressed preference was for non-commercial types (non-treed or non-vegetated; 50% of associations). Early seral stages do serve as a useful index for more mammal species.
- Bird species nesting primarily in shrubs under canopy showed statistical associations with hardwood-leading and mixed wood types >90 years old, likely because these are more open types and permit more abundant understory. While tracking older hardwood and mixed wood provides a broadly practical index of habitat suitability, it is incomplete; presence of shrubs in such stands is not assured.
- Harvesting and other disturbance creates areas of potentially abundant understory. Vegetation management attempts to control this abundance to encourage establishment of crop trees. Because some birds and more mammals prefer open-grown understory, it is important that control not be enduring. Under current harvest rates, the duration of suppression has more influence than the proportion of harvested areas treated.
- Understory often does best in wetter, often riparian areas. The importance of understory to sustaining biodiversity raises two questions, both of which could be addressed by implementation monitoring. First, to what degree is understory maintained in specified Riparian Management Zones? Second, to what degree does harvest influence understory adjacent to small wetland types (unprotected by default riparian regulations) or non-treed types created by wet soil conditions?

Excluding the trumpeter swan (*Cygnus buccinator*), which is treated as a Group 4 species, there are 60 vertebrate wetland and riparian associates in one of the study areas; 51 species show a strong affinity for wetland habitats including small ponds and nine show affinity for riparian areas around larger water bodies (streams, rivers, larger lakes). The distinction is not tidy and many species use both habitats. A further eight cavity nesters prefer riparian and wetland sites, as do 11 understory associates (where it often appears as a preference for NT within the forest cover types).

The majority of wetland and riparian associates are relatively common. Among the 60 species occurring within the study area, 32 occur commonly and 20 are considered uncommon. Eight species are considered casual or transient. Of these species, 13 rank 1 or 2 within the Conservation Framework. Two rank 1 and 10 rank 2 within Goal 2, and the yellow rail ranks 1 in Goal 3. Few show an affinity to particular habitat types because wetlands tend to occur as small inclusions across a broad range of cover types and variants, and rivers flow through many habitat types and variants. Key findings were:

- Northeastern British Columbia contains abundant wetlands and possibly the richest vertebrate wetland fauna in the province, over 80% of which prefer smaller wetlands.
- Because these wetlands are so numerous, so variable in size, and so widely scattered, a general approach is not apparent.
- Implementation monitoring is important to assess the proportion of small wetlands potentially affected by practice; if that proportion is significant it should be followed by effectiveness monitoring of practices potentially affecting wetlands and their inhabitants.

Group 1. Generalists. Canadian lynx, red fox

Group 2. Habitat type. Red squirrel, American redstart

Group 3. Habitat element. Yellow-bellied sapsucker, black-capped chickadee

Figure 3.10 Some of the wildlife species used in the indicator assessment.

Group 4. Localized habitat. Tundra swan, Stone's sheep (*Ovis dallii stonei*)

Group 5. Habitat distribution. Caribou (*Rangifer tarandus*), Fisher

Figure 3.10 (continued)

Group 6. Non-forest. Killdeer (*Charadrius vociferus*), willow ptarmigan (*Lagopus lagopus*)

Localized habitats (Group 4)

Species with highly localized habitats defy coarse-filter analyses. The majority of Group 4 species are designated uncommon or casual within all forest tenures evaluated. Eleven occur on northeastern tenures. Of the 10 in the study area, eight are uncommon and two are casual. Most of these species (8 of 10) rank 1 or 2 within the Conservation Framework. Seven rank 1 or 2 within Goal 2, bull trout ranks 1 within Goal 1, and Nelson's sparrow (*Ammodramus nelsoni*) and the peregrine falcon (*Falco peregrinus*) rank 2 within Goal 3.

Management of these species was addressed by creating an illustrated booklet describing appropriate actions if the species were encountered. For example, designated Ungulate Winter Range and its associated guidelines were in place for Stone's sheep; in addition, drivers were to be alerted to any mineral licks along hauling routes where sheep periodically congregate. For sharp-tailed grouse (*Tympanuchus phasianellus*) the primary guideline was to avoid hauling on roads that pass within 100 m of an active lek during March through May. The peregrine falcon is transient in the area, but, as a precautionary measure, the primary guideline was to avoid harvest or road construction within 500 m of nest sites during breeding season (February through July).

Habitat distribution (Group 5)

Group 5 species are those for which at least one study somewhere has obtained results suggesting that habitat distribution is as or more important than amount of habitat. Within northeastern British Columbia, six vertebrate species meet this criterion and each occurs in all three forest tenures. Among vertebrates, Group 5 species typically are associated with older age classes and are not necessarily rare. Within the study area, three are considered common and three uncommon. Three species rank 2 within one of the Goals of the Conservation Strategy – two within Goal 2 and the fisher within Goal 3.

The Species Accounting System defined the forest cover types preferred by these species. That permitted simple GIS analyses of the patch size and distribution of favoured habitat and, in cases where 20-year harvest plans were available, projection of future distribution. The caribou is addressed by regulation. Major findings for the remaining five species were:

- In northeastern British Columbia, most species showing negative edge effects elsewhere do not exhibit negative edge effects. Golden-crowned kinglet and northern goshawk (*Accipiter gentilis*) are confirmed and probable exceptions, respectively.
- For all five species, evaluation of current patch size distributions and trends in patch size within preferred habitat suggest no negative effects of forest practice.

Non-forested (Group 6)

Within all larger forest tenures some species occur in habitats that are not affected by forest practices, such as alpine, grasslands, and man-made habitats. Examples include bank swallow (*Riparia riparia*), killdeer, eastern phoebe (*Sayornis phoebe*), willow ptarmigan, meadow vole (*Microtus pennsylvanicus*), and hoary marmot (*Marmota caligata*). The peregrine falcon uses non-forested habitat, but is sufficiently localized that it was assigned to Group 4.

Group 6 is included within the Species Accounting System to acknowledge the occurrence of the species within a tenure that rarely occur in forests and are not affected by forest practice. The three northeastern tenures host 29 species; 28 on one of them. There are few strong associations with habitat type, largely because many (18 of 29) prefer man-made habitat which can occur anywhere. An additional seven species use alpine or subalpine, which was not sampled. Like other groups, their relative abundance shows considerable range. On one study area, 28 Group 6 species occur; 13 are uncommon, 10 are common, and five are casual or transient (e.g. Swainson's hawk [*Buteo swainsoni*]). Ten Group 6 species rank 2 within at least one Goal of the Conservation Framework – eight in Goal 2 and one each in Goal 1 and Goal 3. One subspecies relying on non-forested habitats (the wood bison [*Bos bison athabascae*]) is listed as 'threatened' by COSEWIC. This subspecies has expanded its range and abundance by exploiting forage created along new forest roads. That access cannot be controlled, which facilitates illegal hunting and reaffirms that managing to sustain biodiversity is a 'wicked problem' (Chapter 15).

Managing to sustain forest biodiversity will remain challenging, so we must keep learning. Effective monitoring is key to sustained learning. The challenge is to make monitoring cost effective. The Species Accounting System addresses that challenge. In total, 262 vertebrate species are known or believed to occur on the study area, but only 232 are forest-dwelling (Table 3.3). By grouping species into monitoring classes, costs are minimized for 99 species (47%). Where implementation monitoring affirms minimal effect on riparian and wetland species, that total is increased to 159 species (68.5%). Costs are reduced by making as much of the monitoring as possible map-based. Species still have to be monitored to develop credibility for the maps. Simple changes to well-established, relatively cheap monitoring protocols can make them more useful. For example, by adding habitat measures (GIS) to conventional Breeding Bird Surveys, two values are gained: (1) data are in the same format as those used to establish as the major trend indicator in North American birds, and (2) the added habitat or map base permits guided coarse-filter analyses.

Breeding Bird Surveys do not adequately monitor owls (too late in the year), cavity users (many species are quiet during the specified survey period; call playback yields more observations), and wetland and riparian associates (not adequately sampled by roadside counts). Despite these failings, application of the Species Accounting System assists monitoring. An initial step is determining monitoring groups (Groups 1 and 6 are not monitored). Members of Group 2 (habitat type) and Group 5 (distribution, not amount) can be effectively monitored by simple GIS 'table look-ups'. By recording which species are present, the Species Accounting System also guides both coarse-filter analyses and monitoring of Group 3 species, which often are most susceptible to impacts of forest practice (diameters of suitable cavity sites can be translated into appropriate age classes for coarse-filter analysis of suitable habitat). Similarly, the Species Accounting System revealed that many understory associates in the region were responding to shrubs under canopy that are best represented in older hardwood-leading and mixed wood stands, thus guiding coarse-filter analyses of amounts of suitable habitat. For lesser known species the system can guide targeted sampling and has been effective in documenting habitat and the relative abundance of listed warbler species.

Conclusion

For good reasons, sustaining biodiversity will continue to emphasize species. Despite receiving less public scrutiny, underlying processes, such as habitat distribution, remain important. Although we are well past the 'paralysis by complexity' that thwarted management to sustain biodiversity only two decades ago, the challenge of the number of species remains. Processes like the Species Accounting System accommodate both species and processes and help focus limited resources for monitoring and conservation by limiting direct monitoring to those species that most require it and directing specific treatments to where they are most beneficial. It is a near minimalist approach and far less expensive than other approaches, but still credible.

The process of learning while doing is formalized as adaptive management (Chapter 15). In the examples provided here, both practitioners and scientists were involved in the design which facilitated feedback to management and changes in management. Changes derived from the feedback to management included changes in practice (increasing diameter of retained trees, leaving a larger portion of logging residue scattered rather than piled), modifications of normal inventory (diameters rather than volume of logging debris), and focused monitoring (utility of stubbing, duration of undergrowth suppression by vegetation management). In short, for a topic where rapid learning is critical, learning was encouraged, practices were credibly altered, and public meetings indicate that social license was enhanced.

Further reading

Bunnell, F. L. 2013. Social licence in British Columbia: Some implications for energy development. *Journal of Ecosystems and Management* 14(2): 1–16.

Bunnell, F. L., Dunsworth, G. B. (eds.) 2009. *Forestry and Biodiversity: Learning How to Sustain Biodiversity in Managed Forests*. Vancouver: University of British Columbia Press.

Gardner, T. 2010. *Monitoring Forest Biodiversity: Improving Conservation Through Ecologically-Responsible Management*. London: Earthscan.

Hunter, M. J., Schmiegelow, F.K.A. 2011. *Wildlife, Forests, and Forestry: Principles of Managing Forest for Biological Diversity*. Boston: Prentice Hall.

Lindenmayer, D. B., Fischer, J. 2006. *Habitat Fragmentation and Landscape Change: An Ecological and Conservation Synthesis*. Washington, DC: Island Press.

Lindenmayer, D. B., Franklin, J. F. 2002. *Conserving Forest Biodiversity: A Comprehensive Multi-scaled Approach*. Washington, DC: Island Press.

Chapter 4

Productivity of forest ecosystems

Juan A. Blanco, Yueh-Hsin Lo,
Clive Welham, and Bruce Larson

The second criterion under the Montreal Process is the maintenance of the productive capacity of forest ecosystems. This is related to the sustained yield concepts described in Chapter 1, but differs in that it was intended to be extended to the productivity of forest ecosystems as a whole, not just timber. However, over time, revisions to the original wording have resulted in the indicators focusing quite strongly on timber:

2.a   Area and percent of forest land and net area of forest land available for wood production.
2.b   Total growing stock and annual increment of both merchantable and non-merchantable tree species in forests available for wood production.
2.c   Area, percent, and growing stock of plantations of native and exotic species.
2.d   Annual harvest of wood products by volume and as a percentage of net growth or sustained yield.
2.e   Annual harvest of non-wood forest products.

In this chapter, we examine both the factors affecting forest ecosystem productivity and the ways that forest management can influence these. Much forestry training is related to the manner in which productivity can be measured and monitored, but here we assume that readers have this basic level of training, and it is not further elaborated. For those unfamiliar with forest inventory techniques, there is a variety of textbooks available, but many of the field techniques are best learned through hands-on training in the forest.

Environmental controls on plant productivity

The ultimate source of energy for life on Earth is light from the Sun. Pigments in the green tissues of plants absorb light and capture its energy, which is then stored through the manufacture of carbohydrates from simple inorganic compounds – carbon dioxide and water. This energy-trapping process is called photosynthesis. A simplified equation is as follows:

$$\text{Photosynthesis: } 6\ CO_2 + 6\ H_2O + \text{sunlight energy} \rightarrow C_6H_{12}O_6 + 6\ O_2 \quad \text{(Eq. 4.1)}$$

Productivity of an ecosystem is the amount of biomass produced by plant photosynthesis per unit of surface over a given period of time. It can be expressed as energy units, or more commonly, as dry organic mass units (kg ha^{-1} yr^{-1}). The total energy fixed through photosynthesis in a given year is called gross primary production (GPP). Some of this production is used by plants to obtain energy for metabolic activities through the process of respiration:

$$\text{Respiration: } C_6H_{12}O_6 + 6\,O_2 \rightarrow \text{metabolic energy} + 6\,CO_2 + 6\,H_2O \qquad (\text{Eq. 4.2})$$

The difference between GPP and respiration yields net primary production (NPP). This is the actual amount of biomass available for plant growth (Begon *et al.*, 2006).

The rate at which biomass is created by photosynthesis depends on two main plant features: leaf area and the efficiency in which leaves capture and use light. Both features can be limited by available sunlight, carbon dioxide (CO_2), water, and mineral nutrients. These are the resources needed to generate NPP in forests, with temperature as a major factor affecting basic photosynthetic and respiration rates.

Leaf area index (a dimensionless quantity that represents the green leaf area per unit ground surface area) is an important measure of potential productivity because it indicates the extent of site occupation by leaves with respect to light interception. Leaf area index is also affected by the way in which plants allocate NPP between their different organs. As with leaf area, this carbon allocation is influenced by the availability of water, light, and nutrients. NPP is allocated to the root system or to leaf and stem growth in response to which resources are in shortest supply: soil nutrients and water versus energy and carbon (Figure 4.1). Changes in allocation are thought to represent the plant's attempt to maintain optimum ratios between carbon and other nutrients, and between leaf area and the supply of light, moisture, and nutrients to that leaf area.

Leaf area index appears to affect NPP more than photosynthetic efficiency, though both are obviously important. High values for accumulated biomass are associated

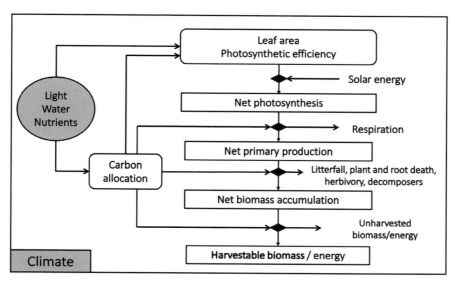

Figure 4.1

The major determinants of economic production (yield) in a forest ecosystem within a particular climate regime. The availability of site resources (light, water, resources) is critically important, both directly and indirectly, through their effect on the allocation of net photosynthesis. (Modified with permission from the original by J. P. Kimmins.)

with high leaf areas, but the efficiency with which the biomass is produced does not show a linear relationship to leaf area index. As leaf area increases, an increasing proportion of the tree biomass is under deeper shade, and therefore contributes less and less to the net production of biomass from photosynthesis. Leaf area index is functionally related to the cross-sectional area of sapwood. Sapwood is the component of tree stems that conducts water and nutrients to the foliage, and therefore is functionally related to leaf area. Basal area (or its surrogate, the number of trees in a stand) is the basic variable that can be manipulated by forest management. Therefore, it is important to understand how forest management affects the different factors that intervene in biomass production by photosynthesis.

Light

Photosynthesis is a process in which the rate of fixation of CO_2 and the capture of solar energy is dependent largely on light intensity. The rate of photosynthesis increases rapidly with light intensity (if other factors are not limiting), but initially there is no net CO_2 fixation (and therefore no biomass production) because the rate of CO_2 generated and lost through respiration is greater than the rate of CO_2 fixation. As light intensity continues to increase, a point is reached at which respiratory losses are balanced by photosynthesis gains. This light intensity is called the compensation point. Above it, the rate of photosynthesis continues to increase rapidly with increasing light intensity, but this relationship is not sustained. With continued increases in light, the rate of increase in photosynthesis diminishes until the saturation point is reached, beyond which further increases in light intensity result in little or no further increase in the net CO_2 fixation rate. At very high light intensities, net fixation may drop because of damage to the photosynthetic apparatus, or for other reasons (Figure 4.2).

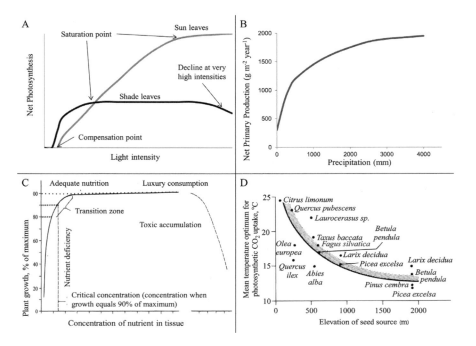

Figure 4.2 Relationship between tree productivity and main growth limiting factors: (A) light; (B) water (annual precipitation, modified from Reichle, 1981); (C) nutrients (modified from Ulrich and Hill, 1967); and (D) temperature. (Modified from Pisek et al. (1969).)

*Figure 4.3 Trembling aspen (*Populus tremuloides*) overstory with an understory of slower-growing white spruce (*Picea glauca*), Dease Lake, British Columbia, Canada.*

Some species can survive in more shaded conditions than others because their leaves can photosynthesize enough at lower light intensities to at least offset respiratory losses. Usually, more light is needed for whole trees than for shrubs and herbs, since the leaves must produce enough carbohydrates to keep stems, branches, and roots alive. The architecture, spatial arrangement of leaves, and respiration rates of stems, branches, and roots are all important in allowing some species to survive at lower light intensities than others. Such species are described as shade-tolerant. The extra sunlight utilized by shade-tolerant species contributes little to growth, although it may enable the species to survive in understory, shaded conditions (Oliver and Larson, 1996).

Forest management can influence light efficiency to enhance biomass production in a stand. For example, having two species coexisting, one fast-growing and shade-intolerant (such as trembling aspen, *Populus tremuloides*), can create an environment where a more valuable but slower-growing, shade-tolerant species (such as white spruce, *Picea glauca*) can survive (Figure 4.3). Later, when the aspen is harvested at maturity, the spruce is released and then grows in direct sunlight until it is ready for harvest (Welham *et al.*, 2002). Other management practices such as thinning can influence light use efficiency and increase stand leaf area by removing the trees which are occluded in the lower parts of the canopy and therefore getting access to less light. Due to this relationship between light and productivity, light has been used as the main factor driving some forest growth models, such as SORTIE (Pacala *et al.*, 1996).

Atmospheric gases (O_2 and CO_2)

Trees require oxygen to obtain energy from respiration. Oxygen is absorbed through bark lenticels, leaves, buds, and roots. Oxygen is usually not a limiting factor for tree production, except when it is not available for roots, as in waterlogged soils. Some species, such as tupelo (*Nyssa sylvatica*), bald cypress (*Taxodium distichum*), or mangroves can survive for several years in standing water, using a physiological method to move oxygen from stems and leaves to the roots (Oliver and Larson, 1996) (Figure 4.4).

*Figure 4.4 Floodplain forest dominated by bald cypress (*Taxodium distichum*) in the Apalachicola National Forest, Florida, USA. The height of the water level is indicated by the black marks on the stems. The spikes in the foreground are pneumatophores, modified roots that rise above the waterlogged ground to enable the trees to breathe.*

Carbon dioxide is absorbed by plants and combined with water to create organic compounds through photosynthesis. Entrance of CO_2 into leaves and its subsequent fixation depend on the concentration gradient between the exterior and the interior of the leaf. This gradient will be higher if atmospheric CO_2 concentrations rise above normal levels, such as in greenhouses or through fossil fuel emissions. CO_2 movement from the exterior to the interior of the leaf needs to overcome stomatal resistance, which is related to the degree to which stomata are open. This resistance is high in species with sclerotic leaves (thick and hard, usually adapted to dry environments). Even with numerous stomata, these species do not open the stomata very much (Terradas, 2001). The degree to which the stomata are open depends on the turgor pressure differential between the guard cells of the stomata and the surrounding epidermis cells. Turgor depends on ionic concentrations inside the cells, and any process that affects this concentration will result in changes to the flow of CO_2. This process is often related to the availability of water.

In a future atmosphere rich in CO_2, plants may become more productive due to the greater availability of carbon for photosynthesis. Historical increases in tree growth in relation to rising CO_2 concentrations within the 20th century have been identified (Graybill and Idso, 1993). However, this 'fertilization' effect will only be maintained if there are no other limiting factors. For example, soil nutrient levels will need to be sufficient to support increased tree growth.

Water

Water is the main resource that limits vegetation growth and survival in terrestrial ecosystems. Water is not only used in photosynthesis, but it acts as the medium in which all life-related chemical processes occur. Water also acts as the transport system permitting nutrient uptake from the soil, and in moving oxygen and metabolites around. Total biomass and NPP are often proportional to water availability. Water can be the limiting factor in many forests, from boreal to tropical ecosystems.

The production of biomass per unit of water used is a measure of how effectively plants use water, and is called water use efficiency (WUE). WUE has been investigated in a wide variety of plants. Many tree features (type of leaves, tree architecture, seasonal growth, etc.) have evolved to maximize WUE. As a consequence, large variations in WUE have been reported between species, sites of different productivity, and under different climatic conditions.

Plants obtain water primarily from the soil through the fine roots. To take up water, plants exert tension through evapotranspiration and are able to draw any water from the soil that is not strongly adhered to soil particles. If plant tension is greater than the attraction tension (matric potential) of the soil, water moves from the soil into the plant. Trees vary in their ability to extract water from the soil and endure conditions of high soil water tension without injury. By opening or closing stomata, trees have the ability to regulate water flows in relation to availability. Most tree growth occurs at soil water tensions of less than −1 MPa. There is a direct relationship among the energy, water, and carbon balances in the leaves, which are also affected by temperature. Any change in temperature changes the water vapour deficit in the air and changes the water tension in the plant, affecting evapotranspiration.

Tree species can use different strategies to endure periods of low water tension (low water availability). The stomata in the leaves can be closed to avoid drought, shutting off transpiration and maintaining high internal moisture levels, but at a cost of impairing photosynthesis. This strategy is typical from species living in arid sites or in sites with seasonal aridity, such as Mediterranean ecosystems or the dry tropics. In these areas, the dry season is usually a time of dormancy for trees.

Excess water can also have negative impact on tree productivity. Flooded soils prevent oxygen from reaching roots. Soil water can also contain dissolved salts, thereby increasing water tension. This makes it more difficult to absorb the water, effectively reducing water availability for plants.

Forest management can influence water availability in different ways. Irrigation can substantially increase productivity by increasing the standing leaf area (Campoe *et al.*, 2013). It is also possible to control the number of plants and trees at a site that are actively removing water from the soil. By reducing tree density, evapotranspiration can be reduced as well as rainfall interception, increasing soil water availability (Bréda *et al.*, 1995). For this reason, thinning can also increase runoff. In semi-arid environments with high solar radiation levels, thinning can increase light incidence to the soil surface, increasing soil temperature and therefore evaporative losses. In saturated soils, forest drainage (ditching) can reduce water content and increase air (and oxygen) content in the soil. However, this practice can have important environmental consequences, such as changes in biodiversity or impacts on forest streams.

Mineral nutrients

A large number of chemical elements can be found in plants. Many occur at very low concentrations (parts per billion) and have no known role in plant metabolism. However, other elements have important roles in plant metabolism, such as calcium (cell walls), nitrogen or sulphur (proteins and nucleic acids), phosphorus (energy-transport molecules), or enzymes controlling chemical reactions (magnesium, potassium, manganese, etc.). Only a few (N, P, K, Ca, Mg, Mn) are found at concentrations that can be expressed as percentages by weight of plant substance.

Most soil nutrients are not in soluble form, though solubility depends on pH. Most of the nutrients needed in higher quantities by trees (macronutrients) exhibit maximum solubility at pH values between 6.5 and 7.5, although metallic ions are generally less available above pH 7. Soil pH is not the only factor determining nutrient availability, however. Clay minerals and organic soil matter are rich in electrical charges, and therefore they have the capacity to store cations. Therefore, soil texture (the relative proportions of clay, sand, or silt) and the amount of organic matter in the soil define the cation exchange capacity (see Chapter 6). How full or empty that pool is depends on the rates of nutrient inputs and outputs. Such flows include the decomposition of organic matter that releases (or sequesters) nutrients into the soil, wet and dry atmospheric deposition, biological fixation, biochemical transformations (volatilization, nitrification, ammonification, and other transformations of soil nitrogen), runoff, erosion, leaching, or combustion of forest soils. These processes define the status of the soil with respect to its ability to supply the nutrients essential to plant growth, or in other words, soil fertility (Soil Science Society of America, 1973).

Once inside the tree, most nutrients are concentrated in the physiologically active organs: leaves, phloem (interior side of the bark), buds, fine roots, and reproductive organs. Trees can resorb some nutrients from foliage before it falls to the ground, reducing demand from the soil. This is done through the removal of amino acids from the leaves; the amino acids are produced by the breakdown of chlorophyll a and b, other ligands and apoprotein. At the same time, an abscission layer forms near the junction of the leaf and the stem, and this restricts the flow of minerals into the leaf, which is actually why the amount of new chlorophyll being produced declines. The loss of the green chlorophyll pigment reveals yellow and orange pigments that were previously masked (xanthophylls and carotenoids, respectively), and some trees start creating red and purple pigments (anthocyanins). Others accumulate tannins, giving the leaves a brown coloration. The rate and intensity of these processes vary geographically, being most apparent in northeast North America and northeast Asia (Figure 4.5). Larches (*Larix* sp.) can also create vivid fall coloration, but similar patterns are generally not seen in deciduous trees found in seasonally dry tropical and sub-tropical climates.

Forest management can influence nutrient availability in many different ways. Any practice that alters the capacity of the soil to retain nutrients or the size of the nutrient

Figure 4.5 Autumnal (fall) colours in Quebec, Canada. These colours are typical of the deciduous forests of eastern North America.

*Figure 4.6 Autumnal (fall) colours in Graubünden, Switzerland. Only the European larch (*Larix decidua*) have changed colour, with the other conifers being evergreen.*

pools will affect availability. Fertilization in forest plantations can increase nutrient availability and therefore tree productivity. On the other hand, slash burning can volatilize nutrients from harvest residues, particularly nitrogen, phosphorous, and sulphur. Tree thinning and harvesting, when removing only stems, exports only small quantities of nutrients from the site as their concentrations in stemwood are small. However, whole-tree harvesting that removes foliage, especially when combined with short rotations, can export an important part of the nutrients in the vegetation, leading to a drop in soil fertility. Some examples of yield declines from conifer plantations can be found in North America (Johnson, 1992), Oceania (Smethurst and Nambiar, 1990), Europe (Blanco *et al.*, 2005), and Asia (Bi *et al.*, 2007). This productivity drop can be relatively fast if slash burning or soil erosion also occur, and depending on the circumstances can reduce soil fertility for very long periods (Blanco, 2012). Forest management can also affect other important processes such as erosion, runoff, and soil leaching (see Chapter 6). If such processes are in constant action through time, they can also lead to losses of nutrient availability and therefore of soil fertility and stand production. Nutrient availability has been used as the main growth limiting factor in forest growth models such as FORECAST (Kimmins *et al.*, 1999).

Temperature

Temperature regulates the speed and activity of chemical reactions. Active metabolic processes for many organisms are restricted to temperatures between 0°C (the point at which water freezes) and 55°C. The respiration rate increases steadily through this range, although the rate of change varies among species. Gross photosynthesis, on the other hand, increases rapidly from the freezing point to a relatively constant 'plateau' between 8°C and 18°C, depending on species. Any further increase in temperature will cause a decrease in net photosynthesis (gross photosynthesis minus respiration) and therefore in tree productivity and growth. Outside this range, trees use different mechanisms to survive temperature extremes. Some species develop deciduous leaves or small and hard evergreen leaves; other species generate thick and isolating bark; and most species close the stomata.

Temperature varies considerably in different parts of a plant. Roots normally assume the temperature of the soil around them. In regions away from the equator, root temperature is lower than shoot temperature in the summer but may be higher in the winter. Shaded stems approximate air temperature under the canopy, although the insulating properties of bark cause stem temperatures to lag behind those of the air. Leaves may have temperatures higher, lower, or the same as the surrounding air. Thin leaves with high transpiration rates can be as much as 15°C cooler than air in summer, whereas thick leaves may be up to 30°C warmer than the surrounding air temperature.

Temperature in forests depends mostly on geotopographic conditions. Slope, aspect, altitude, and latitude affect the amount of solar radiation reaching the forest and influencing temperature. The topographic situation (valley bottom, mountain side or top, leeward side of the dominant winds, etc.) influences air movement and therefore air temperature. As forest management will have very little impact on these conditions, the main management decision related to temperature is to plant species that can survive at the minimum and maximum temperatures of each site, not only the historical ones, but also the ones that are expected in a future under climate change (Wang *et al.*, 2012).

Removing trees through harvesting or thinning changes the amount of solar radiation reaching the soil surface. Storm damage may create similar conditions, especially if the downed trees are salvaged (Figure 4.7). As a result, forest management can have important impacts on microclimate, affecting soil and soil surface temperatures. This has consequences for soil organisms such as decomposers and small plants. It can also impact the reforestation of sites, with the microclimate being so changed in some situations that the re-establishment of trees becomes very difficult.

In boreal and alpine forests, increasing soil temperature usually produces an increase in decomposing rates, as decomposing activity is stimulated. However, in temperate and arid tropical forests, increasing soil temperatures may reduce decomposition due

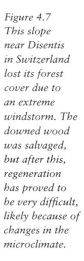

Figure 4.7 This slope near Disentis in Switzerland lost its forest cover due to an extreme windstorm. The downed wood was salvaged, but after this, regeneration has proved to be very difficult, likely because of changes in the microclimate.

to higher evapotranspiration and less moisture available for decomposers (Prescott, 2010). In new plantations, seedlings can also suffer from heat stress as they are close to the soil surface, where the highest temperatures are reached and there is little bark to protect the cambium. This can be prevented by shading all or part of the seedlings, artificially or with nursery crops (Helgerson, 1990). Seedlings and understory can also have a sudden temperature stress if the overstory is removed and solar radiation reaching the soil increases (Childs and Flint, 1987). Temperature is the main growth limiting factor in some forest growth models, such as 3-PG (Landsberg and Waring, 1997).

Ecosystem composition, trophic relationships, and food webs

Ecosystem productivity increases as ecosystem composition becomes more diverse until reaching a maximum at medium diversities. However, productivity decreases if diversity continues to increase. No general model has been derived yet that would explain this relationship. There are several hypotheses to explain this relationship between species diversity and site productivity.

One hypothesis states that species diversity is related to micro-site diversity (Rozenzweig and Abramsky, 1993). In theoretical ecology it is well known that one ecological niche can carry only one species. This theory says that average sites have more niches than very poor or very rich sites. For example, in poor sites (cold, dry, or nutrient scarce), all temperature, moisture, or nutrient availabilities have low values. This is the result of a unique combination of factors that represent a specific site with very low productivity. Similarly, rich sites (warm, moist, and nutrient rich) are also the result of a unique combination of factors. However, sites of average productivity can be reached with a great number of combinations of site factors. Each combination represents a specific niche, which can carry a specific community and hence also have higher species diversity. According to this theory, site diversity is maximum on average sites, and as a consequence species diversity is also maximum on average sites. This theory supposes that productivity is a consequence of resources being used in a more complete way when diversity is high, as different species will be using different portions of the available resources (Terradas, 2001).

A second hypothesis is based on the theory of the limiting factor (Rozenzweig and Abramsky, 1993). This theory states that when site productivity is high all species have the potential to survive. However, a large number of species on the same site leads to strong competition, resulting in productivity reductions. This theory assumes that diversity is a consequence of high productivity and inter-specific competition. Hence, low diversity can be caused by the strong competition of the most competitive and dominant species, which suppresses other species. In forest communities, this phenomenon is easily observed when one dominant tree species dominates the canopy, uses most of the site resources, and suppresses other tree and plant species, reducing species diversity over time.

In forestry applications, the relationship between ecosystem productivity and composition is closely related to the higher production observed in mixed forests compared to monocultures (Pretzsch *et al.*, 2012) (Figure 4.8). This has become an important issue because of pressure to adopt what are viewed as more natural forms of forestry (near-to-nature forestry), and which eschew exotic species and monocultures. In both planted and natural forests, this approach to management results in greater areas of uneven-aged and heterogeneous forest stands, which complicate the use of traditional growth and yield models. Some efforts have been made to create models that would enable the determination of volumes of several mixture types and forms, such as the SILVA model (Pretzsch, 2010).

Figure 4.8
Mixed species
plantation
of Schrenk
spruce (Picea
schrenkiana
var. crassifolia),
Chinese red
pine (Pinus
tabuliformis),
and Hebei
poplar (Populus
tomentosa),
Xining, Qinghai
Province, China.

Trophic relationships and food webs

Forest ecosystem productivity and diversity takes into account not only plant productivity, but also the productivity of herbivores, carnivores, and saprotrophs (decomposers). They use plant production (NPP) as an energy source, but they can also be the source of energy for other organisms. Biomass (and its associated energy and nutrients) flows through the ecosystem following the food (trophic) chains. However, the same organism can be part of different trophic chains, as it can feed on or be food for different organisms. Trophic chains are then organized in trophic webs at ecosystem level.

Understanding how energy and nutrients flow through the different parts of the forest ecosystem is important when developing sustainable forest management plans. Forest management generally involves the production of a crop, whether timber, wildlife, mushrooms, berries, or other materials. Many management practices influence energy accumulated in the forest biomass. For example, maximum aboveground biomass values have been reported to reach up to 422 Mg ha^{-1} and 415 Mg ha^{-1} in mature temperate and tropical forests, respectively, but up to 3,461 Mg ha^{-1} in old growth forests in coastal conifers in California (Kimmins, 2004). Converted to energy at 4,000 cal g^{-1}, the energy contained in this aboveground biomass could equal to 2,000 Mg of coal. This energy could be used for local heat or even electricity production in biomass power plants, but then the energy contained in the forest biomass is not available to flow through the trophic webs, as it would leave the ecosystem through harvesting.

When a forest is harvested there are several important changes in the distribution of energy and how it flows in the ecosystem. First, by removing the overstory canopy, most of the energy entering the trophic web is eliminated, and photosynthesis is temporarily carried on only by understory vegetation. Second, there is a potential increase in the energy flow through the detritus food web. Depending on the type of harvesting (stem-only, whole-tree, or complete-tree), variable quantities of tree biomass (slash) are deposited on the ground. After clearcutting there is characteristically a period of rapid reduction in the thickness and biomass of the forest floor, because biomass inputs to the forest floor through litterfall are greatly reduced. Increased detritus energy flow after

*Figure 4.9
Chinese fir
(*Cunninghamia
lanceolata)
seedling,
Nanping County,
Fujian, China.
In growing
this species,
the organic
matter from the
previous stand is
usually removed,
a practice that
may reduce the
allelochemical
content of the
litter.*

harvesting is the combined result of both increases in the quantity of decomposable organic matter and changes in the condition of the forest floor. Summer daytime temperatures at the forest floor increase after logging. Such changes can enhance decomposer activity in temperature-limited boreal, alpine, and cool-temperate forests, but it can inhibit decomposers in warm-temperate and Mediterranean forests, as warmer soils are usually also drier (Blanco *et al.*, 2011). In addition, removing trees causes the breakdown of mycorrhizal relationships, a reduction in the mycorrhizal fungi biomass, and an increase of free-living saprotrophs.

Loss of forest floor biomass and depth after clearcutting can have several effects on the ecosystem. Forest floor water and nutrient storage capacity may be reduced, which can be undesirable in hot, dry climates and infertile sites, but desirable in cold, humid mineral soils. On sites where the forest floor has adverse chemical properties (too little N in the litter, allelochemicals present, etc.), or where deep layers of low bulk density material pose problems for regeneration (because it dries rapidly in summer), the reduction in depth may favour the reestablishment of a tree crop (Figure 4.9). As the pioneer plant community re-establishes a foliage canopy, the entry of energy into the ecosystem is gradually restored. Litterfall inputs to the forest floor gradually increase, much of which decomposes rapidly. The development of summer shading increases the moisture levels in the slash and forest floor, which promotes decomposition. However, as the tree canopy develops and succession proceeds, the type of litterfall changes, mycorrhizal fungi begin to dominate the soil microflora, and decomposition rates slowly return to pre-logging values.

Accompanying the post-harvesting increase in detritus food web energy flow there is a reduction in energy flowing through the grazing food web. The reduction intensity is proportional to the reduction in leaf biomass available for herbivores. The invasion of the stand by pioneer herbs and shrubs after harvesting rapidly re-establishes the grazing food web, often at a higher level of energy flow than before logging. This is due to the greater physical accessibility, palatability, and nutritive value of the understory in comparison to the mature tree canopy. This can result in more diverse, abundant, and productive animal communities in harvested stands than in the original uncut forest,

*Figure 4.10
The moose
(Alces alces) is
one of a number
of species of
ungulate that
can benefit from
the improved
food supply that
occurs after a
disturbance.*

as long as the wildlife requirements for shelter and winter range are satisfied (Welham *et al.*, 2012) (Figure 4.10).

Productivity of timber and non-timber resources

Producing both timber and non-timber forest products (NTFP) supposes that ecosystem productivity is used to generate both type of values. Each value is associated with different trophic webs. As a consequence, implementing forest management to favour one forest product could reduce the energy and materials available to produce another product. For example, reducing tree cover to increase understory production would increase production of game animals by diverting energy into the herbivore trophic web. However, it also lowers the potential for timber production. Similarly, favouring forest floor development could increase production of mushrooms (fruiting bodies of saprotrophic fungus in the decomposers trophic web), but reduce nutrient availability for tree production as nutrients get sequestered in the litter layer.

Maintaining forest structure associated with NTFPs production is generally acknowledged as being positive, contributing to forest environmental functions like carbon storage, nutrient cycling, erosion control, and hydrological regulation. Moreover, forests managed for NTFP production can retain large amounts of plant and animal biodiversity (Michon and de Foresta, 1997). On the other hand, NTFP harvesting results in direct and indirect pressures on the forest, due to competition between humans and animals for some forest foods (Boot and Gullison, 1995). There are consequently concerns about just how benign the harvesting of NTFPs is, and whether these arguments have been overstated. NTFP depletion or removal can rapidly influence such forest characteristics as composition and structure of vegetation. For example, extraction of bark can lead to the death of the tree, while the harvesting of fruits and flowers may have negative results for the reproduction of the population. Some species are better able to sustain continuous off-take than others. In the case of vegetation, those species exhibiting abundant and frequent regeneration and rapid growth will prevail.

Table 4.1 Estimated trade-offs between the effects of certain management options on selected ecosystem goods and services.

Management options	Tree productivity	Understory productivity	Biodiversity	Carbon	Water	Amenity values
Stand level						
Structural retention	–	(–)	+	?	+	+
Use of native species	–	?	+	(–)	(+)	+
Mixed-species stands	+	–	+	+	–	+
Long rotations	(–)	–	+	?	+	+
Short rotations	(+)	+	–	?	–	–
Thinning	0	+	+	(–)	+	(+)
Whole-tree harvesting	(+)	+	–	–	+	–
Site preparation	+	–	–	?	?	–
Herbicides and fertilizer	+	–	–	?	–	–
Landscape level						
Riparian buffers (of native vegetation)	–	+	+	(+)	+	+
Retaining patches of native vegetation	–	+	+	(+)	+	+
Connectivity between plantations and native forests	0	(+)	+	0	0	+
Maintaining landscape heterogeneity (different land-use types, special places, etc.)	–	(+)	+	?	+	+

Note: + = positive effects, – = negative effects, 0 = neutral effect, ? = unknown or uncertain effects, brackets indicate that the effect may not be so clearly positive or negative depending on other factors not captured here.

Modified from Bauhus *et al.* (2010).

Likewise, rodents, ungulates, and other animals that have broad niches and prolific reproductive strategies are more able to withstand heavy hunting (Fa *et al.*, 1995).

As can be seen in Table 4.1, most of the measures that benefit biodiversity impact negatively on tree productivity, both at the stand and the landscape level. However, at the landscape level, it is important to separate between effects that have impact on the tree production per unit of forest land and on the overall landscape. The synergies and trade-offs between biodiversity and productivity at the landscape level depend largely on the forest policy context. If plantation establishment is directly related to and dependent on the area of native forests set aside for conservation, there can be strong synergistic effects (Paquette and Messier, 2010; see also the later section, 'The TRIAD approach to the management of forested landscapes').

As with any form of natural resource management, there is no form of plantation management that can provide a maximum of all timber and non-timber products. It is simply not possible to maximize wood production, carbon sequestration, conservation of biodiversity, and social and cultural benefits in the same forest stand (Bauhus *et al.*, 2010), and attempts to do so will always result in one or more values suffering while others benefit.

Changes in site productivity as they relate to semi-natural forests and plantations

Plantations are forests established from the intentional planting or sowing of propagules (seed, seedlings, or cuttings) to generate an even-aged stand, with the aim of producing particular forest products, or for protective purposes (Schuck *et al.*, 2002). Geometric plant spacing is a classic feature of these stands, in addition to artificial regeneration with either native or non-native species. Intensive management techniques and protection measures are employed initially, and sometimes on an ongoing basis, to maintain growth trajectories and stand composition. Stands established as plantations but which have been without intensive management for a significant period of time are likely to revert to a semi-natural condition. Semi-natural forests include tree species that occur naturally in the stands, either from ingress or through artificially assisted regeneration. Hence, they show similarities to natural forests in terms of age class distribution, composition, and structure.

Conversion of natural forests to plantations can occur after harvesting (live and salvage logging), following a stand-replacing fire, or when the existing stand is killed by pathogens or a severe climatic event, such as a drought. Regardless of the disturbance type, site preparation is undertaken in most cases to improve planting conditions and ensure successful establishment of the crop species. This can include slash removal, prescribed burning, mechanical treatments, and in some cases, drainage and irrigation (Smith, 1962). Natural forests build up considerable stocks of dead organic matter and propagule banks over generations (seed stock, for example). This historical legacy can confer substantial resilience to disturbance in terms of the ability of a system to return to equilibrium stability (see Holling and Gunderson, 2002) following a perturbation event. Productivity of the crop in the next rotation is therefore likely to be at least equivalent to the prior cohort, assuming disturbance is not too severe (see Bi *et al.*, 2007). Productivity could even be enhanced if a previously limiting factor can be mitigated through management (i.e. insufficient stocking, nutrient availability, poor drainage). Conversely, tree growth could be reduced if the disturbance event and/or site preparation significantly impair key processes driving ecosystem recovery.

A fundamental tenet of sustainable forest management is to maintain forest capacity to provide a characteristic suite of ecosystem goods and services. Although disturbance alters structure, composition, and function, given sufficient time a forest will restore the original balance in these ecosystem attributes. A pattern of non-declining change is thus sustained over the long-term as the disturbance-renewal cycle is repeated. A characteristic of many plantations, however, is that in an effort to generate higher economic returns, the renewal time between harvesting events (i.e. the rotation age) is often much reduced over what occurs in the natural forest. If the rotation age is too short, not all ecosystem attributes will be completely restored, leading to an eventual decline in productivity. One example is dead organic matter (DOM). In northern coniferous forests, DOM is an important source of nutrient inputs (see Laiho and Prescott, 2004) and long-term nutrient storage and release. Depletion of DOM stocks can occur because branches and foliage are often separated from the log bole at roadside, where they are then burned in slash piles. As part of site preparation, DOM is often cleared from the site to facilitate planting (Figure 4.11). Finally, shorter rotation times interrupt the mortality patterns associated with stand development in natural forests, which does not allow for sufficient re-accumulation of DOM (Harmon *et al.*, 1986). Because of inherent resilience, the impacts associated with the degradation of DOM stocks may not become apparent until several rotations later.

*Figure 4.11
Slash derived
from harvesting
piled up rather
than distributed
through the
forest, western
Tasmania,
Australia.*

Maintenance of both forest area and forest quality

Sustainable forest management needs to provide the resources that society is demanding from the forests, but in a way that both current and future demands can be met. This requires the maintenance of both forest area and productivity (forest quality).

Maintaining forest area depends largely on forest policies, forest types, and how nutrients, energy, and water are circulated in the ecosystem. For example, tropical forests growing on extremely nutrient-poor soils, such as weathered soils in the Amazon basin and in northern Australia, or the peat forests of Sumatra, are of low productivity. These low-productivity forests lack the species diversity, multilayered canopies, and tree heights associated with tropical forests growing in more fertile sites, such as alluvial soils in the floodplains of tropical rivers, or young volcanic soils (Kimmins *et al.*, 1999). Where forests on nutrient-poor, low-quality sites are cleared for agriculture or other land uses, soil tends to degrade very rapidly, and such areas are usually abandoned within a few years. If they are abandoned before soil fertility or soil structure has been severely damaged, an impoverished secondary forest will rapidly invade. Over time, trophic webs and energy flows will be gradually restored, and the original forest could substitute the secondary forest over time. This is the usual course of events under the traditional practice of shifting cultivation (Figure 4.12).

On the other hand, if the soil is used for longer times and to produce more intensive harvests, the soil becomes too impoverished to allow the development of secondary forest. Pioneer grass or shrub species with the ability to survive in very poor sites can dominate the area and prevent trees from establishing. Burning the site may remove the understory and improve site conditions temporarily as some of the nutrients stored in the plants are released into the soil. Generally, however, the result is a loss of soil organic matter and nutrients (Bi *et al.*, 2007). This is because most of the nutrient pools in low-quality tropical forests are in the vegetation. Once the plants are removed, this pool is lost and nutrients become more limiting, even if water, light, and temperature are favourable. The resulting low site quality forces the farmers and loggers to clear more forest land, and deforestation then spreads. Forests on fertile, high-quality sites,

Figure 4.12 Forest cleared during shifting cultivation, Crocker Mountains, Sabah, Malaysia.

*Figure 4.13 A plantation of hybrid Eucalyptus (*Eucalyptus urophylla x grandis*) on Hainan Island in southern China. The trees, which are 30 months old, have an average height of 15 m and average dbh of 12 cm. The trees will be harvested for pulp at an age of about 36 months. The spacing is 4 m × 1.67 m. Great care will be needed on this site to ensure that productivity is maintained.*

in contrast, are much more resilient, and have soils with appreciable nutrient reserves. Such soils are capable of supporting sustainable forestry if managed in a way that conserves their soil organic matter, nutrient reserves, and structure and avoids erosion. Intensive forestry in plantations established in such sites is possible, if the site quality is preserved (Figure 4.13).

In temperate and boreal forests, the same considerations for sustainable ecosystem management apply. To preserve the existing forest area, productivity should be maintained through time so no new areas need to be brought under management. It is therefore necessary to avoid overexploiting site resources. In many temperate forests, and most boreal sites, soils are relatively young and mineral nutrient pools are usually adequate to support tree productivity. Species diversity is lower in temperate and boreal forests than in the tropics. Management practices that create a mosaic of forest and non-forest landscapes can increase species diversity at landscape scales. In temperate regions of Europe, the abandonment of low-productivity agriculture and pasture lands has led to an expansion of forest area during the last decades, thereby contributing to the landscape mosaic.

In terms of forest quality, practices used to increase short-term tree productivity such as short rotations, whole-tree harvesting, and intensive site preparation can result in nutrient and organic matter exports exceeding natural inputs. This will eventually lower site quality and cause a reduction in productivity. This decline is often cumulative and non-linear across rotations. This phenomenon has been documented in conifer plantations around the world (Bi *et al.*, 2007). Natural nutrient inputs (by atmospheric deposition, biological fixation, mineral weathering, etc.) can be supplemented by fertilization. To maintain growth rates, fertilizer doses usually need to be increased every rotation. However, increasing the amount of fertilizer can have environmental consequences, such as increased nitrogen leaching to watercourses and groundwater (Wei *et al.*, 2012).

Managing the forest stand to grow under longer rotations enables the development of biological structures such as snags, logs, canopy structure, and so forth, which mimic some features of natural forests and could even reach some of the attributes of old growth forests, provided that the rotation is long enough (Franklin *et al.*, 2002).

A sustainable basis for the intensification of production

Advantages and disadvantages

As discussed in Chapter 1, the intensive culture of woody crops is an activity dating back many centuries. The modern concept of the short rotation woody crop (SRWC) can be defined as a silvicultural system based upon short clear-felling cycles that employs intensive cultural techniques such as fertilization, irrigation, and weed control, and utilizing genetically selected planting material (Drew *et al.*, 1987). A key feature of successful SRWC culture is density regulation; plantations are typically established at densities well exceeding 10,000 stems ha^{-1} to minimize weed competition and ensure that site resources are fully utilized. Dickmann (2006) added the proviso that SRWC systems often rely on coppicing, but this is only appropriate for some species. Extensive selection and biotechnology programs have also played a critical role in producing high-yielding tree varieties and efficient clonal propagation methods. SRWC systems are used as sources of fuel, bioenergy feedstock, and bioremediation, and to promote carbon sequestration.

For obvious reasons, intensive production has the benefit of generating the desired products or objective in the shortest time possible. Establishing a new plantation is relatively easy because many species can be propagated asexually, or seed can be quickly produced in large amounts (see Dickmann, 2006). Following harvest, coppice regeneration often ensures vigorous regeneration and biomass accumulation, although clonal vigour often declines over subsequent rotations and a new plantation must be re-established. Many eucalypts regenerate freely through coppicing. Jarrah (*Eucalyptus marginata*) is particularly recognized for this ability, but some eucalypt species important for timber, including mountain ash (*E. regnans*), alpine ash (*E. delegatensis*), and rainbow gum (*E. deglupta*) do not regenerate in this way. There is evidence of improvements in soil organic matter, structure, and fauna when formerly agricultural sites are converted to plantation forestry (Makeschin, 1994a, b).

Challenges to realizing the potential of intensive production and then ensuring that it is sustained can be grouped into two broad categories: issues with plantation performance and site-related features (see Makeschin, 1999). Weed competition is a major factor limiting plantation establishment and growth. Effective weed control depends on intensive site cultivation usually in combination with herbicides. Fertilization is often required to promote early growth and then sustain productivity over the long term (both within and across rotations). This can introduce potential issues with nutrient leaching and runoff, thereby impacting ground and surface water quality. As with any monoculture, pests and diseases can be problematic. Protection of young seedlings from herbivore browsing may be a necessity. Insect and fungal pest control often requires the application of chemical pesticides and fungicides. Selection of disease-resistant clones and materials well suited to prevailing site and climate conditions are prerequisites to avoiding problems and realizing the sustained yield potential.

Site-related features include the inherent characteristics of a given site (biodiversity, soil), and how repeated cropping impacts soil quality. Experience has demonstrated that the potential of intensive production systems can best be realized when sites possess adequate nutrient and moisture regimes at the outset. As discussed in Chapter 6, soils that are either poorly or excessively drained, infertile, stony, shallow, highly acid or alkaline in pH, or degraded through erosion, compaction, or salinization simply will not support the productivity needed to justify an investment in intensive forestry. Wet soils prohibit the use of mechanized equipment because of soil compaction, while steep slopes introduce the risk of soil erosion. Conversion from a natural forest to intensively managed single-species plantations results in a severe reduction in biodiversity. This is not only in terms of overstory composition but also in the understory, whose abundance must be severely controlled to eliminate competitive effects. It is likely that the soil biota is also affected (Makeschin, 1999). In the long term, intensive plantation management will likely lead to losses in soil organic matter, with associated reductions in available nutrients. This will require the use of fertilizers if production is to be maintained (Merino *et al.*, 2005). Alternative management options include selecting tree species with higher nutrient use efficiencies, lowering the planting density, increasing the rotation length, and reducing the harvesting intensity.

Even-aged versus uneven-aged management

Traditionally, forest management is divided into two categories: even-aged and uneven-aged management. The categories are quite administrative in nature, but do represent different biological approaches to timber production and the progressions of stand structures through time. In even-aged management, a stand regenerates after a disturbance which removes most of the preceding stand. This disturbance could be natural, such as a hot fire or intense windstorm, or anthropogenic, such as a clearcut. In either case trees regenerate during a short period after the disturbance and then grow as a cohort of trees quite similar in age (Oliver and Larson, 1996). In uneven-aged management, the stand is not replaced after a large disturbance, but rather regenerates periodically after partial disturbances which only remove part of the stand. Trees regenerate after each disturbance, creating a stand with trees of multiple ages.

Uneven-aged management can be carried out either with a small amount of volume removed in very short cycles, such as every five years, or by larger harvests on a longer cycle, such as 20–25 years. The idea with light but very frequent harvests is that regeneration will occur continuously and no distinct cohorts will exist. If the harvests are heavier, but less frequent, trees will invade until the growing space is reoccupied and then there will be no new regeneration until the next cut. The stand will consist of multiple distinct cohorts. The silviculture system using the periodic light cuts is called single tree selection because the openings are very small. The system using periodic larger cuts and therefore bigger openings is referred to as group selection. The former results in an all-age stand and the latter in a multi-cohort stand. Both are variations of uneven-aged management.

Theoretically there should be no difference in biomass production between the various systems. Leaf area per hectare is a function of the species and the amount of available water and reaches equilibrium relatively quickly after each harvest, as soon as the growing space is reoccupied (a point often referred to as canopy closure). Differences in productivity on a given site, therefore, are a result of how much time the site is not fully occupied and the species composition. Shade-tolerant species tend to be more productive than shade-intolerant species, but often the shade-intolerant species are more commercially valuable. The lighter the harvest, the more difficult it is to establish regeneration, particularly for shade-intolerant species. This has important implications, as it means that even a very light harvest operation can significantly change the composition of a forest, moving it towards dominance by shade-intolerant species. In group selection cutting in the temperate latitudes the openings can be spaced so that light entering the stand through one opening will shine diagonally below the live crowns of the uncut trees and allow regeneration to become established throughout the stand. When light is entering through multiple openings there can be enough light for more shade-intolerant species to establish.

The greatest impact of different types of management on productivity is the frequency of stand entries with heavy ground-based harvesting equipment such as skidders. Frequent entries will compact the soil throughout the stand and reduce productivity over time unless permanent extraction trails are established and used for each entry. Even with frequent entries for thinning, it is easier to utilize a permanent trail system in even-aged stands, somewhat more difficult in group selection, and very difficult in single tree selection (Figure 4.14). Where single tree selection has been practiced for some time, a dense network of permanent trails may develop: this could be seen as an adverse consequence of single tree selection systems. Logistically, the harvest system will largely dictate the economic feasibility of other silvicultural activities, and this will

Figure 4.14 Lightweight tractor being used to extract logs at Saihanba Tree Farm, Hebei Province, China. Use of smaller equipment reduces the risk of problems associated with soil compaction.

affect productivity. For example, the costs of the management of competing non-tree vegetation are more a function of the area to be covered than the amount of vegetation; therefore it is usually only practiced in even-aged management since the treatments would occur infrequently (usually only once a rotation).

As well as the major effect of species composition on productivity, within-species genotypic variation should also be considered. Although planting can be used in uneven-aged management, there is usually reliance on natural regeneration. Natural regeneration maintains the genotypes on the site, but genetically improved stock can be used in planting, as can genotypes more suitable for the site (given that climate change may have changed the growing conditions of the site since the establishment of the previous stand). As a result, even-aged plantations using improved stock can have much higher productivity of a target tree species than an uneven-aged stand.

Short rotation versus long rotation

Controlling rotation length is one of the common strategies in forest management. Rotation length is defined as the period between the establishment of a stand of trees and the time when that stand is ready for the final cut. Rotation length depends on the growth performance of the tree species, its provenance, phenotypic variations, market needs, financial concerns of the landowner, and other factors (e.g. mitigation of climate change effects). For example, in Finland, a short rotation length for Norway spruce (*Picea abies*) is about 30–50 years and a long rotation length is about 60–80 years (Pyörälä *et al.*, 2012), while for Chinese fir (*Cunninghamia lanceolata*) in China, a short rotation length is about 10 years and a long rotation length is about 50 years (Xin *et al.*, 2011).

Rotations can be calculated in a number of ways. For example, a technical rotation is the period required to produce a specified type of forest product. An economic rotation is the period over which mean annual return on investment is maximized. An ecological rotation, on the other hand, would be the time required for a site to return to the pre-harvesting ecological condition (Kimmins, 2004). Intensive silvicultural practices, such as site preparation and vegetation control, reduce initial competition from woody species and thus permit the persistence of early successional species, increasing overall diversity (Jeffries *et al.*, 2010). Therefore, whenever a forest is harvested, the site is reverted to an earlier stage of the ecological succession. This can be both ecologically sound and economically desirable in order to favour a particular species in the ecosystem – for example, Douglas fir (*Pseudotsuga menziesii*) in a predominantly western hemlock (*Tsuga heterophylla*) area, willow (*Salix*) or poplar (*Populus*) in a predominantly oak (*Quercus*) area, or Scots pine (*Pinus sylvestris*) in a predominantly Norway spruce (*Picea abies*) or silver fir (*Abies alba*) area.

However, if the disturbance is excessive it can create problems for future forest productivity. For example, the more frequent the disturbance is (or in other words, the shorter the rotation), the more prolonged are the early stages in the succession, which are usually dominated by less desirable or less productive vegetation. An alternative could be a moderate degree of disturbance (intermediate rotation length), which can be a desirable compromise between economical, technical, and ecological rotations. However, if repeated at intervals shorter than the time required for complete successional recovery, such a practice can result in a gradual retrogression, ultimately reaching the same non-productive condition produced by the high degree of disturbance. A good example of this situation is the management of Chinese fir plantations in China (see the section 'Chinese fir plantations for furniture and structural timber').

In terms of the renewability of resources, site nutrient capital is tightly linked to the influence of forest management on nutrient flows. As discussed earlier in this chapter, each forest has a certain capital of these nutrients which exists as a dynamic equilibrium between inputs from the atmosphere and soil weathering, and losses in stream water and other pathways. Harvesting inevitably results in some depletion of the site nutrient capital through losses in harvested materials (Blanco *et al.*, 2005), as a result of disruption of the nutrient flows and retention mechanism, or as the result of other post-logging site treatments such as slash burning. These losses are replaced in time by natural inputs. However, for a given loss of nutrients on a given site, there will be a given nutrient recovery period, and if this period is shorter than that needed to replenish the nutrient pools, a reduction in site fertility will develop over time. This period can be referred to as the nutrient recovery rotation. In a sustainable forest management scheme, rotation length should be at least as long as the period to recover nutrient capitals. Otherwise, artificial nutrient inputs such as fertilization may be needed to accelerate recovery and allow for shorter rotations without endangering site productivity. However, fertilization carries risks such as eutrophication. The minimum sustainable rotation length can vary markedly among different ecosystem types, depending on the size of the nutrient pools and their flows through the trophic chains.

Rotation length also depends on the management objective. For example, for a given site, the annual mean timber production could be increased by using longer rotation lengths (60–80 years in the case of Norway spruce). However, if the same stand is managed to produce biomass for energy, this objective could be maximized by using shorter rotations of 30–50 years (Pyörälä *et al.*, 2012). In this sense, the use of short rotations is becoming an increasingly promising tool in willow and poplar plantations for biomass. However, conflicting long-term effects have been reported: soil bulk density decreased, soil porosity increased, and organic matter accumulated at a study site in northeast Germany (Kahle *et al.*, 2007), whereas in the United Kingdom long-term negative impacts on mammals, tree productivity, and diversity have been reported (McKay, 2011). Such differences point again to the need to assess the sustainability of long or short rotations in the context of specific forest ecosystems.

International examples of intense production systems

The proportion of the world's industrial wood sourced from intensively managed forest plantations is estimated at just over a third today and is increasing. This increase is having major implications for planted forests around the globe, and it will have major consequences on the sustainability of intensive forest management. Some examples are provided in the next three sections.

Loblolly pine plantations in southern USA

In North America, industrial forest harvesting is highly mechanized. Most operations are similar to those found in boreal forests and are driven by concerns about profitability, cost-efficiency, and operational control. Clearcutting is frequently used to harvest mature softwood stands. In sensitive areas, various types of selective harvesting systems are practiced, both in coniferous and deciduous forests (except, until recently, in harvesting operations along the Pacific coast). Harvesting is usually carried out by companies, contractors, or small-scale loggers, regardless of the type of forest or forest ownership.

In the southern USA, loblolly pine (*Pinus taeda*) is one of the most important commercial species, occupying more than 13 million ha. Within its native range loblolly pine encompasses 15 southern and mid-Atlantic states from Delaware to Florida and west to southeastern Oklahoma. Nearly 90% of the forest land in US southern states today is in private ownership (Wicker, 2002). Loblolly pine is managed in plantations at a range of intensities (extensive to very intensive) and is used for the production of both roundwood pulp and saw logs. Although most of the pine harvest comes from non-industrial private forest landowners (NIPFs), intensively managed plantations are only found on industrial land, usually held in the form of land management corporations. Plantations owned by the NIPFs are managed at a much lower intensity.

Intensive management practices include deployment of genetically improved seedlings, cloned from fast-growing varieties. Each plantation usually includes three to five clonal varieties to maintain genetic diversity at the local level. Because the land is very flat, site preparation can be intensive (using both prescribed burning and chemicals) and planting can be mechanized, using bare root nursery stock, usually at densities ranging from 1,200 to 1,500 stems ha^{-1}. Competing vegetation is aggressively controlled using an application of herbicide along each row at the time of planting and then broadly either by air (usually a helicopter) or by machine (usually a skidder) at least twice in the first five years. Rotations on the most intensive plantations may be 15 years.

Later stand management operations generally include occasional or frequent thinnings. Fertilization is also often applied, especially in industrial plantations.

Short-rotation willow plantations for bioenergy in Europe

In Europe, timber harvesting is prohibited throughout substantial forest areas because of their special importance for biodiversity, recreation, soil, and water protection. Only 75% of the total forest area (around 139 million ha) is considered to be available for wood supply in the 39 countries of the European Economic Area (EEA) region. However, maintaining the economic viability of forest production remains a challenge for many European forest owners, and to cope with this, forestry is expected to develop new and more intensive methods.

Increasing demands on forests as a resource for bioenergy and other products is already affecting the plantation area in Europe. The area of planted forests in the EEA region has increased over the last 15 years from 10.9 to 13.3 million ha (almost 8% of the total forest area). This legitimate need does not necessarily conflict with biodiversity and ecosystem conditions. However, recent EU policy developments have reinforced the need for renewable energy and may lead to several countries further promoting biomass from forestry. Measures to increase forest production and/or more intensive use of forests could potentially conflict with biodiversity protection.

In areas prone to desertification, such as the Mediterranean basin, forest plantations have been promoted to combat land degradation. In northern Europe, many new plantations have been established on former abandoned or low-productivity agricultural lands. In Sweden, there have been efforts over the past 20 years to develop intensively managed willow (*Salix*) plantations to produce biomass for energy generation while maintaining sustainable yields. Willows grow in areas with high evapotranspiration rates and high nitrogen retention rates. In addition, the ability of willow to take up heavy metals makes it suitable for multipurpose forestry, in which biomass production can be combined with water purification or as a receptor of municipal sludge (Perttu, 1999).

Willow is a pioneer species, with high light demands, and therefore competition with the understory for light can be critical. Successful establishment is usually achieved without competition with other vegetation. After establishment, willows can attain a leaf area of 6 m^2 m^{-2} or more, effectively suppressing understory growth through shading (Lindroth *et al.*, 1994). Soil pH should be in the 5 to 7 range, and although it is very efficient at using water, the high levels of biomass that it can achieve makes water availability a critical issue. As a consequence, light, shallow soils with low water retention capacity should be avoided. Understory, especially perennial grasses, can be controlled by growing cereal in the field in the season before planting the willows, or by applying herbicides. Understory competition must be controlled during the establishment year. Mechanical weeding may be needed as many as three times during the first year, after bud burst and early shoot formation, taking advantage of the flexibility of willow shoots, which makes them less prone to mechanical damage.

Typical plantation densities are 10,000–15,000 trees ha^{-1} (Figure 4.15). Rotation lengths are usually in the range of three to five years in a coppice system undergoing multiple cutting cycles, although the development of faster growing clones is enabling reduced cutting cycles of two to four years. In southern Europe, cutting cycles can be shortened to as little as one year, but this is only possible when water and nutrients are provided in sufficient quantities to ensure the replenishment of growth reserves every year, and a three- to four-year cycle is more typical. Frost can reduce biomass production by 50%, and therefore sites exposed to late spring frosts should be avoided. For planting, one-year-old monoclonal cuttings are used, with clonal mixtures being favoured to avoid pathogen adaptations and the spread of diseases (McCracken *et al.*, 2003). Planting is usually carried out mechanically, with 20 cm cuttings being taken from 2 m long willow rods; these are then pressed into the soil until only 2 cm tops protrude above the soil surface (Verwijst *et al.*, 2013).

To fully exploit the growth potential of willows, soil fertility must be comparable to that of agricultural fields. To maintain growth in the long term and avoid yield decline, dry sites are avoided and nutrients are added at a rate that balances the removal of nutrients though harvesting. Adequate fertilization is needed to maintain this form of intense forest management. Plantations using modern willow clones should be fertilized with at least 220 kg N ha^{-1} during the second and consecutive cutting cycles. This can be achieved with mineral fertilizers, farm animal residues or municipal organic sludge.

Harvesting is undertaken mechanically, taking into account the capacity of the soil to support heavy machinery. In northern Europe, harvesting occurs in winter, when

*Figure 4.15 Willow (*Salix sp.*) being grown as short-rotation coppice for use as biomass, Uppsala, Sweden.*

the frozen soil can support the machinery used for harvesting, chipping, and hauling the willow chips directly to where they can be consumed for heating production. This avoids the need for storing the chips. Winter harvesting also avoids the presence of leaf biomass in the harvested material, reducing nutrient exports and resulting in a lower water content in the biomass. Expected productivities in Great Britain and northern Europe, if fertilization and understory control are done properly and fast-growing clones used, can reach 5–12 Mg ha^{-1} of oven-dry biomass (Verwijst *et al.*, 2013). In southern Europe, productivity ranges from 10–15 Mg ha^{-1} of oven-dry biomass (Vega *et al.*, 2010).

As a final consideration, water use by short-rotation bioenergy plantations needs to be studied to determine the long-term sustainability of this forestry system. The relationship between water and energy is critical for planning energy and water policies in the future. The interrelation between water, energy, and CO_2 emissions exposes the complex trade-offs that arise from the use of tree energy crops. While bioenergy plantations represent an effective and viable energy source compared to fossil energies, especially in relation to the mitigation of climate change (see Chapter 8), the associated water consumption is much higher than non-renewable sources of energy. For example, poplar and willow plantations in southern Europe should be restricted to areas of high water availability (Sevigne *et al.*, 2011). In such situations, woody energy crops should be applied as a complement to other types of biomass such as forest and agricultural residues.

Chinese fir plantations for furniture and structural timber

China, Indonesia, and Malaysia contain over half of the forest land of Asia. All three have struggled with deforestation, and their governments are actively encouraging the development of plantation forests, which are usually oriented towards wood production, not conservation. Many such plantations are intensively managed to gain productivity. A rapidly increasing human population, demand for forest products, the increasing designation of protected forest areas, and the depletion of natural forest resources are all behind this increase in plantation investment. Competition for forest land in Asia is intensifying as a range of industries such as agriculture, palm oil, and mining seek to expand their activities in forested areas. Demand for forest products is also on the rise with increasing urbanization and rising income levels (Cheng and Le Clue, 2010).

China accounts for about one-third of the plantations established globally and has seen a significant increase since 2000. A major plantation species in southeast China is Chinese fir (*Cunninghamia lanceolata*), a subtropical coniferous tree species. It is a fast-growing species that on good sites can produce 450 m^3 ha^{-1} at a final harvest age of 25–30 years. The timber of Chinese fir is straight and decay resistant, and has a long history as an important construction and furniture material. There has been a steady increase in its use in plantations in the past few decades, making it one of the most important timber tree species in China, accounting for 60%–80% of the total area of timber plantations in southeast China, and for 20%–25% of the national commercial timber output (Lu *et al.*, 2015).

Chinese fir occurs naturally as a component of mixed subtropical evergreen broad-leaved forests. It is a species of moderate shade tolerance, but grows best in full sunlight. The species is moderately nutrient demanding, and in unmanaged natural forest it normally grows on moist and fertile sites. The response of Chinese fir to forest fertilization varies with stand age. Growing on yellow-red earth soils, the greatest response of young Chinese fir is to P and then to K, with little response to added N. In contrast, the greatest growth response in mid and late rotation is to N (Li *et al.*, 1993).

Traditionally, Chinese fir plantations were established after native evergreen broad-leaved forests were harvested and the slash burned. Sloping sites were sometimes terraced and intercropped with food plants before canopy closure (see Figure 4.16). Plantation sites were generally abandoned after one or two rotations and allowed to regenerate naturally by stump sprouting and natural seeding to mixed species stands that acted as a fallow period to restore the site. However, since the 1950s, the plantation area of Chinese fir has been enlarged, and this species has been repeatedly planted on the same sites without intercropping or fallow periods. Farmers have generally used a 25-year rotation, varying from 20 to 30 years depending on site quality. However, some plantations in Fujian Province are being harvested as young as 17 years or even less, a trend driven by the increasing demand for timber. With recent restrictions on the harvesting of natural forests in China, this pressure is increasing. At present, most Chinese fir plantations are in the second or third rotation on the same sites; some are thought to be in even later rotations, but this is difficult to confirm due to a lack of documentation.

Farmers have reported yield declines in multi-rotation Chinese fir plantations since the 1960s, but few scientifically rigorous investigations were undertaken until the late 1970s and the 1980s. A variety of competing hypotheses on the causes have been proposed, including soil nutrient depletion caused by nutrient removals in harvested materials and/or nutrient losses by slash; physical degradation of the soil; toxic substance accumulation, including allelopathic effects; slow decomposition of Chinese fir litter and consequent slow nutrient cycling and consequent reduced biological activities of the soil (Bi *et al.*, 2007). Nutrient depletion caused by harvest removal has traditionally been considered to be one of the main factors responsible for plantation yield decline. Successive short rotations of Chinese fir on a site will deplete the soil nutrient pool and soil carbon stocks. Over the same time period, short-rotation harvesting removes more nutrients than long-rotation harvesting because of differences in the nutrient content of sapwood and heartwood, in the proportion of sapwood and heartwood in the harvested materials, and the different quantities of nutrients removed in foliage and branches. Short rotations result in a higher frequency of site preparation by burning

Figure 4.16 Chinese fir (Cunninghamia lanceolata) plantation at Xi Hao, Nanping County, Fujian, China. Note the terraces, indicating that this plantation has been established on previously cultivated land.

*Figure 4.17 Chinese fir (*Cunninghamia lanceolata*) plantation at Xi Hao, Nanping County, Fujian, China. The site has been burned following harvesting, resulting in the loss of the organic layers. Burning on a steep slope such as this can make the soil very prone to erosion.*

and the associated loss of nutrients and organic matter, and deterioration in soil physical, chemical, and biological properties (Tian *et al.*, 2011).

Based on experience, farmers assert that Chinese fir will not survive and grow well unless the site is burned prior to plantation establishment. However, there is increasing evidence that slash burning is the ultimate cause of soil degradation and site yield decline because of accelerated loss of nutrients and topsoil. One of the most serious effects of slash burning is an increase of soil erosion (Figure 4.17). In a three-year study conducted in Youxi, Fujian Province, Sheng (1992) concluded that erosion on slash-burned study sites was about 37 times greater for mineral material, 10 times greater for organic matter, and eight times greater for the total loss of soil nutrients than that for an unburned control site. Such losses via burning are often much higher than those through harvest removals.

The slash burning of harvested Chinese fir sites has traditionally been conducted largely to reduce competition from herbs and shrubs that can cause plantation failure if not controlled. Competition from minor vegetation has been identified as one of the main factors contributing to yield declines in second and third rotation plantations of Chinese fir. The main problem from weeds is the vigorous sprouting of rhizomatous shrub and herb species. Sprouts of non-crop species grow much faster than their seedlings, and therefore exert much greater competition for light, moisture, and nutrients. Competition from weeds is exacerbated by declines in soil fertility and tree growth, both of which reduce the competitive strength of Chinese fir (Bi *et al.*, 2007).

The Triad approach to the management of forested landscapes

Traditionally, the maintenance of social, economic, and ecological values in a forested area has been through multipurpose forest management. In other words, the same working forest should provide all the values. However, in recent years a new option is gaining in popularity: dividing the forest into a number of zones for different but complementary uses (Seymour and Hunter, 1992).

Forest production depends on the spatial scale at which it is measured. With increasing spatial scale (moving from a single forest or plantation stand to the watershed or landscape level), it becomes increasingly easier to reconcile conflicting or non-complementary management objectives. In addition, in any landscape setting there will be a range of different interest priorities with regard to natural resource management represented by different stakeholders. Adopting a 'landscape approach' has become increasingly fashionable, and is currently considered to be essential to ensure that forestry is placed in the context of other forms of land use. If the landscape is divided into different zones, each one dedicated to different management objectives, there can be clear, specific, and effective management directions for each unit, which reduces conflicts between stakeholders by establishing a hierarchical order of uses within each zone (Côté *et al.*, 2010). Zoning can also help concentrate harvesting activities in the landscape, thus minimizing anthropogenic fragmentation and the extent of the road system and optimizing economic benefits (Beese *et al.*, 2003).

One of the most cited zoning strategies proposed in North America is the TRIAD (or three-zone) approach proposed by Seymour and Hunter (1992), in which three different zones are established with three different sets of objectives and priorities. To maintain the ecological integrity of the forest, one zone is usually dedicated to conservation, as a network of reserves. To counterbalance the decline in wood available for harvest in the conservation zone, a wood production zone has been proposed (MacLean *et al.*, 2008). The third zone is devoted to ecosystem management designed to emulate natural disturbance such as fire or windthrow. In this zone various harvesting techniques may also be applied to emulate old-growth attributes of structural and functional diversity, maintaining trees of different ages (including those over 100 or 200 years old), different types of deadwood, and more late-successional species (Gerzon *et al.*, 2011).

The TRIAD concept has largely been ignored by the logging industry. It immediately represents potential losses by designating a conservation zone, and there may also be losses in the potential to harvest in the ecosystem management zone. The proposed gains in the intensive production zone are not being met as logging companies are unwilling to make the necessary investments.

Conclusions

About one-third of the Earth's land surface is covered by forest, the vast majority of which has developed naturally. Once established, forest cover tends to be self-perpetuating, although the composition and structure of the forest will change over time. Nutrients within these forests are cycled, sometimes tightly. For example, the majority of nutrients in a tropical rainforest are in the trees and roots – very little is stored within the soil. Consequently, when the forest is converted to agricultural crops, the nutrients are quickly exhausted.

Many problems in managed forests can be related to disturbances to the nutrient cycle; others are related to the poor choice of species for a particular site. A key skill for a forester is to decide on a regeneration system that conserves the nutrients, and which also promotes the species that best meet the management objectives for the site. In some cases, those management objectives may be inconsistent with the site capability, and an adjustment will be necessary if forestry is to be successful. An added complication for all site-based work is that site conditions are changing. For years, it was assumed that site conditions were relatively static, but it is now evident that the climate is changing, and in some cases, the soil is changing due to the atmospheric deposition of nitrogen. This presents significant challenges to foresters, as it is no longer appropriate to rely on reproducing the forest present on a site prior to harvesting. The increasing prevalence of invasive weed species adds a further level of complexity.

Foresters need to be aware of these challenges, but in some parts of the world, the official responsibility of a forester ends when regeneration has reached a 'free to grow' stage (defined as a young crop of trees that is no longer impeded by competition from other species). As a result, there is no requirement to practice what in other places would be described as silviculture, namely the stewardship of a growing forest. These differences are most apparent between areas where a forest manager has a vested interest in the future forest (e.g. through private or corporate ownership) and those areas where reforestation is simply an onerous requirement associated with the (temporary) right to harvest trees in an area, usually in the form of a logging concession.

Further reading

Kimmins, J.P. 2004. *Forest Ecology: A Foundation for Sustainable Forest Management and Environmental Ethics in Forestry*. 3rd edition. Upper Saddle River, NJ: Prentice Hall.

Minang, P.A., van Noordwijk, M., Freeman, O.E., Mbow, C., de Leeuw, J., Catacutan, D. (eds.) 2015. *Climate-Smart Landscapes: Multifunctionality in Practice*. Nairobi: World Agroforestry Centre. Available at: http://asb.cgiar.org/climate-smart-landscapes/

Oliver, C.D., Larson, B.C. 1996. *Forest Stand Dynamics*. updated edition. New York: John Wiley & Sons.

Peh, K.S.-H., Corlett, R.T., Bergeron, Y. (eds.) 2015. *Routledge Handbook of Forest Ecology*. London: Routledge.

Pretzsch, H. 2010. *Forest Dynamics, Growth and Yield*. Berlin: Springer Verlag.

Chapter 5

Forest ecosystem health and vitality

Richard Hamelin and John L. Innes

Criterion 3 of the Montreal Process requires the maintenance of forest ecosystem health and vitality. Two indicators are used, with one being the area and percent of forest affected by biotic processes and agents (e.g. disease, insects, invasive alien species) beyond reference conditions, and the other being the area and percent of forest affected by abiotic agents (e.g. fire, storm, land clearance) beyond reference conditions. While seemingly clear, the wording for this criterion and its associated indicators presents many problems for the forest manager. For example, there is no clear definition of 'forest ecosystem health and vitality'. In addition, both indicators refer to a state known as 'beyond reference conditions'. This is extremely difficult to define, as it represents a static view of the environment that simply is not present. As an example, there are no reference conditions for plantations of a tree species that has never before been grown in plantations.

In this chapter, we examine the various ideas surrounding forest ecosystem health and vitality, and how it can be managed, if appropriate. We consider the concept of biotic and abiotic disturbance, showing how some disturbances are essential for the maintenance of values such as biodiversity. We also examine how a forest ecosystem can be affected by external agents, such as climate change and invasive species.

The evolving concept of forest health

Forest health is affected by many biotic and abiotic disturbance factors that influence species composition, nutrient cycling, biodiversity and water quality as well as economic activities related to the forest. The concept of forest health has evolved in recent years from a simple anthropogenic view to a much broader concept that depends upon context and circumstances.

What is a healthy forest?

Everyone probably has an idea of what a healthy forest is. However, the concept of forest health is complex and depends on several factors. Defining a healthy forest is challenging, and it is virtually impossible to come up with a single definition of a healthy forest that will satisfy everyone. Forest protection textbooks often list multiple definitions, illustrating the complexity of this concept (Edmunds *et al.*, 2000). Forest

Figure 5.1 Lodgepole pine (Pinus contorta) forest killed by the mountain pine beetle (Dendroctonus ponderosae), Barkerville, British Columbia, Canada.

health has a different meaning to a forest manager planning to harvest the timber and to a conservationist with an objective of increasing biodiversity.

Forest health is often considered from a narrow scale, both spatially and temporally. But the spatial scale matters in the evaluation outcome; evaluating tree mortality at local and landscape scales can produce very different assessments. A localized infection centre can be considered a threat at the local level but may not be considered important at the landscape level, unless it is a problem such as *Phytopthera* that has the potential to spread rapidly from a stand to a landscape scale. Perspectives can also change: when in the early 2000s patches of lodgepole pine (*Pinus contorta*) in British Columbia, Canada, were killed by the mountain pine beetle (*Dendroctonus ponderosae*), forest managers failed to recognize that this would develop into an epidemic killing 17 million ha of forest, something that occurred because of a combination of warmer winters and the availability of large areas of over-mature pine created by years of fire suppression. The temporal scale also matters. Tree mortality can be viewed as a threat at the beginning of a rotation, especially for biotic disturbance agents that can become epidemic, but less of a risk towards the end of the rotation. A more appropriate approach to defining a healthy forest should clarify both the scale and the perspective of the assessment.

Forest health is also intrinsically viewed from an anthropogenic perspective, with some economic expectations of the resources present in the forest. Traditionally this meant measuring the volume of wood available in a stand at the time of harvest. But the increase in the number of stakeholders who benefit from the forest has forced a re-examination of this view. Single components of the system, whether the trees, the wildlife, the plants or the microbes, cannot be considered in isolation. Instead, it is important to recognize the need for whole-ecosystem approaches.

It is always tempting to focus on a few of the components of the forest ecosystem, especially when a pest causes an outbreak. The concept of a biotic agent being a pest is always an anthropogenic bias, just as the concept of some plants being weeds is equally biased. A pest is only a pest when it threatens resources that have a value for

102

humans, or when it threatens the survival of species that we care about. It is actually possible for an insect, a fungus or a plant to be a pest in one setting but beneficial in another. Eucalyptus rust (caused by *Puccinia psidii*) was used as a biocontrol agent of an invasive weed in the USA in the 1980s, but is now considered a global threat to eucalypts worldwide.

Can we completely eliminate anthropogenic bias from a definition of forest health? Humans probably cannot and should not be taken out of the equation of forest health. Human activities now form an integral part of most forest ecosystems. Forests can instead be viewed as dynamic social-ecological systems with interactions and feedbacks between management and disturbance agents, including human, and shaped by climate.

Ecosystem function and forest health

Independent of scale or stakeholder, there is a convergence for ecosystem processes and sustainability to be at the core of the definition of forest health (Raffa *et al.*, 2009). One approach is to consider forest health as a function of the extent to which ecosystem processes are functioning within natural historical boundaries and using appropriate modifiers to specify the scales and/or human expectations. The challenge is then to define natural historical boundaries.

The concept of planetary boundaries has been used to determine the levels of disturbances that are within a safe range for the planet (Rockstrom *et al.*, 2009). It was proposed that the maintenance of functional biodiversity and redundancy could help improve resilience and prevent ecosystems from tipping into undesired states. Ecosystems with keystone species providing critical functions, such as whitebark pines (*Pinus albicaulis*) in alpine ecosystems, are particularly vulnerable to disturbances, such as disease and insect outbreaks, and at a greater risk of tipping into undesired states. The double threat of white pine blister rust (*Cronartium ribicola*) and mountain pine beetle (*Dendroctonus ponderosae*) are now threatening the survival of whitebark pine. This is having cascading effects on these fragile ecosystems, affecting wildlife such as grizzly bear (*Ursus arctos*) and Clark's nutcrackers (*Nucifraga columbiana*), and impacting processes such as erosion and snowmelt. Clearly, the boundaries of that ecosystem have been crossed by this double attack. It has resulted in whitebark pine being listed on the IUCN Red List as an endangered species.

A useful forest health definition should then include concepts of sustainability, resilience and ecosystem functions, with humans and their activities as an integral part of the equation, but without anthropogenic or economic expectations. Anthropogenic

Figure 5.2 Grizzly bear (Ursus arctos) and Clark's nutcracker (Nucifraga columbiana), two species that are suffering from the loss of whitebark pines (Pinus albicaulis) in the mountains of western North America. The nutcracker is particular important in ensuring the continued survival of whitebark pine.

expectations will almost certainly be met if the forest is resilient, in a sustainable state and functioning within ecosystem boundaries.

Forest biotic disturbances

Living organisms that infect or infest trees cause some of the most important disturbances in forest ecosystems. Epidemics caused by the mountain pine beetle (*Dendroctonus ponderosae*), white pine blister rust (*Cronartium ribicola*), ash dieback (caused by *Hymenoscyphus fraxineus*), spruce budworm (*Choristoneura* spp.) and sudden oak death (caused by *Phytophthora ramorum*) illustrate how biotic agents can cause ecosystem-wide outbreaks that sometimes result in irreversible changes. Forest pest outbreaks can have wide-ranging impacts, from severe economic hardship in communities affected to impacts on biodiversity and ecosystems processes.

Trees can be attacked by a wide variety of living organisms, including fungi, insects, nematodes, bacteria, viruses, mammals and parasitic plants. These biotic agents can be highly specialized and attack only a narrow range of tree species, or they can be generalists and attack a multitude of tree species. Both specialists and generalists can be found at equilibrium with their tree hosts in native ecosystems. Outbreaks occur when conditions make it possible for an insect or a pathogen to increase their populations beyond a baseline level. In some cases, this may result in a new form of equilibrium, such as with the bush mopane (*Colophospermum mopane*) ecosystem in parts of Zimbabwe that is maintained by artificially high populations of elephants (*Loxodonta africana*). Understanding the conditions that lead to outbreaks and trying to predict them is challenging but is one of the most important tasks of a forest health specialist.

Figure 5.3
*Mopane (*Colophospermum mopane*) woodland that is maintained at bush height by elephant browsing pressure. The photo was taken in Gonarezhou National Park, Zimbabwe, at the end of a particularly dry 'dry season', so the trees are without foliage. If the browsing pressure was released, this bush mopane would develop into mopane woodland.*

Understanding the multiple drivers of outbreaks

Biotic disturbance agents do not act in isolation. They are part of complex multilevel interactions that comprise multiple drivers. The disease triangle concept provides a useful framework to study the complex interactions between the environmental factors, the host and the biotic agents. A pathogen or an insect attack will occur only when there is a combination of a susceptible host and a virulent or aggressive biotic agent in the presence of environmental conditions that are favourable to the development of an outbreak (Figure 5.4).

This concept is simple when applied to agro-forestry situations where a single host species is planted in even-aged homogeneous environments, but becomes far more complex when considering biotic disturbances in natural forest stands. Forest stands are generally heterogeneous, comprising different tree species or various age classes. Trees are exposed to biotic agents over rotations that generally last several decades, and different biotic agents occur at different times during a rotation. Biotic disturbance

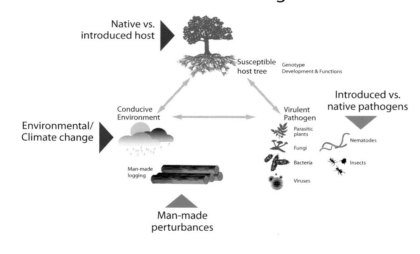

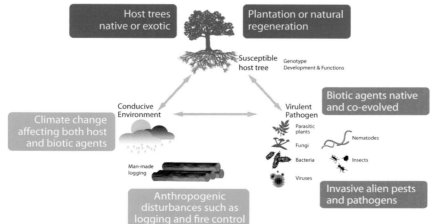

Figure 5.4 Multiple drivers affect the outcome of forest biotic agent outbreaks. The disease triangle illustrates how different forms of stress interact to create a condition.

agents rarely act alone and there are often multiple factors acting in concert to cause outbreaks. Anthropogenic influences such as logging practices and fire control affect the outcome, often over long rotations. Climate change is another important modifier of the disease triangle, affecting not only the conditions for infection but also the host condition and even composition. The simultaneous effects of multiple stressors act as additive deleterious effects and reduce the tree's capacity for a defensive response (Desprez-Loustau *et al.*, 2006). For example, root diseases often act as agents that weaken conifer tree defences and enhance bark beetle attack success, especially during the endemic phase of outbreaks (Lewis and Lindgren, 2002). The very nature and complexity of forest ecosystems should therefore serve as a warning against oversimplification. One important aspect to better understanding forest pest outbreaks is to study their role in natural forest ecosystems.

Still, the disease triangle can be useful for understanding forest disturbances and can be applied to forest pest management. Management strategies can focus on modifying any side of the triangle to affect the outcome. For example, thinning of a plantation can be used to increase light in a stand and reduce relative humidity, thereby making the environmental conditions less conducive to foliar diseases. This can be combined with planting mixtures of hosts and non-hosts to reduce the risk of spread of pests, thereby affecting the overall outcome of an outbreak.

The role of biotic disturbances in forest ecosystems

Insects and pathogens do not always have a negative impact on forest ecosystems. In fact, native biotic disturbance agents form an integral part of natural forest ecosystems. They often play essential roles in cornerstone forest ecosystem processes, such as nutrient cycling, regeneration, succession, biodiversity and the creation of wildlife habitats. Understanding the roles of these biotic agents in balanced ecosystems can help recognize patterns that transform innocuous agents in ecosystems into causal agents of large outbreaks with landscape-scale impact on ecosystems. This can provide some clues to finding better ways to predict and potentially prevent outbreaks. Comparing the biological attributes and population dynamics of species that are at equilibrium with their host with those that undergo rapid population expansion can provide a useful framework. Identifying factors that distinguish forest ecosystems that are at equilibrium and contrasting them with those that are unbalanced is a fundamental question in forest health studies.

Forests need a healthy amount of diseases

Manion (2003) proposed that diseases are part of the natural forest ecosystems and that they are part of a balanced ecosystem. Some biotic disturbance agents, in particular root diseases, create gaps in the forest that are important to several processes. Because of their biology and epidemiology, root pathogens attack and kill trees in infection centres, thereby creating canopy gaps. This mortality increases light penetration into forest stands, promoting regeneration and the growth of shade-intolerant species, with a variety of consequences for the forest ecosystem. Some biotic disturbance agents cause different levels of mortality in different tree species. For example, laminated root disease, caused by *Phellinus sulphurascens*, attacks Douglas-fir (*Pseudotsuga menziesii*) and western hemlock (*Tsuga heterophylla*) in the Pacific Northwest of North America. This favours regeneration by non-susceptible species such as western red-cedar (*Thuja plicata*) or deciduous trees. These disease-created gaps increase heterogeneity, diversity and resilience in the stand by creating patches composed of different species and age classes.

Wildlife trees as animal hotels

Wildlife biologists consider that dead, dying and decaying trees play a vital role in maintaining biological diversity and ecological processes within forest ecosystems. They provide shelter, nesting sites and food for a wide range of organisms. Root diseases and decay fungi are particularly important players in generating wildlife trees. In some cases, keystone species appear at the centre of a community web. In the interior of British Columbia, Canada, decayed and dead quaking aspen (*Populus tremuloides*) are overwhelmingly favoured by cavity-nesting birds. The majority of those trees are decayed by a fungus, *Phellinus tremula*, and are central to a nest web that starts with northern flickers (*Colaptes auratus*) as the keystone excavator, followed by a community of secondary nesters (Martin *et al.*, 2004). But the situation is different in the tropics, where a high diversity of non-excavating bird and mammal species interacts with different species of trees, fungi or avian excavators for cavity production.

Such considerations indicate the importance of understanding these patterns. In some cases, healthy trees can be artificially transformed into wildlife trees that are propitious for excavation and nesting by primary cavity nesters such as woodpeckers. Several approaches are possible, including inoculation of decay fungi using rifles to shoot the inoculum through the bark, or simply girdling or topping trees to encourage natural infections. Most of these approaches are still experimental but point to a new era where stewards of the forest deliberately infect trees to improve wildlife habitats.

Pathogens, saprophytes and symbionts

Biotic agents often form a continuum that ranges from mutualism to parasitism, and there is often a fine line between pathogens and innocuous close relatives. Within a genus, it is possible to find some species that are pathogenic and others that are saprophytic, endophytic or even symbiotic. These represent different evolutionary paths to develop adaptation and obtain food and nutrients from hosts. Endophytes grow non-parasitically on trees and plants and can be beneficial or neutral to the host. In some

Figure 5.5 Northern flicker (Colaptes auratus) *at a nest hole, in this case in a Ponderosa pine* (Pinus ponderosa).

*Figure 5.6 Indian-pipe (*Monotropa uniflora*). This plant, a species within the Ericaceae family, cannot photosynthesize and instead forms a symbiosis with* Armillaria *and other species to obtain nutrients from tree roots. It occurs in Russia, Asia, North America and northern South America.*

instances the presence of endophytes is associated with increased protection against insects and pathogens. Symbionts, such as mycorrhiza, provide mutual benefits to both the host plants and the fungi by providing the fungi with carbohydrates in exchange for access to nutrients. Saprophytes derive their nutrition from dead organic matter, while pathogens attack their hosts to derive their nutrition.

Some closely related species can have different interactions with their hosts, and comparing their biology, epidemiology and attributes could highlight what underlies those differences. *Phellinus sulphurascens* and *Armillaria ostoyae* are two of the most important root diseases in the Pacific Northwest of North America. However, the genus *Phellinus* comprises several decay fungi that do not have the ability to attack roots; and although the genus *Armillaria* comprises several root pathogens, one species of *Armillaria* can colonize orchids, but in a reversal of the roles, the fungus is the host and the plant (*Gastrodia elata* and *Galeola septentrionalis* from Asia, and *Gastrodia cunninghamii* from New Zealand) is the parasite. Dissecting such interactions could reveal the essence of what differentiates a very successful pathogen from a non-pathogenic close relative.

Role of anthropogenic activities in forest pest outbreaks

Although balanced forest ecosystems comprise native diseases and pests, some biotic disturbance agents can cause severe outbreaks that affect entire ecosystems. Anthropogenic actions are often responsible for some of the worst forest pest outbreaks. The following factors are commonly responsible: (1) introduction of new invasive alien species; (2) management practices that are conducive to insect or pathogen reproduction, survival or spread; and (3) climate change.

Forest invasive alien species

Invasive alien species have increased in incidence over the last century, driven by globalization (Figure 5.7) (Meyerson and Mooney, 2007). This trend will continue to accelerate with increasing international trade, travel and transport, leading to the mixing of biota from across the world. Indeed, a study looking at the causes for this increase in European countries found that the only significant predictors of invasive species abundance are national wealth and human population density. There is a direct correlation between the number of invasive alien species discovered and the economic activity of a country (Lambdon and Desprez-Loustau, 2010).

Invasive alien pests and pathogens are usually introduced accidentally by various pathways. For insects, a common pathway of introduction is via transport of

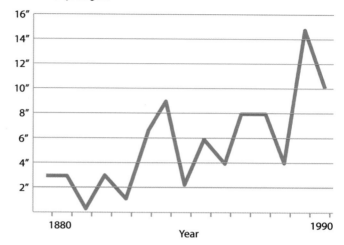

Figure 5.7 Increase in the number of alien pathogens during the last two centuries. (Redrawn from Lambdon and Desprez-Loustau (2010).)

material that carries eggs or larvae directly on imported goods, on wood packing material or by the transport of firewood. For pathogens, the main pathway of introduction of invasive species is via the 'plants-for-planting' industry. Indeed, pathogens have the ability to infect trees and plants while delaying symptom appearance. This complicates the tasks of phytosanitary inspectors since asymptomatic infected plants look identical to healthy plants, except that they are carriers of an exotic pathogen.

Invasive alien species tend to be successful in their new environment because they have not co-evolved with native biota. This generally results in the absence of natural enemies, competitors or natural resistance in the host species in the new ecosystem. In extreme cases invasive alien forest pests can change the ecosystems they invade and affect forest structure, species composition, food webs and vital processes such as nutrient cycling (Loo, 2009). Chestnut blight (*Cryphonectria parasitica*) provides one such example of irreversible ecosystem damage. This fungal pathogen introduced accidentally from Asia to North America has severely reduced the abundance of American chestnut (*Castanea dentata*), formerly a dominant or co-dominant tree species in forest communities in eastern North America. The removal of chestnut has caused dramatic changes in the eastern forests. Trees such as oaks (*Quercus* spp.) have now replaced chestnut as the dominant tree species. Additional impacts on the ecosystem include a reduction in leaf-litter processing and decreased abundance of cavity-nesting birds.

Since there is no built-in defence in the ecosystem against invasive alien forest pests, their management can be challenging. Clearly, preventing their introduction in the first place is a priority in all countries with valuable natural resources. The best way to prevent outbreaks of invasive alien forest pests is by increased vigilance, including quarantines, inspections and citizen involvement. Some novel tools and approaches are promising to help in the efforts to reduce risks of invasive species introductions. Genomics offers a number of tools, such as DNA barcoding and DNA-based diagnostics, that are revolutionizing the way we identify, name and detect pests and pathogens. Genetic and genomic profiling can also help identify the sources of invasive species, an important step when trying to prevent future introductions.

The increased availability of social networks to connect and mobilize citizens with common interests could also change the way invasive pest surveillance is done. For example, the Ashtag program (https://www.ashtag.org/about/) enables citizens to help monitor the progress of ash dieback in the United Kingdom. The program offers citizens the possibility of tagging and geo-referencing ash trees, tracking their health progress over time and posting photos, thereby becoming a 'steward' of that tree. In another instance, Google Street View was used to map the pine processionary moth (*Thaumetopoea pityocampa*) (Rousselet *et al.*, 2013). More citizen involvement and new tools are crucial to wage the war against invasive species.

Impact of forest management practices on forest pest outbreaks

Another important factor that can lead to biotic disturbances is the (mis)management of forests. In particular, tree harvesting, reforestation and fire control can be major drivers of biotic disturbances.

An outbreak of the mountain pine beetle (*Dendroctonus ponderosae*) is a good example of an indigenous insect undergoing a major outbreak, driven at least in part by anthropogenic activities. The beetles are a component of some natural forest ecosystems, with some forest ecosystems, and some of these pine-dominated ecosystems are dependent on disturbances, particularly wild fires, for regeneration. A combination of logging a century ago, overstocking with pines, better control of forest fires, and climate change has resulted in a perfect storm that has created an unprecedented outbreak covering 35 million ha in 19 US states and two Canadian provinces in western North America. Unusually hot summers combined with much milder winters have contributed to the expansion of the beetle far beyond its historical range.

Because of the magnitude of this outbreak, there can be consequences even beyond forest ecosystems. It was estimated that the beetle outbreak converted the forest from a net carbon sink to a large carbon source (Kurz *et al.*, 2008), with a peak of carbon emissions associated with the outbreak in 2006 and 2007. However, as the trees are killed but not destroyed, the carbon emissions are fairly limited, and much less important than fire. Over time, the carbon in the trees will be gradually released into the atmosphere through natural decay processes, but this may be compensated by regeneration of trees in the affected areas.

Root pathogens also benefit from tree harvesting. In particular, the stumps that are left after logging provide a unique opportunity for some root pathogens. A freshly cut stump offers a point of entry for Annosus root disease caused by *Heterobasidion annosus*, from which it can colonize the roots and eventually spread via root-to-root contacts to attack neighbouring healthy trees. Selective harvesting and pre-commercial thinning are clearly beneficial to root pathogens. Selectively removing trees in a plantation allows root pathogens such as *Armillaria ostoyae* to colonize stumps and spread from infested stump to healthy tree via root-to-root contacts. Annosus and Armillaria root diseases have flourished in second-growth forests and thinned plantations in conifer monocultures in various parts of the world (Otrosina and Cobb, 1989).

The homogeneity that follows reforestation is another factor that can promote pest outbreaks. The extreme case is a tree plantation with a single planted species, often from a narrow genetic base. This is compounded by homogeneity of age and spacing. A common practice in plantation forestry is to have high-density stocking at the beginning of the rotation, followed by thinning. This can facilitate tree-to-tree spread of pathogens and increases the risk of outbreaks. In addition, high-density plantations often result in competition stress that can incite weak parasites to turn into pests. An example is *Sirex noctilio*, a wood wasp, and its fungal symbiont *Amylostereum aerolatum*. This pest complex rarely causes forest damage in Europe, where both the wasp and the fungus are indigenous. However, this complex was introduced in the Southern Hemisphere where it causes severe losses in intensively managed pine plantations. Stressed pines in overstocked plantations are particularly at risk. In the Southern

Hemisphere, both the pine hosts and the *Sirex* wood wasp complex are non-native, resulting in an interaction without the necessary population control. To remedy this situation, a nematode (*Beddingia* [*Deladenus*] *siricidicola*) that parasitizes the wood wasp larvae is used as a biocontrol agent.

Even anthropogenic activities that aim to protect forests can be unwittingly beneficial to forest pests. The success of controlling forest fires has impacted, directly or indirectly, several forest pests. In western North America, reducing the occurrence of forest fires has resulted in large areas of even-aged pines that contributed to the mountain pine beetle outbreak. Fusiform rust of pines (*Cronartium quercuum* f. sp. *fusiforme*) has also exploded with the reduction in forest fires in the southern USA. This disease was previously endemic in the southern USA, and depends on two hosts to complete its life cycle: oaks and pines. Under a natural fire regime, oaks would be at low density following fires and the rust life cycle would be interrupted. But the incidence of fusiform rust increased rapidly with the intensification of extensive planting of slash (*Pinus elliottii*) and loblolly pine (*P. taeda*) well beyond their natural range, and was enhanced by fire control that resulted in large oak populations building up. The impact of dwarf mistletoes (*Arceuthobium* sp.), particularly in western conifers, increased as fire control and cutting practices left low-value deformed and infected trees as sources of seed to regenerate the forest and to serve as mistletoe inoculum.

The history of disturbance and the increase of forest plantations often upset pre-existing balances between trees and their microbial associates. As a result, pests and pathogens that are functioning within normal forest ecosystem boundaries and providing a positive role can become significant biotic disturbance agents that upset the ecosystem balance.

Climate change

Changes in climatic conditions have the potential to modify the outcome of the disease triangle in ways that are difficult to predict. Climate change will certainly affect the distributions of tree species. In the shorter term, assisted migration has been proposed

*Figure 5.8 Torreya tree (*Torreya taxifolia*), Torreya State Park, Florida, USA.*

to ensure that trees that will grow for rotations lasting several decades will be adapted to the future climate. However, assisted migration is controversial, and vigorously opposed by some groups who fear unintended ecosystem consequences in the areas where the trees are being established.

An example of assisted migration specifically in response to climate change is the movement of Florida torreya (*Torreya taxifolia*) from its current highly restricted range in northern Florida to a range of sites in the northern Midwest of the USA. This is a special case: the species is a relict population dating back to the last ice age. Its chances of survival in a changing climate are very limited, and it appears best adapted to a climate far to the north of where it currently occurs. The movement of the species is a private effort, rather than being undertaken by government agencies.

Pest and pathogen ranges will also expand in response to climate change. They are more mobile than their hosts and can more rapidly change locations in response to a changing climate. Although there are still only a few well-documented examples of climate change–driven forest pest outbreaks, epidemics of the mountain pine beetle and the Dothistroma pine needle blight have both been clearly impacted by changes in the climate.

The Dothistroma pine needle blight epidemic, though much smaller in scale and impact than the mountain pine beetle, is no less significant in terms of its signal of unforeseen climate change implications to forest health. Dothistroma needle blight is believed to have been part of the natural forest ecosystems in British Columbia, Canada, for centuries. However, a recent unprecedented outbreak has drawn attention to the patterns observed (Figure 5.9). What sets the current epidemic apart from earlier events is both its extent and severity in pine stands and the fact that the foliar pathogen is killing mature native hosts, a previously unreported event (Woods, 2011).

Is predicting the impact of climate change on pest and disease outbreaks possible? Modelling the incidence of forest pests under different scenarios is an important endeavour. For some forest biotic agents, modelling could reveal future trends and affect management. The current mountain pine beetle epidemic in British Columbia was arguably predictable given the strong link between the insect's vulnerability to very

*Figure 5.9 Young lodgepole pine (*Pinus contorta*) infected by Dothistroma needle blight, British Columbia, Canada.*

cold winter temperatures and the recent marked increase in winter minimum temperatures. But it is difficult to believe that anyone would have predicted the unprecedented nature of the outbreak.

For others, notably pathogens, it will be more challenging. One reason is that the occurrence of pathogen outbreaks, in particular those that infect the aboveground portion of trees, are dependent on rainfall. Unlike the consistency of global circulation model scenario projections for increased winter temperatures that were used to model the mountain pine beetle increase in survival, the potential utility of the rather variable projections of changes in precipitation is less clear (Woods, 2011).

Abiotic disturbances in forests

All forests are subject to abiotic disturbances, which refer to any disturbance not directly caused by another organism. They include geological disturbances such as volcanoes, earthquakes and tsunamis; geomorphological disturbances such as landslides, avalanches, floods and blowing sand; soil-related problems, such as infertility, salinization and waterlogging; climate-related problems, such as windstorms, droughts, extreme heat, extreme cold and aseasonal frost; fire; anthropogenic problems such as air pollution; and complex issues such as desertification. Interactions between these occur, such as the increased susceptibility of trees to insect pests after they have been weakened by exposure to sulphur dioxide or ozone.

The extent of damage caused by an abiotic disturbance is quite variable. For example, large volcanic explosions, such as occurred at Krakatau in Indonesia in 1883 or Mt. St. Helens in Washington State in 1980, can have devastating effects over large areas. Such events can cause massive, catastrophic disruption to forests and other ecosystems, completely destroying the existing system. In the case of Krakatau, the majority of the island was simply destroyed. At Mt. St. Helens, in addition to the blast, a significant part of the mountain collapsed. However, the apparent devastation in the blast zone was not complete, and the first stages of ecosystem recovery were almost immediate. The greatest effects were at higher altitudes, so recovery has been slow.

Figure 5.10 Early recovery within the blast zone at Mount St. Helens, Washington State, USA. Although trees are present, their growth is slow as this is the sub-alpine zone, and the trees are without the benefit of any form of shelter.

Figure 5.11
Fresh growth
on Sitka spruce
(Picea sitchensis)
in Wales, UK,
killed by a late
spring frost.

At the other end of the scale, many abiotic disturbances are relatively minor. For example, wind can result in whipping, fraying branches and foliage loss, but the damage generally quickly recovers. A late frost can damage a tree's fresh young foliage, but it is usually able to replace it, sometimes in the same season. Summer drought may induce premature discoloration and loss of foliage, but the tree generally recovers the following year, although repeated droughts may have much more severe impacts.

Disturbances associated with abiotic events may be aggravated by biotic disturbances. For example, windthrow may create the ideal conditions for an outbreak of a pest that benefits from a supply of fresh downed wood, such as the European spruce bark beetle (*Ips typographus*). Drought may significantly weaken a tree, allowing a pest that would not normally be capable of killing a tree to cause mortality. These interactions may be extremely complex, and some have still to be fully explained. In all cases, it is important to remember that disturbance is a natural part of a forest ecosystem, and that some level of disturbance is usually required to maintain forest ecosystems (Pickett and White, 1985).

Some anthropogenic activities can also be classed as abiotic disturbances, particularly those associated with pollution. A variety of chemicals can cause problems for trees and forests, with the most important gases being sulphur dioxide, the oxides of nitrogen, hydrogen fluoride, ozone and ammonia. Particulates, especially heavy metals such as copper, lead and zinc, may also be important, although the effects are generally localized around smelters and often occur through impacts on the soil and soil biota. At high exposures to some pollutants, the trees can be killed. At lower exposures, the trees can be weakened, making them more susceptible to other forms of disturbance, including both biotic and abiotic agents. The effects depend on the pollutant involved, the exposure, the sensitivity of the tree to that pollutant, and a variety of mitigating environmental factors. For example, trees may be less susceptible to high ozone concentrations during drought conditions because their stomata are closed, and uptake is severely reduced. This has led to an emphasis being placed on the dose received by a tree, rather than relying on exposure.

*Figure 5.12
Damage to
a stand of
Norway spruce
(Picea abies)
in Muotatal,
Switzerland. The
initial damage
was caused by
a windstorm,
and the downed
stems then
provided an
opportunity for
an infestation
by the European
spruce bark
beetle (Ips
typographus) to
develop.*

The complexity surrounding the interactions between abiotic and biotic disturbances has given rise to a model known as the disease spiral (Manion, 2001). This recognizes three types of factors that can lead to the decline of individual trees and whole forests. Predisposing factors create the conditions that can result in other agents causing a tree to start declining. Inciting factors push the tree into the degree spiral, and contributing factors may ultimately be responsible for the death of the tree. The different factors all push the tree inwards towards the final state: death.

Climate- and weather-related effects

While climate change is sometimes invoked as a cause of damage to forests, most forest disturbances are related to events that are better associated with extreme weather conditions, such as hurricanes, typhoons, ice storms and droughts. Weather and climate are related, but it is very difficult to attribute a particular weather event to climate change. Increasingly, it is possible to relate the probability of a weather event to climate change, but this is still an emerging science. Climate change is more important in relation to adaptation. In particular, if a disturbance occurs, what trees should be planted to replace those lost in the event? While many foresters would argue that natural regeneration should be allowed to happen, the trees present on a site reflect those that were able to establish some time ago, perhaps under quite different environmental conditions. For example, if a 150-year-old stand is burned, the natural regeneration will be from trees that established successfully 150 years ago, when we know the climate was considerably cooler. Will these trees still be suitable, and will they survive for the next 150 years, when we know the climate will be considerably warmer? Obviously, the longer the rotation, the greater this problem will be, and forest managers may be required to intervene to accommodate the effects of anthropogenic changes to the climate. In some cases, this may involve assisted migration, the movement of trees geographically, although this remains very controversial among some environmentalists.

Figure 5.13 Damage caused to tropical rainforest on the Atherton Tablelands of Queensland, Australia, by Cyclone Larry in March 2006.

Figure 5.14 Pit-and-mound topography caused by repeated windthrow in a subalpine forest at Beatenberg, Switzerland.

Windstorms

Windstorms take various forms and are given a variety of names. The most important are generally related to cyclones, an atmospheric circulation phenomenon that is normally associated with very strong winds. In the tropics, they are frequently referred to as typhoons, whereas they are often referred to as hurricanes in temperate regions. Damage to forests may occur from strong winds that do not reach hurricane force (which is defined on the Beaufort scale by wind speeds in excess of 117 km hr^{-1}). During a particular storm, gusts may occur that are significantly stronger than the typical wind speed during the storm, and these can be responsible for particular windthrow events. The degree of turbulence is also important as it creates a range of stresses on a tree.

During a storm, trees can suffer various forms of damage. The most extreme is when the trees are uprooted (windthrow), although the effects may be similar when trees are snapped at a point on the lower stem (wind snap). Repeated storms involving uprooting can result in the development of hummocky terrain known as pit-and-mound topography (Oliver and Larson, 1996), which can have significant effects on regeneration (Figure 5.14). More often, damage is limited to the occasional uprooted or snapped tree, and damage to limbs and crowns.

In addition to the powerful winds associated with cyclonic events, strong winds can also be associated with a number of other conditions. Tornadoes are common in some areas, and are associated with very strong, swirling winds that can be particularly destructive to forests because of the forces that are created. Tornadoes generally move along paths, with the greatest damage occurring where they touch down. The paths vary in width from about 30 m to a maximum of about 1 km. Derechos are periods of strong winds associated with thunderstorms, and the damage occurs over much larger distances, with the swath of damage generally exceeding 400 km in length. Wind speeds can be very high, exceeding 160 km hr^{-1}.

Many parts of the world are exposed to frequent, lower intensity wind. For example, in the Northern Hemisphere, the west coasts of North America and Europe are subject to persistent winds. These areas are often considered particularly suitable for forestry, because of their mild and humid climates, although wind can limit what can be grown. In Scotland, Sitka spruce (*Picea sitchensis*) plantations in exposed situations on the west coast are at increasing risk of windthrow beyond about 40 years, and final harvesting takes this into account. Wind exposure also influences the sites chosen for afforestation in Scotland and has affected silvicultural techniques, with late mechanical thinning no longer being done on vulnerable sites. Windthrow can limit the effectiveness of a number of silvicultural treatments in such situations. For example, in the Pacific Northwest of North America, windthrow can affect small stands left as part of variable retention silvicultural designs.

A wide range of factors determine the susceptibility of a tree to windthrow or wind snap. These include the nature of the wind (direction, strength, variability), the nature of the tree (species, shape, height, presence of foliage, presence of decay pockets), the nature of the tree's surroundings (proximity to other trees, recent thinning in the stand) and the nature of the soil (particularly the soil moisture content but also soil texture). The resistance of the tree to wind flow is particularly important.

Persistent wind can also shape tree crowns (termed flagging) and restrict growth, although this is generally localized. The effects are particularly obvious in coastal situations, where both wind and saltwater carried by the wind may contribute to this effect.

*Figure 5.15 Typical windthrow damage in a Norway spruce (*Picea abies*) stand following a severe wind storm, southern Sweden.*

*Figure 5.16
Wind-flagging in
the Marlborough
Sounds area of
New Zealand.
In situations
such as this,
regeneration
of trees may be
prevented by
wind.*

Saltwater carried by wind can penetrate quite far inland, and is sometimes responsible for the appearance of salt injury to foliage.

Windthrow can be managed through a number of strategies. These include a hazard assessment of a site (especially prior to afforestation) and various silvicultural interventions (Quine, 1995). The nature of any stand thinning may have a significant effect on the wind firmness of a stand, although the responses are complex and difficult to predict. Any treatment that opens up the stand to significant eddying may result in an increase in windthrow hazard, as will any treatment that leaves more susceptible trees exposed. Reducing the volume of the crown by pruning can increase wind firmness, but is expensive and therefore only practiced where the consequences of damage may be substantial yet there is still a desire to retain the tree (such as in many urban situations).

The practice of clearcutting often results in windthrow of the remaining trees. There are a number of reasons for this. First, trees growing in the middle of a stand may be much less wind-resistant than edge trees, although over time they will have an opportunity for adaptive growth which will increase their wind firmness. Clearcutting also changes the movement of wind across the landscape, potentially creating turbulence and funnelling the wind. Consequently, steps should be taken to avoid increasing the wind hazard; Harris (1989) provides some useful examples.

Ice, hail and snow
In temperate and boreal zones, and in the case of hail occasionally in sub-tropical zones, ice and snow can create substantial problems. One of the most dramatic forms of damage is associated with freezing rain that can build up ice on the foliage and branches of trees. Eventually, the weight of the ice results in limbs breaking off and, if accompanied by wind, stems can also break. A similar situation can arise when freezing fog impacts on foliage, resulting in the development of a thick layer of rime ice. Such storms are quite common in some areas, particularly eastern North America, and while damage may be extensive, it rarely results in tree mortality unless the damage is particularly severe (involving >75% branch damage) (Manion *et al.*, 2001).

Figure 5.17 Windthrow on the edge of a Sitka spruce plantation, Scotland, UK.

*Figure 5.18 Snow damage to a stand of European alder (*Alnus glutinosa*), Val Bavona, Switzerland.*

Snow has a similar effect to ice, with the problem being the additional pressure placed on branches and crowns by the weight of the snow. In some cases, this will break off limbs, whereas in others it will result in stems being bent over (Figure 5.18). In areas that normally experience high snowfall, many trees, particularly conifers, have adopted a growth form involving a very narrow crown that prevents the accumulation of large amounts of snow.

In some regions, a number of different forms of winter injury can occur. Snow cover insulates the soil from extreme air temperatures, which in the boreal can drop to below −30°C or −40°C for prolonged periods of time. Cold-adapted trees can survive this, although the roots are more sensitive than the aboveground parts of the tree. If these temperatures are accompanied by a lack of snow, root damage can occur. Conversely, if the ground is already frozen, and the air temperature warms sufficiently, evergreen trees may start transpiring and then suffer drought stress because of the lack of available soil moisture. This particular form of damage is recorded quite frequently on the leeward slopes of mountains such as the Rockies in North America, resulting in what has been termed 'red belt' damage. It is associated with warm, dry winds during winter, and the reddening is caused by needle desiccation. Another form of injury occurs when a warm period in later winter or early spring results in de-hardening, but is then followed by freezing temperatures. Such conditions resulted in severe needle browning in northwestern Ontario in spring 2012, affecting more than 250,000 ha of forest (Man *et al.*, 2013).

Drought and excessive heat

With severe, multi-year droughts being recorded in several places, including California, USA, and Victoria, Australia, in the last couple of decades, there are growing concerns about the potential impacts of drought on forests. The incidence of droughts is expected to increase with global warming, especially in continental interiors, and may present significant challenges for forest managers.

The direct effects of drought include foliage wilting and loss, dieback and mortality. Foliage discoloration may be evident, but this can be brought about by many different agents, of which drought is only one. Often a tree can survive a drought in a particular year (although the timing of the drought is critical), but droughts in subsequent years can cause increasing levels of mortality as the tree's capacity to recover is eroded.

Trees tolerate drought through two main mechanisms: stomatal control and developing extensive (and deep) root systems. In Australia, a country prone to drought conditions, eucalypts tend to adopt the latter strategy, having developed a deep rooting ability. In general, the rooting depth of eucalypts increases as annual rainfall decreases. However, there are marked differences between species, and river red gum (*Eucalyptus camaldulensis*) has a much deeper root system than blue gum (*E. globulus*). Eucalypts have a number of other properties that enable them to withstand drought conditions better than other genera, including an ability to rapidly absorb and transpire water when it becomes available and to shut down rapidly at the onset of drier conditions (Bell and Williams, 1997).

Eucalypts can also tolerate drought through the osmotic component of water potential in the leaves. Essentially, the trees either maintain or build up a high concentration of solutes in the leaves, and this helps to maintain turgor and productivity during periods of drought, as well as enabling faster recovery after the drought stress is reduced. Spotted gum (*Corymbia maculata*), bull mallee (*Eucalyptus behriana*) and grey box (*E. microcarpa*) appear to be among the species that are particularly adept at this, although only a small number of species have been tested, and it is likely that many other species have similar responses. Another mechanism is inter-cellular water storage, recorded in manna gum (*E. viminalis*), but it is unclear how common this is.

Excessive heat is also a problem for some trees in some situations. The greatest problems are usually related to management activities, but can also be associated with other forms of disturbance. For example, when a thin-barked species such as European beech (*Fagus sylvatica*) is suddenly exposed to direct sunshine (as might happen after a thinning operation), the heat on the bark can cause severe damage, killing the underlying cambium.

*Figure 5.19 Examples of sun scald on the bark of European beech (*Fagus sylvatica).

Waterlogging

There are very few tree species that can withstand prolonged periods of waterlogging, although some species that exist through special adaptations. These are most obvious among mangrove species, but this represents a special case as the forests are flooded on a daily basis, rather than continuously for long periods. Bald cypress (*Taxodium distichum*) in the southeast USA and Montezuma bald cypress (*Taxodium mucronatum*) in central Mexico are frequently found in wet areas and are able to survive because, like most mangrove species, they have aerial roots known as pneumatophores. A variety of other species are able to survive waterlogged conditions, with willows (*Salix* sp.) and alders (*Alnus* sp.) being common in Europe. Some of the best adaptations occur in the seasonally flooded forests of South America, including the várzea forests of the Amazon basin (seasonally inundated by whitewater rivers) that cover 180,000 km², and igapó forests (seasonally inundated by blackwater rivers). These forests can be flooded for up to seven months each year, and all the species that are present have adapted to such conditions (Wittmann *et al.*, 2006).

In Australia, many paperbark (*Melaleuca* spp.) can survive waterlogging. Experiments with the swamp paperbark (*Melaleuca ericifolia*) have found that seeds can germinate and the cotyledons emerge while underwater. Seedling growth is increased by waterlogging, and root growth is stimulated. Not all species behave in the same way. For example, adult trees of *Melaleuca halmaturorum* (also known as swamp paperbark) can survive permanent waterlogging, but seedlings cannot. Melaleuca species appear to be able to survive waterlogged conditions by having dense mats of fine roots at the soil surface. These have a large surface area, and can scavenge oxygen effectively from the water.

To survive waterlogged conditions, root systems must be able to cope with very low amounts of oxygen and high levels of other gases, including carbon dioxide. They must also be able to cope with the anaerobic environment, which is associated with very low soil redox potential and with toxic forms of microelements. Very few tree species have the ability to do this and, as a result, flooding can result in considerable mortality.

*Figure 5.20
The interior of
a* Melaleuca
*forest near
Perth, Western
Australia.
This forest has
permanently
waterlogged
soils, and the
trees have special
adaptations to
cope with these
conditions.*

*Figure 5.21
Mortality
associated with
flooding near
Quesnel, British
Columbia,
Canada. The
height that the
water table
reached is
clearly evident.
The cause of
the flooding
was a beaver
dam, which was
subsequently
breached,
reducing the
height of the
water.*

Fire

Fire is the disturbance that forms the greatest challenge for many forest managers, especially in drier forests and woodlands. On the one hand, fire represents a potentially catastrophic disturbance that can destroy all the values that a forest manager is trying to maintain. On the other hand, fire is a natural process, and its exclusion from some types of forest can result in changes to the forest that are not only undesirable but which can also increase the risk of a catastrophic fire. Increasingly, it is being

*Figure 5.22
The endangered
monkey puzzle
tree (Araucaria
araucana) has
a thick bark
that enables it
to withstand
fire. Its natural
range covers an
area of Chile
and Argentina
characterized by
volcanic activity,
with eruptions
being a frequent
if unusual source
of fire ignition.*

recognized that fire is a necessary process in many forests, and that allowing it to happen is an important part of maintaining the health of the ecosystem.

To a certain extent, fire can be regulated through management actions. For example, prescribed fires can be set to reduce the amount of fuel in the forest. Specific operations can be undertaken to achieve the same end result, while at the same time reducing the chances of a fire moving from the ground to the canopy through laddering. Species that are more fire-resistant can be favoured – those with thick barks seem particularly capable of withstanding fire.

Plants in ecosystems have a number of adaptations that can help them withstand fire. Bark thickness has already been mentioned. Another mechanism is the presence of serotinous cones that only open after they have been heated to a certain temperature. Lodgepole (*Pinus contorta*) and jack pine (*P. banksiana*) are well-known examples from the boreal forest, but a number of other conifer families also exhibit serotiny, including *Cupressus*, *Picea* and *Sequoiadendron*. It is also common among species of Proteaceae, *Eucalyptus* and *Erica* in the Southern Hemisphere. However, there is substantial variation in the degree of serotiny, both between and within species, and populations from areas with more frequent crown fires tend to exhibit higher levels of serotiny. The simplest forms of serotiny involve the melting of waxes that seals the fruit or cone scales. However, some genera such as the Australian *Banksia*, which occurs in dry woodland environments characterized by relatively frequent fire, have more sophisticated forms of serotiny that first require fire to open the follicles and then moisture that enables the cone scales to expand and reflex. This means that seeds are not released immediately after a fire, when the environmental conditions might not be suitable, but instead are released after there has been some rainfall, resulting in ideal germination conditions.

Other forms of adaptation include the release of seeds prior to a fire that are capable of surviving in the soil seedbank until their hard seed coats are cracked by a fire. This strategy is adopted by many *Ceanothus* species in California, USA. Another strategy

is to be able to sprout after a fire – this can occur from epicormics buds set deep in the bark of the stem and branches, or from basal stems, roots and horizontal rhizomes (Clarke *et al.*, 2013). Eucalypts show many different forms of adaptation, including serotinous seeds, epicormic buds and lignotubers, although the strategy to cope with fire varies from species to species. Most eucalypt species are capable of resprouting, but whether or not they do depends on a number of factors, particularly the extent of the fire damage to the bark and lignotuber. However, three important timber species, namely karri (*Eucalyptus diversicolor*), flooded gum (*E. grandis*) and mountain ash (*E. regnans*) lack lignotubers and rely on seed for regeneration after a fire. Both alpine ash (*E. delegatensis*) and the ridge-fruited mallee (*E. incrassata*) produce large amounts of seed after a fire, improving the likelihood of rapid ecosystem recovery. An illustration of the potential complexity of fire interactions with regeneration is provided by two eucalypt species (swamp gum [*E. ovata*] and white ironbark [*E. leucoxylon*]), which regenerate best after fire because the fire reduces seed predation by ants.

Fires need fuel, heat and oxygen for combustion, the so-called fire triangle. Fire suppression normally concentrates on removing one of these, for example by creating a fireline that removes fuel, or by dropping water or fire retardant on the flames, which removes heat and oxygen. Firefighting has become increasingly sophisticated, and many forest managers are unlikely to be directly involved. However, initial fire attack is often done by local forest staff, and in large fires, all available staff may be involved in firefighting operations.

The traditional response of forest managers to fire has been immediate suppression. In many parts of the world, this is still the case, as the immediate concern is the loss of economic value, especially in the timber. Increasingly, there are also concerns about the wildland-urban interface: as people build more and more houses in fire-prone areas, the risk of wildfires causing damage to human structures has increased rapidly. This is particularly true of western North America but is also an issue in other parts of the world, such as Australia. Suppression remains the instinctive response, and now that concerns about carbon emissions associated with fires are growing, the pressure to suppress fires is increasing.

*Figure 5.24
Vigorous
resprouting from
epicormics buds
among eucalypts
after a severe
fire in Kinglake
National
Park, Victoria,
Australia.*

*Figure 5.25
Part of the area
that burned in
February 2009
in Victoria,
Australia.
The town of
Marysville was
located in the
valley on the left:
it was almost
completely
destroyed in
the fire.*

At the same time, there is increasing recognition that fire suppression has seriously affected the health of forests, especially in western North America. Fuel loads have increased and, as a result, when a fire does occur, there is a much greater risk of it developing into a catastrophic crown fire. Such fires can be extremely difficult to contain. On occasion, they can reach such a scale that containment is impossible, as with the 'Black Saturday' fires in Victoria, Australia, on February 7, 2009. These occurred on a day that recorded some of the hottest temperatures ever recorded in Victoria, with Melbourne recording 46.4 °C. The very high temperatures were accompanied by strong winds in excess of 125 km h^{-1}. The fires that occurred under these conditions were so powerful that they were impossible to control.

Fire management normally involves addressing risks, hazards and values. Again, this can be viewed as a triangle. Risk involves an assessment of the likelihood of an ignition. The hazard is normally assessed in terms of the fuels that will be burn if an ignition occurs. Values involve the assessment of the net change in resource condition if a fire occurs, and includes damage to the forest itself and to other aspects of the ecosystem, including human infrastructure. The actual response to a specific fire will depend on a range of factors, including the severity of the fire, the capacity to undertake firefighting operations and prevailing policies towards fire suppression.

In some areas of the world, a century or more of fire suppression has created a highly artificial forest. These forests are much denser than they would be under natural conditions, have higher rates of pests and diseases and are more susceptible to catastrophic fires. Arno and Fiedler (2005) provide an interesting account of some of these forests, indicating the extent of the problem faced by forest managers. They advocate a return to more natural disturbance patterns, which they argue would reduce the adverse effects of fire suppression. However, reaching this stage will be difficult and costly. If steps are not taken to improve forest health, it is likely that some of the values that are so cherished will be lost, at least in the short term. The danger was illustrated dramatically in 1988, when fires burned over 300,000 ha of the iconic Yellowstone National Park, directly affecting 36% of the land area of the park. Moving to a more natural fire regime will involve recognizing the different types of fire frequency and severity that characterize western forests, although some of these, such as the stand-replacing fires that occur at intervals of more than 500 years, are unlikely to be encouraged. They may however affect the response to fires. For example, the Paradise Fire in the temperate rainforest in the Queets River drainage of the Olympic National Park in 2015 was suppressed where necessary but also allowed to burn in order to 'benefit the ecosystem'.

The recovery of the forest after fire depends on a range of different factors, including the vegetation present before the fire, the fire intensity and the environmental conditions following the fire. The nature of the forest prior to the fire will determine the seedbank in the area, but if this is destroyed, as in a severe fire, the forest will have to be re-established through recolonization or through planting. Some of the most intensive

*Figure 5.26 Twenty-five years after the fires, a grizzly bear (*Ursus arctos*) moves through an area of burned forest in the Ninemile section of Yellowstone National Park, USA. The lack of regeneration is an indicator of the intensity of the fires in this area. While many studies of post-fire recovery have focused on the timber, studies in Yellowstone have examined how the entire ecosystem responds to severe fires.*

studies of fire recovery have taken place in Yellowstone National Park following the severe fires of 1988. Since the fires, the recovery of the vegetation has been studied in detail (Wallace, 2004), greatly enhancing our understanding of how the whole forest ecosystem responds to severe fire.

Air pollution

Another form of disturbance is air pollution. This can occur naturally, for example around sulphurous hot springs and springs with high carbon dioxide emissions, but such occurrences are relatively rare. Air pollution resulting from anthropogenic activities is much more frequent, although the distribution of air pollution has shown some interesting changes over the past 100 years. The effects of air pollution can be broadly divided into acute and chronic. Acute effects are associated with short-term exposures to high concentrations of pollutants and consequently high doses received by organisms. They are generally characterized by visible foliar injury and, in severe cases, mortality. Chronic effects are associated with long-term exposures to lower concentrations although the actual doses may be quite high. Chronic effects are often characterized by hidden injury. There may be no visible injury, even when the organs within the foliage are being damaged. The effects may be subtle, such as reduced responsiveness to physiological stress and reduced growth, making diagnosis difficult.

Acute effects were widespread around a variety of different types of industrial plants in Europe and North America in the 19th and 20th centuries. In some cases, the pollution was so severe that all vegetation was killed, creating an 'industrial desert' (Kozlov *et al.*, 2009). Some of the most severe effects were seen around smelters, such as at Nikel and Monchegorsk in Russia, and Sudbury and Trail in Canada. Such environmental disasters may not be entirely the result of pollution: often the smelters required large amounts of wood, and at least some of the ecosystem loss in some cases can be attributed to the felling of the forest. Many such sites have seen significant recovery, primarily because air quality legislation (associated with human health, not ecosystem health) required that the smelters improve air quality (which they often did by

Figure 5.27 Industrial desert in the vicinity of a major smelter at Nikel, in the Kola Peninsula of northern Russia.

increasing the height of the smoke stacks rather than reducing the volumes of sulphur dioxide being emitted). Brick kilns were and some cases still are another significant source of localized pollution, with the toxic gas being hydrogen fluoride. Hydrogen fluoride has also been a problem around aluminum ore reduction plants, phosphate fertilizer plants and glassworks, with textbook examples of hydrogen fluoride injury being recorded around Spokane, Washington, in the 1950s and later around Columbia Falls in Montana, USA. A good description of the damage around the smelter at Sudbury, Canada, and the early stages of its subsequent recovery, is provided by Gunn (1995).

In areas with concentrations of heavy industry, the damage extended beyond the immediate vicinity of a point source. The industrial heartlands of England had widespread pollution damage in the 19th and early 20th centuries. Similarly, the Ruhr region of Germany was also badly affected. In the second half of the 20th century, attention was drawn to the Ore Mountains (Erzgebirge), an area that straddled the borders between Poland and the former East Germany and Czechoslovakia. Here, a major episode of decline and mortality of trees was recorded, particularly among conifers such as Norway spruce (*Picea abies*) and silver fir (*Abies alba*). European beech (*Fagus sylvatica*) was also affected, although seemingly not to the same extent. The damage was attributed to exposure to very high concentrations of sulphur dioxide, especially during winter, which was released by the concentration of industrial plants burning brown coal (Lomský *et al.*, 2002).

Increasingly stringent air quality regulations in the developed world have led to much cleaner environments, and today injury caused by sulphur dioxide and hydrogen fluoride is relatively rare. However, significant problems still exist in developing countries, although the extent of the problem remains largely unknown.

While the damage around point sources was relatively easy to identify (by looking downwind of the plant), ecosystem impacts caused by gases such as ozone can be more difficult to identify (and mitigate). Ozone is formed by the interactions between oxides of nitrogen (car exhausts are a major source) and hydrocarbons in the presence of sunlight. One of the classic areas where this was recognized this was the Los

*Figure 5.28 Dead Norway spruce (*Picea abies*) in the Krkonoše Mountains of the northern Czech Republic. The mortality has been attributed to high exposures to sulphur dioxide.*

*Figure 5.29 Dead Ponderosa pine (*Pinus ponderosa) *in the San Bernardino Mountains, California. Mortality was likely due to a combination of weakening by ozone, followed by attacks by bark beetles.*

Angeles basin. Oxides of nitrogen emitted by the large numbers of cars and industries were sucked inland by the local wind systems, enabling ozone to form. This caused widespread foliar injury to Ponderosa pines (*Pinus ponderosa*) in the San Gabriel and San Bernardino Mountains, and even to trees in areas as far east as the Sierra Nevada (Miller and McBride, 1999). Trees were seriously weakened, making them susceptible to mortality induced by bark beetles and other pathogens. Similar conditions have been identified in southern Switzerland, where pollutants from the area centred on the city of Milan in Italy are drawn up into the Alps. Another area of notable ozone injury is the Desierto de los Leones, on the slopes surrounding Mexico City, where various species, including sacred fir (*Abies religiosa*) have been severely impacted.

Attempts have been made to identify the relative sensitivity of different species to particular types of pollutants, with some species being classed as tolerant and others as sensitive. This however fails to take into account the very significant variation in sensitivity that can occur within a species. For example, black cherry (*Prunus serotina*) is a North American species that is widely considered to be sensitive to ozone. However, some genotypes are actually quite tolerant and will fail to show symptoms even when neighboring trees have severe foliar injury.

The symptoms associated with injury by air pollution are quite varied. Fortunately a number of pictorial atlases exist, and these can aid in the diagnosis. A general one is Flagler (1998), which deals with a range of different forms of pollutant-induced injury. A more specific guide deals in detail with the injury caused by ozone to broadleaved species (Innes *et al.*, 2001). While such guides can provide important clues as to the cause of a particular symptom, and can provide an indication of what to look for when pollutant injury is anticipated, great care is needed when making a diagnosis, as many other damaging agents can cause similar symptoms.

There has been a huge amount of speculation about the role that 'acid rain' may have caused in the mortality of trees observed in Europe and North America in the late 1970s and 1980s. Great care needs to be taken when trying to interpret this literature, as it is characterized by poor observations and a failure to correctly diagnose many common pathogens (Skelly and Innes, 1994). For example, damage to Fraser

*Figure 5.30
Dead Fraser fir
(Abies fraseri)
at Mount
Mitchell, North
Carolina, killed
by the invasive
balsam woolly
adelgid (Adelges
piceae). Many
scientists initially
attributed by
the mortality
to pollution
(specifically acid
rain), without
conducting
any proper
diagnoses.*

fir (*Abies fraseri*) at higher elevations in the southern Appalachian Mountains of the eastern USA was initially blamed on air pollution, but was subsequently shown to be the result of an introduced insect pathogen, the balsam woolly adelgid (*Adelges piceae*). In hindsight, it is likely that drought played a major part in triggering some of the symptoms, either directly or through its effect on biotic pathogens. There is still speculation, particularly in North America, that soil acidification is involved, but this remains controversial.

While attention has focused on trees because they are highly visible, iconic and have clear economic value, other components of the forest ecosystem are actually more sensitive to pollution, and may have suffered much more. In particular, many lichen species are extremely sensitive to pollution. In the United Kingdom, many lichen species were severely impacted by the high levels of sulphur dioxide that were emitted during the Industrial Revolution, and are only now showing signs of recovery. Other ecosystem components were also likely affected, but the lack of baseline information and the small numbers of scientists with an interest in such matters makes it difficult to determine how extensive any such effects are.

One last aspect of pollution deserves mention. This is the eutrophication of ecosystems, particularly freshwaters. In most cases, the pollution is associated with agricultural runoff, particularly when nitrogen fertilizers have been used. However, fertilizers such as urea can be volatilized, releasing ammonia, which is then transported by air and subsequently deposited as ammonium. This became a particular problem in the Netherlands in the 1980s, primarily because slurry from intensive animal production units was directly applied to fields. Because of the associated problems, such slurries are now injected into the soil, dramatically reducing ammonia emissions. However, nitrogen oxides emitted from vehicular exhausts is still an important source of eutrophication in some areas. Ecological impacts are most evident in freshwater ecosystems, including streams and lakes in forested landscapes, but some terrestrial ecosystems have also been adversely affected. Management responses have been quite dramatic, including the complete removal of the organic layer of the soil.

Figure 5.31 Experimental treatment at Kootwijk in the Netherlands, where the organic layers were removed in an attempt to counter the impacts of eutrophication. The grass on the right of the picture reflects the edge of the treatment area.

Links between ecosystem health and human health and well-being

Unlike outbreaks in agricultural crops, which affect mostly growers and suppliers, forest pest outbreaks can affect a large number of stakeholders: communities depending on jobs, companies whose business model depends on the availability of wood, Indigenous peoples for whom forests are part of their cultural heritage, outdoors enthusiasts, and the public whose quiet enjoyment of nature is affected. As a result, forest pest outbreaks rarely go unnoticed, and governments, companies and other stakeholders devote large amount of resources to managing these outbreaks.

Both biotic and abiotic agents can cause damage to trees and forests that has immediate effects on people. Among the most obvious is the loss of timber supply associated with major events, such as large-scale fires and major pest outbreaks. However, impacts extend much further, and can even affect urban populations. For example, the loss of street trees due to a pathogen can reduce the value of houses in the street, and the amenity values of parks and other recreational resources can be severely affected by pest outbreaks.

There are also direct effects to contend with. Some important tree pests can directly affect humans: the oak processionary moth (*Thaumetopoea processionea*), the pine processionary moth (*Thaumetopoea pityocampa*) and the gold tail moth (*Euproctis chrysorrhoea*) all have hairs that can cause severe irritation to humans. The hairs of the processionary moth are only 0.1–0.2 mm in length, so are almost invisible, and can be carried by the wind for distances of several hundred metres. Tree health in urban environments is also important to human health. Medical studies have found a direct link between tree mortality caused by an invasive alien beetle and human health outcomes in North America. The air pollutants that affect trees also affect humans; in fact, most air pollution regulations are based on human impacts rather than ecosystem impacts. There is also the possibility that air pollution has subtle effects on trees, such as changing the proteins in pollen, which may be affecting allergic reactions of humans to pollen from trees affected by air pollutants.

Considerations such as these have increasingly led to a belief that there are strong links between ecosystem health and community well-being. Forest health is no longer about the impacts of pests and diseases on the production of timber. Instead, all agents that cause disturbances in forests need to be seen in the context of the whole ecosystem, which not only involves trees but also the whole ecosystem, and especially the humans that depend on those ecosystems for their livelihoods.

Salvage

Faced with a natural or human-induced forest disaster, there are several strategies that can be adopted. In areas being managed for conservation, the ecosystem may be left alone, with no or minimal management interventions. More often, there is a need to try to recover at least some of the economic value in the damaged timber. This may be both dangerous and costly. Windthrow is particularly difficult to salvage because modern harvesting equipment is not designed to work with fallen trees, thus requiring manual operations. The unpredictable tensions in fallen trees make any work with a chainsaw dangerous, and salvage operations after major storms are often accompanied by increased mortality and injury rates in the workforce. In other cases, the presence of dead trees and damaged limbs adds to the danger of working in the forest, with the problem increasing over time as natural decay sets in to the damaged and dead trees.

The immediate management response to a large-scale forest disturbance is often to initiate salvage logging. However, Lindenmayer *et al.* (2008) argue strongly that salvage operations generally need a lot more care and preparation than they are generally accorded. The ecosystems may be very sensitive after a disturbance, and salvage operations may further damage them. The opportunity to diversify the landscape is often missed, and the focus on salvaging the largest and most valuable timber may result in the selective removal of some of the most valuable wildlife habitat.

Disturbed forests should, then, be seen as both a challenge and an opportunity.

Conclusions

The maintenance of forest health presents a major conundrum for a forest manager, mainly because of the differing concepts over what constitutes forest health. In some traditional systems, particularly those associated with the maximization of timber yield, any unhealthy tree had to be removed immediately, especially if there was a risk of others being infected by whatever biotic agent was present. Today, natural disturbances are seen as an essential part of the forest ecosystem, and there is less emphasis on eliminating the agents that cause the disturbances. However, there are clearly boundaries to this tolerance, and the tolerance decreases as management becomes more focused on timber yield. Consequently, in plantations designed for timber production, the tolerance for agents of disturbance is low, whereas in natural forests with no obvious commercial value, natural disturbances may be much more tolerated.

There are, however, some cases where there is zero tolerance. These are mainly related to highly contagious diseases such as *Phytophthora* or cases where invasive alien species could cause irreversible ecosystem changes. In such situations, immediate control is called for, although in many cases, by the time a problem is discovered, the authorities notified and a decision taken over the appropriate action, it is too late. The destruction caused by the emerald ash borer (*Agrilus planipennis*) in eastern North America is a good example. In another case, that of the Asian long-horned beetle (*Anoplophora glabripennis*), early detection and aggressive eradication have successfully eliminated the pest in some areas. There are also major difficulties associated with some plantation species. For example, *Acacia mangium* in Sarawak, Malaysia, is encountering a number of problems, including a red root rot disease associated with *Ganoderma philippii* and a phyllode rust associated with the fungus *Atelocauda digitata*. Meanwhile, another plantation species, *Falcataria moluccana*, in Sabah, Malaysia, is being attacked by species of the rust *Uromycladium*, probably *Uromycladium tepperianum*. Such problems are likely to increase in the future.

There is a need to recognize the complexity of the interactions between trees and the biotic and abiotic stressors. Understanding this complexity will be crucial in the face of climate change. For forest managers, it is critically important that regular monitoring of the forests under their management is undertaken so that early actions can be taken. This involves both aerial survey and ground investigation of any identified problems. Decisions need to be taken about the extent of any interventions, balancing potential economic losses against a whole suite of other values. Some of these require disturbance, and if forests are to be managed in a more natural way, some economic losses are inevitable.

Further reading

Arno, S. F., Fiedler, C. E. 2005. *Mimicking Nature's Fire: Restoring Fire-Prone Forests in the West.* Washington, DC: Island Press.

Edmonds, R. L., Agee, J. K., Gara, R. I. 2011. *Forest Health and Protection.* 2nd edition. Long Grove, Illinois: Waveland Press.

Innes, J. L., Haron, Abu Hassan (eds.) 2000. *Forests and Air Pollution in Rapidly Industrializing Countries.* Wallingford, UK: CABI Publishing.

Karnosky, D. F., Scarascia-Mugnozza, G., Ceulemans, R., Innes, J. L. (eds.) 2001. *The Impact of Carbon Dioxide and Other Greenhouse Gases on Forest Ecosystems.* Wallingford, UK: CABI Publishing.

Schoenenburger, W., Innes, J. L., Fischer, A. (eds.) 2002. *Vivians's Legacy: Windthrow Research in Switzerland Following the Vivian Storm.* Birmensdorf, Switzerland: Swiss Federal Institute for Forest, Snow and Landscape Research.

Wainhouse, D. 2005. *Ecological Methods in Forest Pest Management.* Oxford: Oxford University Press.

Wylie, F. R., Speight, M. R. 2012. *Insect Pests in Tropical Forestry.* Wallingford: CABI Publishing.

Chapter 6

Maintenance of soils

John L. Innes

Criterion 4 (Conservation and maintenance of soil and water resources) has a rather limited range of indicators related to the management of forest soils. A general indicator deals with the area of forest within a jurisdiction whose primary management goal is the protection of soil or water resources, and then there are two status indicators dealing specifically with soils:

4.2.a Proportion of forest management activities that meet best management practices or other relevant legislation to protect soil resources.
4.2.b Area and percent of forest land with significant soil degradation.

Soils are fundamental to forests: without soils, there would be no substrate for trees to grow in. While it is always possible to find trees growing in extreme situations, apparently without any soil, forests can only grow where there is an adequate soil. Maintaining this soil is an important part of forest management, as incorrect practices can lead to soil damage and even soil loss. Soils and trees interact in many ways – trees obtain their nutrition and water largely through the soil, and roots anchored in soil provide a means for trees to achieve the heights that they do. In fact, soils provide three of the five primary resources that all vegetation requires: water, mineral nutrients and a medium for growth (the other two are radiant energy and carbon dioxide).

Soils develop over time, with that development being influenced by interactions between the parent rock, the climate, the vegetation and the topography, and being very much influenced by time. This development is always occurring, and soils should be seen as a dynamic part of the forest ecosystem. Their structure and composition is determined by a range of physical, chemical and organic processes, all of which interact with each other. The nutrients that all plants require are largely derived from the minerals in the soil, but there are also a range of organic nutrients that are present, mainly as a result of the recycling of plant materials.

Soil development

While soil development (termed pedogenesis) may seem to be outside the scope of forest management, an understanding of the processes involved is important when interpreting soils, and it is becoming increasingly important to understand soil developmental processes when considering land reclamation. In reclamation, a manager

Figure 6.1
Pines growing
on minimal soils
at Zhangjiajie
National Park,
Hunan, China.

may be faced with establishing a new soil at a site, and while a natural soil might take hundreds or even thousands of years to develop, it is possible to establish a new soil capable of growing trees, something that is especially important in urban forestry.

The basic source of soil is the parent material. This may be derived from bedrock or consist of unconsolidated material deposited there by a geomorphic agent such as water, wind or ice. Examples of the latter include floodplain gravels, loess and till (also known as boulder clay). Bedrock breaks down through a process known as weathering. This takes many forms, but can be broadly divided into physical, chemical and biological processes.

Physical weathering involves the disintegration of parent material and the reduction of its grain size. The initial failures may be related to structures within the bedrock, such as bedding planes and faults, or may involve the separation of individual particles within the rock. Failure is generally caused by pressures within the rock or pressures created by foreign substances within voids in the bedrock, and occurs most rapidly where there are no confining pressures. Consequently, failures are concentrated towards the surface of rocks. Several different processes of physical weathering are recognized, and these have been broadly summarized in Table 6.1. In general, physical weathering results in the gradual comminution of material down to sand and silt-sized particles (i.e. >2 μm), although such particles can also be created by various forms of erosion.

In contrast to physical weathering, chemical weathering results in the decomposition of the parent material, involving a chemical change. Water is by the far most important agent in chemical weathering. In nature, water is never 'pure' – it always contains ions and other materials derived from the atmosphere or organic materials picked up while passing through the soil. This has led to a division between geochemical weathering (involving inorganic processes) and which leads to the production of 'rotten' rocks (known as saprolites), and pedochemical weathering, which involves the breakdown of the saprolite by both geochemically and biologically controlled phenomena.

Another common division is into congruent and incongruent dissolution. Congruent dissolution involves the transfer of a mineral's outer mass into solution without any

135

Table 6.1 Processes of physical weathering.

Process	Explanation
Thermal expansion	This process involves stress caused by differential expansion, either of surface layers or of individual mineral grains. It is particularly apparent during fires, but diurnal temperature fluctuations may be sufficient to cause cracking, especially if water is present in the rock.
Unloading	This occurs when pressures within a rock are released by the removal of pressure from above or from the side. This may occur due to erosion, but is particularly apparent following glaciation (the ice not only removes rock, releasing pressure, but acts as a support as long as it there). Artificial rock cuts can generate this process.
Hydration and swelling	This starts as a chemical process when water is absorbed by minerals (hydration), but the failure is caused by the resultant swelling, and is a physical process. It is particularly apparent with clay minerals, especially those containing layers of OH⁻ or H_2O. On a large scale, the process is known as exfoliation, whereas at the scale of an individual boulder it is termed spheroidal weathering.
Growth in voids	This involves the widening of spaces within the rock, and can be caused by various agents. In temperate latitudes, the freezing of water in rock spaces is particularly important. In boreal latitudes, there may be fewer freeze-thaw cycles, reducing the effectiveness of the process. The hydration of salts within cracks may also be important in some regions.

Based on Ritter *et al.* (2011).

precipitation of other substances, such as the dissolution of limestone. With incongruent dissolution, some of the ions are transferred to the solution, while others are incorporated into new minerals. Incongruent dissolution also involves a range of processes that break the mineral grain apart. A number of chemical processes may be involved, outlined in Table 6.2. One of the important results of chemical weathering is the development of secondary minerals, which are the result of the alteration of the primary minerals found in the bedrock. Secondary minerals primarily result from the alteration of feldspars, ferromagnesium minerals and micas, and are usually clay-sized (<2 μm). The very small size of secondary mineral particles results in a significant increase in surface area and is accompanied by the development of an electrical charge on the surfaces of the particles. There are always negative charges, and these attract cations. The capacity of a soil to attract such cations is termed the cation exchange capacity. Calcium, magnesium, potassium and ammonium, all of which are important plant nutrients, are all cations. Secondary minerals can also have positive charges (and hence an anion exchange capacity), but this is dependent on the pH of the soil. The soils of temperate and boreal regions are almost always dominated by negative charges (with the exception of some volcanic soils), whereas many tropical soils have a positive charge.

Biological weathering is frequently mentioned in geomorphological textbooks, but in practice this form of weathering is relatively minor in comparison to the processes of physical and chemical weathering. It involves both mechanical processes, such as when plant roots widen cracks within rocks, and a range of chemical processes caused by the presence of organic materials in soil solutions.

Weathering alone does not make a soil, and a complex series of physical, chemical and biological interactions is involved. These were described by Jenny (1941) as:

$$S \text{ or } s = f(cl, o, r, p, t \ldots) \tag{Eq. 6.1}$$

where S = soil, s = any soil property, cl = climate, o = biota, r = topography, p = parent material, and t = time, with the dots representing external factors that can influence soil development locally. These factors all interact, and the final product (S or s) may be the result of multiple processes over time. In particular, specific combinations of climate and vegetation, called pedogenic regimes, can result in important soil-forming processes.

One such regime is podzolization, which involves the removal of iron, aluminium and organic matter from the A horizon and their accumulation in the B horizon. It is particularly common in cool, humid-temperate environments dominated by forests. The process is increased under coniferous forest, as coniferous litter tends to produce acidic soil solutions, and cations such as Al^{3+} and Fe^{3+} are more mobile under acidic conditions.

Another pedogenic regime is termed laterization. This develops under conditions of high rainfall and temperature, intense leaching and oxidation and is therefore most common in humid tropical regions. There is very little organic accumulation, and in contrast to podsolization, silica is leached from the soil, leading to very high concentrations of hydrated oxides of iron and aluminium.

Table 6.2 Chemical reactions involved in the decomposition of rocks and minerals.

Oxidation and reduction	Involves the transfer of electrons from one element to another. Oxidation occurs above the water table (i.e. in the presence of atmospheric oxygen); reduction occurs below the water table.
Solution	This involves the transfer of mass from a solid to an aqueous phase. The solubility of minerals is very variable, and depends on the form that they are in.
Hydrolysis	Involves the reaction between mineral elements and the hydrogen ion of dissociated water. Involves the reaction between a salt and water to produce an acid and a base.
Ion exchange	Involves the substitution of ions in solution for those held by mineral grains.

Based on Ritter *et al.* (2011).

Figure 6.2 A typical podzol at Okstindan in northern Norway. The very sharply defined horizons are typical of this type of soil.

A third pedogenic regime is calcification. This involves the accumulation of carbonates within a particular layer within the soil, generally in the B horizon. The carbonates are derived from the upper levels of the soil profile and move down through the soil to the accumulation zone. The source of the carbonate is most often wind-blown dust (known as loess). The process is found in sub-humid to very arid climates where there is insufficient precipitation to carry dissolved carbonates down to the water table.

Much of the emphasis on soil development has been on the physical and chemical status of the soil, with relatively little emphasis placed on the biological state. Yet, the fungi and fauna of a soil play a critical role in the decomposition and recycling of organic matter and in the acquisition of mineral nutrients by trees. Over time, many organisms will colonize a new soil, but in terms of forest establishment, mycorrhizas are a particularly important component. These take a number of forms (seven in total), with the most important being the arbuscular mycorrhizas (AM), the ericoid mycorrhizas (EcM) and the ectomycorrhizas (EM).

Arbuscular mycorrhizal fungi form associations with a wide range of plant species, and include the families Glomaleaceae, Acaulosporaceae and Gigasporaceae. These grow symbiotically with plant cells; they derive nutrition from the host plants while at the same time providing nutrition (particularly phosphate ions). They also have hyphae that extend to a maximum of 6–10 cm into the soil. The AM, which are members of the Phycomycete fungi, are the most widespread mycorrhizae, and have the greatest species diversity.

The ericoid mycorrhizas are restricted to a single plant family: the Ericaceae. This widespread family of plants dominates some Mediterranean-type ecosystems, and may also be the dominant understory in some temperate and boreal forests. The EcM are predominantly Ascomycota, with several species forming mycorrhizae, although some Basidiomycota may form EcM with species of Ericaceae.

The third group is the ectomycorrhizas. These are an important group for trees, particularly in high-latitude forests, although they are also found in tropical forests. The diversity of EM increases with increasing latitude, reaching a maximum in temperate forests. There are exceptions to this, and EM are for example relatively rare in New Zealand, colonizing only three genera there. In contrast to the AM, the ectomycorrhizae develop between cells in plant roots, and they also develop what is known as a mantle and Hartig net on the exterior of the root. Hyphae can extend several metres into the surrounding soil and are an important source of a number of nutrients, including nitrogen and phosphorus.

In contrast to many other soil processes, and the development of a soil biota, the development of mycorrhizal communities in soils seems to be much more random (Dickie *et al.*, 2013). Colonization appears to depend on a range of factors, and there is no orderly progression through time as a soil develops. As a result, improved growth of trees can be obtained on mycorrhiza-deficient soils through artificial means, primarily inoculation.

Soil description

Soil is generally defined as a natural surface layer that contains living matter and supports or is capable of supporting plants (Strahler and Strahler, 1992). This definition is preferable to that commonly used by engineers, which embraces all unconsolidated mineral material and is used to distinguish from bedrock. Soils contain minerals, organic matter, gases and water. They contain both inorganic and organic fractions, and the organic fraction contains both living and dead material. In fact, the greatest species diversity in a forest is generally found within the soil. With increasing depth, the proportion of organic material decreases, and where it ceases to exist, the soil is termed regolith or sediment.

Soils are generally divided into a number of soil horizons. These horizontal layers have differing proportions of organic and inorganic materials, and may differ substantially in their physical and chemical composition, even within a particular soil. The smallest unit of soil is termed a polypedon. A polypedon contains a unique set of properties, and is composed of pedons, a soil column that extends from the surface of the soil to its base. This is generally viewed as a hexagonal-shaped column, such that the surface of the land is divided into pedons. Each of the six faces of a pedon displays a set of horizons, known as the soil profile. Horizons dominated by inorganic material are designated by a set of capital letters, with the uppermost being the A horizon, and the regolith being the C horizon. In between there is usually a B horizon, and there may be others as well (such as an E horizon). The organic layer is generally designated as the O horizon, and is often defined as having at least 20% organic carbon (dry weight). Organic layers may also be designated as L, F or H, with L referring to a layer of relatively undecomposed organic matter (litter), F being a layer of partly decomposed organic matter in which some of the original structures can still be made out, and H signifying a layer of highly decomposed, predominantly colloidal matter (Pritchett, 1979). F and H layers are generally absent in tropical soils, and the L layer may be very thin. The A, E and B horizons constitute the soil solum and represent the horizons in which roots are active. The horizons are summarized in Table 6.3.

All parts of the organic matter in the soil are critical as, ultimately, the organic matter represents an important source of mineral nutrients. The organic matter breaks down through a process known as humification, which involves both biotic and abiotic processes of chemical transformation. While water and oxygen are important in breaking down simple and accessible organic molecules, the most important role is played by the soil microorganisms. These rely on extracellular digestion and the secretion of a wide variety of enzymes which break down complex chemicals into simpler, water-soluble compounds. Decomposition starts with the sugars and proteins, followed by the polysaccharides, hemicellulose, cellulose, lignin, waxes and resins

Table 6.3 Commonly used nomenclature for soil horizons.

Horizon	Characteristics
O	Dominated by organic material, although the required amount of organic matter differs between classifications. The required organic matter may be greater if the mineral fraction is dominated by clay minerals.
A	A mineral horizon formed at the surface or below the O horizon. It is a mixture of humic organic material and mineral material.
E	A mineral horizon characterized by the loss of silicate clay, iron or aluminum, leaving a concentration of sand and silt particles.
B	A mineral horizon characterized by broken-down bedrock and accumulated amounts of clay minerals, carbonates, and sesquioxides of iron and aluminum.
C	Mineral horizon above the hard bedrock and lacking the properties of O, A, E or B horizons.
R	Hard bedrock underlying the soil.

Based on Ritter *et al.* (2011).

(Lukac and Godbold, 2011). The end result is a relatively stable form of organic matter, known as humus. Two different forms of humus are widely recognized: mull and mor. The classification does not work readily in tropical regions. Mull tends to form under broadleaved trees that produce readily decomposable litter, and contains a thin (0–5 cm) litter layer, with the underlying A horizon being dark in colour and showing evidence of soil organism activity (such as earthworms). Mor humus is more acidic, and tends to form under coniferous trees. It has a thick (10–20 cm) litter layer that is sharply divided from the A horizon, and there is usually a very pale mineral horizon. Mull and mor represent opposite ends of a continuum, and most temperate forest soils are intermediate; their form of humus is referred to as moder or duff mull. The term duff is often used in North America to refer to the layer of decomposing organic material just below the litter layer and immediately above the mineral soil. It is made up of humus but refers to a specific layer, whereas humus can occur throughout the soil profile.

Soils are generally described by their horizons, which are best exposed by digging a soil pit. One of the fundamental characteristics is the colour of a soil horizon, and this will often indicate the dominant processes active in the soil. Black generally indicates the presence of organic matter, whereas red indicates the presence of sesquioxide or iron. While seemingly subjective, the description of colour has been standardized through the use of standard colours. These are known as the Munsell soil colours, and similar standardized colours are now available for describing foliage colours.

A second major descriptor is the texture of the soil. As with the classification of soils, there are differences between the boundaries used to classify different particle sizes within soils, but the system developed by the US Department of Agriculture is widely used (Table 6.4). The relative percentages of sand, silt and clay determine the texture of the soil. The texture of the soil plays a big part in determining its water-holding capacity and the rate at which water can pass through the soil. It also affects the stability of the soil and how easily it can be eroded.

Table 6.4 Classification of soil and sediment particles.

Grade name	Diameter limits
	2.0 mm
Very coarse sand	
	1.0 mm
Coarse sand	
	0.5 mm
Medium sand	
	0.25 mm
Fine sand	
	0.10 mm
Very fine sand	
	0.05 mm
Silt	
	0.002 mm (2 microns)
Non-colloidal clay	2 to 0.01 microns
Colloidal clay	Less than 0.01 microns

US Department of Agriculture.

Soil classification

There are different soil classifications in operation around the world. Different disciplines approach soils from differing perspectives, and so what an engineer considers to be soil is not necessarily the same as what an ecologist would consider to be soil. Most of these classifications have been developed for a particular region, and as a result they cannot be readily translated into one another. Many ecological textbooks adopt the US Department of Agriculture system, simply because the textbooks have been written for American students. Other classification systems used in the USA include the Unified Soil Classification System (USCS) of the American Society for Testing Materials, and the classification developed by the American Association for State Highway and Transportation Organizations (AASHTO). Internationally a classification known as the World Reference Base for Soil Resources has been developed (Table 6.5).

Table 6.5 The 32 reference soil groups of the World Reference Base for Soil Resources.

Group	Soil type	Abbr.	Description
1	**Soils with thick organic layers**		
	Histosols	HS	
2	**Soils with strong human influence**		
	Anthrosols	AT	Soils with a long and intensive agricultural use
	Technosols	TC	Soils with many human artefacts
3	**Soils with limited rooting due to shallow permafrost or stoniness**		
	Cryosols	CR	Ice-affected soils
	Leptosols	LP	Shallow or extremely gravelly soils
4	**Soils influenced by water**		
	Vertisols	VR	Alternating wet-dry conditions, rich in swelling clays
	Fluvisols	FL	Floodplains, tidal marshes
	Solonetz	SN	Alkaline soils
	Solonchaks	SC	Salt enrichment upon evaporation
	Gleysols	GL	Groundwater affected soils
5	**Soils set by Fe/Al chemistry**		
	Andosols	AN	Allophanes or Al-humus complexes
	Podzols	PZ	Cheluviation and chilluviation
	Plinthosols	PT	Accumulation of Fe under hydromorphic conditions
	Nitisols	NT	Low-activity clay, P fixation, strongly structured
	Ferralsols	FR	Dominance of kaolinite and sesquioxides
6	**Soils with stagnating water**		
	Planosols	PL	Abrupt textural discontinuity
	Stagnosols	ST	Structural or moderate textural discontinuity
7	**Accumulation of organic matter, high base status**		
	Chernozems	CH	Typically mollic
	Kastanozems	KS	Transition to drier climate
	Phaeozems	PH	Transition to more humid climate
8	**Accumulation of less soluble salts or non-saline substances**		
	Gypsisols	GY	Gypsum
	Durisols	DU	Silica
	Calcisols	CL	Calcium carbonate

(Continued)

Table 6.5 (Continued)

Group	Soil type	Abbr.	Description
9	**Soils with a clay-enriched subsoil**		
	Albeluvisols	AB	Albeluvic tonguing
	Alisols	AL	Low base status, high-activity clay
	Acrisols	AC	Low base status, low-activity clay
	Luvisols	LV	High base status, high-activity clay
	Lixisols	LX	High base status, low-activity clay
10	**Relatively young soils or soils with little or no profile development**		
	Umbrisols	UM	With an acidic dark topsoil
	Arenosols	AR	Sandy soils
	Cambisols	CM	Moderately developed soils
	Regosols	RG	Soils with no significant profile development

IUSS Working Group on the World Reference Base (2006).

Unfortunately, a list such as presented in Table 6.5 cannot be easily translated into national or regional soil classifications. This presents considerable difficulties when moving between jurisdictions, and knowledge of the local soil classification is generally important when working within a particular management context.

Biogeochemical cycles

The physical, chemical and biological nature of the soil often provides important information about the processes that may be present or have been active at some stage in the past. At any given time, the forest represents a snapshot of what is happening. Nutrients are present in a number of compartments within the ecosystem, and are always being transferred between them. This is not a closed system: nutrients are continually being added to and being lost from the forest ecosystem. Most work on biogeochemical cycles has been done on carbon (addressed in Chapter 8) and on nitrogen, since nitrogen tends to be the most limiting of the nutrients used by forests.

Nutrients are added to the ecosystem through the weathering of the bedrock and minerals within soils, through biological fixation and through atmospheric inputs. Atmospheric deposition represents an important source of nutrients. Some of this is natural, representing dust, volcanic aerosols and marine salts. However, pollution is also an important source of atmospheric deposition, especially for sulphur and various forms of nitrogen. The deposition may be dry (solid particles), wet (dissolved in precipitation) or occult (deposited directly from cloud and fog particles). The composition of the water reaching a forest floor may be significantly different from the water that hits the canopy, as nutrients are washed off the leaves and trunks, leached out of leaves, or even taken up directly by leaves. Water dripping off the foliage is known as throughflow, whereas water running down tree trunks is termed stemflow (see Chapter 7 for a discussion of these processes). Both throughflow and stemflow tend to show much larger concentrations of nutrients than precipitation above the canopy.

Nitrogen can also be derived directly from the atmosphere. This occurs through the symbiotic relationships that exist between bacteria and tree roots, through

Figure 6.3 Cycas armstrongii, a cycad found in sclerophyll woodland in the Northern Territory of Australia. Cycads are unusual in that they form a symbiotic relationship with cyanobacteria, enabling them to fix atmospheric nitrogen. The bright green foliage is typical for this species after a fire.

free-living bacteria in the soil and through epiphytic lichens living on trees. Nitrogen can be fixed by a number of species of bacteria, and also by actinomycetes and cyanobacteria. The cyanobacteria can fix various forms of nitrogen. In forest ecosystems, they are found in a range of forms, including free-living (e.g. *Azotobacter* and *Clostridium*) and as symbionts in, for example, some lichen species (e.g. *Lobaria* sp. and *Peltigera* sp.) and some plants (including cycads, *Gunnera* sp. and mosquito ferns [*Azolla* sp.]).

The majority of nitrogen-fixing bacteria are found as symbionts in the roots of plants. One of the most widespread and diverse of the plant families, the Leguminosae, is characterized (with a few exceptions) by having root nodules containing the nitrogen-fixing *Rhizobium* bacteria. Two sub-families of the Leguminosae, the Caesalpinioideae and the Mimosoideae (which includes the acacias), are particularly well represented by sub-tropical and tropical trees, and these undoubtedly play an important role in maintaining the fertility of tropical forest soils. The only non-legume known to host *Rhizobium* bacteria are the five species of *Parasponia* (Cannabaceae), a tree genus found in Malaysia and the islands of the West Pacific.

A number of other species are able to fix nitrogen through an association with *Frankia* bacteria in root nodules. This appears to be the last vestiges of an ancestral trait that has almost been lost. About 25 genera among eight families are known to be involved: such species are termed actinorhizal plants. The families and genera involved include Betulaceae (*Alnus*), Cannabaceae (*Trema*), Casuarinaceae (*Allocasuarina, Casuarina, Ceuthostoma, Gymnostoma*), Coriariaceae (*Coriaria*), Datiscaceae (*Datisca*), Elaeagnaceae (*Elaeagnus, Hippophae, Shepherdia*), Myricaceae (*Comptonia, Morella, Myrica*), Rhamnaceae (*Ceanothus, Colletia, Discaria, Kentrothamnus, Retanilla, Trevoa*), and Rosaceae (*Cercocarpus, Chamaebatia, Dryas, Purshia*). While this list is likely incomplete, not all species within the listed genera are actinorhizal.

143

Figure 6.4 Examples of actinorhizal plants. From left to right: Colletia spinosa *(Brazil and Uruguay)*, Discaria chacaye *(Argentina)*, Casuarina glauca *(Australia)*, Alnus rhombifolia *(North America)*.

Soil fertility

The nutrients that forests require can be broadly divided into the macronutrients (relatively large amounts needed) and micronutrients (generally small amounts being needed). The macronutrients include nitrogen (N), phosphorus (P), potassium (K), calcium (Ca), magnesium (Mg) and sulphur (S). The micronutrients include iron (Fe), manganese (Mn), zinc (Zn), boron (B), copper (Cu) and molybdenum (Mo).

Ironically, air pollution has in many areas resulted in excess deposition of nitrogen and sulphur. This has created imbalances in ecosystems: an excess of nitrogen is termed eutrophication (see Chapter 5). When the nutrients in a soil are in balance, there is unlikely to be a problem for plants, but an excess of one or a shortage of another can lead to disruption (Marschner, 2012). Trees will absorb the nutrients, and if more are available, absorption will be greater and growth will increase. However, as soon as one of the nutrients is in short supply, deficiencies will appear, and growth will be slowed. In extreme cases, visible symptoms of nutrient deficiency may appear, and ultimately trees may die. This process is particularly important for nitrogen, which may often be in abundant supply due to anthropogenic sources.

More often, one or more nutrients may be in short supply. This is very much related to the weathering processes in soils (which result in the release of nutrients), and the movement of water through the soil (which can wash out nutrients in a process known as leaching). In some soils, nutrients may be present, but very tightly controlled. In many tropical soils, macronutrients such as nitrogen, sulphur and phosphorus, and micronutrients such as zinc, manganese, iron and copper, are incorporated into the organic matter. As this decays, any nutrients released are quickly taken up by the fine roots of the vegetation. Soil nutrient declines in the tropics following deforestation are closely linked to the loss of this soil organic matter (Tiessen *et al.*, 1994), rather than the soils being inherently nutrient deficient, as once thought.

A special case involves chemicals released by the trees themselves, a process known as allelopathy. This generally occurs through the litter (leaves, needles and small branches that accumulate under tree canopies). It is widely thought that this prevents competition, especially for water, but the process may be more complex. An important example of this is provided by Chinese fir (*Cunninghamia lanceolata*), a species widely grown in southeast China. This species is also known to be subject to soil-induced losses in productivity (Bi *et al.*, 2007). Other trees species known to be allelopathic include black walnut (*Juglans nigra*), many eucalypts, and the tree of heaven (*Ailanthus altissima*). In some cases, the chemicals only affect certain plant species, whereas in others they are more generic (Sasikumar *et al.*, 2004).

*Figure 6.5 Crimson bottlebrush (*Callistemon citrinus), an Australian species known to produce allelochemicals.*

145

Soil nutritional disorders can be recognized in several ways. The first indication is generally the appearance of the trees, which may be growing more slowly than expected or, in severe cases, the foliage may show colour changes. While some of the changes in colour are distinctive, others are not, and diagnoses generally have to be accompanied by chemical analysis of the foliage. Species vary in their response to nutritional deficiency, and other factors, such as soil water availability and disease may play a role. In conifers, deficiencies may first appear in older needles, but this should not be confused with the breakdown of chlorophyll in older needles prior to them being shed.

A second approach is to undertake chemical analysis of the foliage. This method is not fail-safe: it does not provide any indication of the cause of a particular deficiency, and the normal range of levels for a particular species may be uncertain. In addition, care is needed in cases of nitrogen enrichment, as a tree may show signs of mineral deficiency even though the soils have adequate amounts of nutrients (under normal conditions and without the nitrogen enrichment). Consequently, rather than absolute amounts, in some cases the ratios of nutrients may be more useful as a measure of imbalance.

The most expensive approach is conduct chemical analyses of the soil itself. This technique is widely used when assessing agricultural fertility, but is less useful in forestry because of the complexity and spatial variability of the soil. Soil analyses usually involve assessments of pH, organic carbon, and total nitrogen, but may also involve more detailed assessments of the other nutrients.

The effects of forest management activities on soil fertility are discussed later in this chapter in the section 'Changes in soil chemistry and nutrient availability'.

Soil water

Soil is a complex ecosystem in itself, and like all ecosystems, water is an essential component. Water is required for most of the reactions involved in chemical weathering and is of course required by plants. The physical nature of the soil affects how much water it can hold, and also determines how quickly water can pass through the soil.

Water can occur in soil in three phases: solid, liquid or gaseous. In its solid form (ice), it is unavailable to plants, and if trees are transpiring when the ground is frozen, desiccation damage can occur (see Chapter 5). As water freezes, it expands, and this can be disruptive to soils, causing frost heaving. However, frozen ground is able to bear the weight of heavy machinery, and forestry operations undertaken when the ground is frozen may have less of an impact than when it is wet. Similarly, if road surfaces are frozen, there is less likely to be damage to the road surface during forestry operations and log haulage, and consequently less sediment will be washed off the roads.

Liquid water may occur in pore spaces within the soil, especially after precipitation or snowmelt, or may be more tightly bound in the form of capillary water or adsorbed water. During rainfall events, water enters the soil until the soil becomes saturated, or the rate of water reaching the surface exceeds the infiltration capacity of the soil. When surface water accumulates, it may form ephemeral streams: such runoff can cause considerable erosion.

The relative humidity of the soil atmosphere is generally very high (98%), even when the gravimetric water content is as low as 2% (Coleman and Crossley, 1996). This enables a range of soil biota to exist, and these are essential for most processes occurring in the soil, particularly the breakdown of organic matter and the recycling of nutrients.

The relationships between forests and water are examined in Chapter 7.

Soil as a resource

All forests in their natural state have some form of soil. As discussed earlier, these are highly variable, depending on climate, bedrock, topography, biota and time. However, the one factor that has not been discussed so far is the disturbance of soils caused by human activities. In many parts of the world, areas now important for agriculture were once covered by forests. The soils in these areas have been heavily altered, and now bear very little resemblance to their original state. Forest soils can also be heavily altered: in much of the southeastern USA, forests have again developed on what was for a time agricultural land. Elsewhere, changes in forest composition have resulted in changes in the soil, with a well-known example being the podsolization of many European forest soils brought about by changes from broadleaved forest to plantations dominated by conifers, especially Norway spruce (*Picea abies*).

The alteration of soils by management activities can be both obvious and very subtle. More subtle changes include gradual changes to the nutritional status of the soil brought about by repeated harvesting of trees. This is often assumed to be sustainable, but if the rate of removal of nutrients in forest products exceeds the rate of natural replenishment, then the system will only be sustainable through the addition of fertilizers. This is a particular problem on some organic substrates, such as peat, which rely on very small nutritional inputs from the atmosphere. When planted with trees such as Sitka spruce (*Picea sitchensis*), as has happened in parts of Scotland and Wales, nutritional deficiencies can quickly develop if artificial fertilization is not adopted.

To a certain extent, this problem can be relieved by leaving the foliage and branches at the site, since the majority of nutrients are contained in these parts of the tree. However, with the increased interest in biomass as a fuel, the demand to remove as much of the tree as possible is growing. A considerable amount of research has been done on the effects of removing all parts of a tree (whole-tree harvesting), and this is informing some of the debate on biofuel removal from forests (Berger *et al*., 2013). As mentioned in Chapter 4, the development of short-rotation coppice presents a particular problem, since relatively large amounts of nutrients may be lost, especially

Figure 6.6 A pile of debris left after a logging operation in coastal British Columbia, Canada. In the past, this would have been burned, resulting in the loss of any nutrients to the atmosphere. Increasingly, there is interest in using this material as biomass fuel, with the same end result for the ecosystem, namely the removal of nutrients from the site.

potassium, calcium and magnesium. One possibility is to return the ash to the forest, but this is expensive and the ash does not replace all the nutrients that have been lost. In addition, returning the ash is really only effective when nitrogen is non-limiting, as the wood ash is generally low in nitrogen (Pitman, 2006). The application of wood ash will, however, help offset any soil acidification that may be associated with whole-tree harvesting (Stupak *et al.*, 2008).

Soil degradation

A number of different processes can result in the physical, chemical or biological degradation of the soil. Allelopathy has already been mentioned – this can be considered a form of biological soil degradation in cases where species with allelopathic properties have been introduced into ecosystems where they were previously absent. Physical degradation generally involves damage to the soil structure, with the worst case being complete removal of the soil (soil erosion). This can occur through the action of running water or wind. Chemical degradation involves either loss of nutrients from the soil or the addition of potentially toxic chemicals.

Compaction, rutting and puddling

The physical structure of the soil can be significantly changed by forest management operations. Structure, porosity, density, strength, pore size distribution, aeration, water retention, infiltration capacity and hydraulic conductivity can all be affected. The extent of this damage is highly variable and ranges from light to catastrophic. It is generally associated with forest machinery and may be direct, such as through bulldozing the soil, or indirect, caused by the weight of vehicles passing over a soil surface.

Many of the most severe problems are associated with the construction of the roads needed to access forest areas, especially when these are in wet areas with steep slopes. In such situations, roads are generally built using cut-and-fill, where the upslope side of

Figure 6.7 Oversteepened slope associated with a forest road in Gippsland, Victoria, Australia. Such slopes may fail or be a source of sediment as a result of small-scale erosion. Both processes are evident here.

*Figure 6.8
Poor road
surface
conditions
caused by an
interruption to
water movement
in a relatively flat
area, Williams
Lake, British
Columbia,
Canada.*

the road is oversteepened, and the material cut from the slope is spread on the downward side. The upward side is then subject to erosion, and may fail because the slope has been oversteepened. The downslope debris may be subject to erosion, spreading the debris over a large area. Without adequate management, surface and sub-surface water flows may be disrupted, and if water is diverted in ditches, erosion may ensue. Erosion of the road surface itself may also occur, depending on the nature of the materials, and the water management procedures in place.

Disruption of surface and subsurface water movement by roads is not restricted to steep slopes, and may occur in flat areas such as peatlands. In such cases, the problems may be associated with restrictions in water flow caused by the road, resulting in waterlogging and even flooding. This can not only result in damage to the road caused by vehicular traffic on a waterlogged road surface, but can delay or even prevent regeneration of forest in areas on either side of the road affected by waterlogging.

Any unmetalled road on a slope can be a focus for erosion. Management of this involves minimizing the amount of water running down a road by directing off the road as quickly as possible. This can be achieved using a suitable camber or by the use of wash bars.

Soil compaction occurs when a heavy weight is applied to a soil, destroying soil pores and breaking down surface aggregates. The primary cause of soil compaction within the context of forest management is machine traffic. Much of the equipment used in modern forestry is heavy, and its passage over soil results in compression. This is because undisturbed forest soils tend to have low soil bulk density and high macroporosity. The extent of compaction varies with the type of soil and the amount of traffic, and in some cases it can be severe. The majority of compaction occurs the first time a vehicle passes over a soil, or after only a few passes (Froehlich and McNabb, 1984). Compaction takes a great deal of time to recover naturally, and it may be necessary to rip areas such as log landings that have been subject to severe compaction, but this is rarely done for skidding trails (also known in Australia as snig tracks). Estimates of the extent of soil disturbance differ between studies, with a review by Grigal (2000)

Figure 6.9 Gullying developing on forest roads in Fujian, China (left), and Victoria, Australia (right). If not rectified, these gullies could develop rapidly, making the roads impassable to any form of traffic and potentially contributing sediment to rivers.

Figure 6.10 Rehabilitating a log landing at a fire salvage site by breaking up the compacted soil, Gippsland, Victoria, Australia.

estimating that there is substantial alteration of soil properties over 25%–50% of the harvested area, which may result in a loss of site productivity of as much as 10%.

While the most intense physical disturbance occurs on roads and landings, skid trails can also be a source of compaction and rutting, and equipment may also cause problems as it passes over a soil during a harvesting operation. There is clear evidence that even a single pass can cause considerable compaction, especially in a forest soil

Figure 6.11 Rutting caused by repeated machine traffic during harvesting, Otway Ranges, Victoria, Australia.

that has not previously been disturbed, and that this compaction takes a long time (up to 100 years) to recover (Rab, 2004).

Soil compaction is important because it can reduce the infiltration capacity of the soil, leading to surface runoff and erosion (see Chapter 7). It can also prevent roots from developing adequately in the upper soil layers, leading to lower intakes of water and nutrients. Seedling mortality in the next generation of trees may be increased, and growth of both seedlings and any remaining trees may be reduced. There may also be a shift in the ground cover, from one dominated by interior forest species to a community dominated by pioneer species. Other impacts include adverse effects on the soil fauna and flora. In particular, there may be significant effects on soil bacteria, which is important as the bacteria play a critical role in a wide range of ecosystem processes, including carbon mineralization, nitrification and nitrogen fixation. The effects are related to the severely reduced drainage and gas permeability associated with compaction (Frey *et al.*, 2009).

A number of steps can be taken to reduce the degree of compaction and rutting. A commonly used method is to strip the branches from stems at the bole, leaving the slash on the ground for machines to pass over. This is commonly used only for sensitive soils, since it more difficult than skidding trees to the roadside and removing the slash there. There have been questions over its effectiveness, with a 60 cm thick cover having little amelioration effect for heavy equipment (Horn *et al.*, 2007), but other studies suggest a significant benefit of using a brash mat (e.g. Hutchings *et al.*, 2002). The technique has been extended to skid trails and even roads, using small diameter wood, and archaeological evidence suggests that the technique has been in use for thousands of years to enable wheeled (and foot) traffic to pass over areas subject to rutting and puddling. Various wheel, tire and track designs on vehicles are intended to reduce the amount of compaction; the impact of wheeled vehicles can sometimes be further reduced by lowering the tire pressures.

Another process of concern when operating on wet soils is puddling. Puddling affects soils rich in silt and clay. It occurs when they lose their structure due to the reorientation of soil particles, and is usually caused by pressure being placed on the

*Figure 6.12 Logging slash left onsite and used to protect the soil in an upland alpine ash (*Eucalyptus delegatensis*) thinning operation in Gippsland, Victoria, Australia.*

Figure 6.13 Examples of tracked and wheeled vehicles used in forestry. Top left: tracked line skidder with blade (Canada); top right: grapple skidder with small blade and chained tires (Canada); bottom left: bulldozer with large blade (Malaysia); bottom right: grapple yarder (Canada).

soil surface when the soils are wet – either by animals such as cattle or by machinery (Peth and Horn, 2006). It results in increased density of the soil, and may inhibit the establishment of seedlings in the future. In many jurisdictions, problems associated with wet soils have resulted in laws and best practices that limit forestry operations when the soils are wet. This limits the amount of damage, but obviously restricts the time available for harvesting, and consequently the total amount of timber that can be harvested.

*Figure 6.14
Skid trail
protected by
brash mat
(foreground)
and corrugating
(laying small logs
at right angles to
the direction of
traffic), British
Columbia,
Canada.*

Changes in soil chemistry and nutrient availability

An important aspect of soil management in forest lands is the effects of the trees themselves on soil chemistry and nutrition. A number of concerns exist, with the greatest being that repeated harvests of wood will deplete certain soil nutrients faster than they can be replenished through the weathering of minerals or inputs from atmospheric deposition (mainly in the form of dust). In the late 20th century, there were also concerns that forests could capture pollutants and transfer them to the soil, causing soil acidification and the acidification of surface waters.

Intensive plantation forestry is a particular concern, especially when it involves the conversion from one forest type to another. This has happened for centuries in central Europe, where the native broadleaved forests and mixed forests of European beech (*Fagus sylvatica*) and silver fir (*Abies alba*) were replaced by monocultures of Norway spruce (*Picea abies*), often from a restricted range of provenances. More recent examples include the replacement of tropical rainforests with monocultures of Caribbean pine (*Pinus caribaea*), gamhar (*Gmelina arborea*) and *Eucalyptus* sp. in Brazil, the replacement of native vegetation with Chinese fir (*Cunninghamia lanceolata*) in southeast China and rubber (*Hevea brasiliensis*) in tropical southwest China, and the replacement of native tree species with radiata pine (*P. radiata*) in Australia, New Zealand, Portugal and Chile.

For most forms of extensive (i.e. natural and semi-natural) forest management, a principle that would minimize the changes to the chemistry of forest soils is 'near-to-nature' forestry. This practice is used to varying degrees in many parts of the world, although it has perhaps received greatest attention in places such as Germany, where many natural forests have been replaced by plantations (e.g. Bode, 1997). This approach to forest management seeks to have a forest cover that resembles as closely as possible the original forest cover of the area. The original forest cover is estimated based on the geographical position of the site and clues such as the ground flora, and is the basis for many silvicultural practices in central Europe. Variations of this approach have been adopted in places such as British Columbia, Canada. It is formalized through

Figure 6.15
Gmelina
arborea, *a
species native to
southeast Asia
that is being
increasingly
utilized in
monocultures in
tropical regions.*

Figure 6.15 Gmelina arborea, *a species native to southeast Asia that is being increasingly utilized in monocultures in tropical regions.*

measurements of hemeroby, which is measured on a scale designed to indicate the extent of human impact on a site, and which includes not only the botanical composition of the flora but also the naturalness of the restocking process, the amount of dead wood, the size of felling coupes and the current stand structure (Grabherr *et al.*, 1998). Similar ideas have emerged in discussions about forest quality (Dudley *et al.*, 2006), with hemeroby and authenticity being similar concepts.

Such approaches tend to assume that forest soils (and forest communities) are relatively stable in the long term. For example, in central Europe, near-to-nature forestry frequently involves attempts to restore ecosystems to their pre-disturbance state. However, this fails to account for long-term changes in stand conditions – it assumes that there is a climax vegetation for a site, and that this is fixed. Forest ecosystems are actually much more dynamic, and with climate change occurring, the idea of returning an ecosystem to a state present several hundred years ago is very questionable. This problem is examined in the concluding chapter of this book, since it affects all the biophysical indicators used in sustainable forest management.

When a forest is harvested, a large number of changes occur to the nutrient flows. Gross primary productivity declines, the soil surface warms and the rates of organic matter decomposition increase. Soil water contents increase, as much less water is being transpired and there is no evaporation of water from canopy surfaces. The low rate of nutrient uptake combined with the higher rates of decomposition can lead to temporary nutrient losses from the ecosystem. Nutrient losses associated with the removal of harvested biomass are generally quite low, and seem to be less than the rate of replenishment. The amounts of nutrients that are removed vary with tree species and the nature of the material removed. For example, removal of only the bole results in fewer losses than whole-tree harvesting. Generally, atmospheric deposition of nitrogen and sulphur is generally sufficient to replace losses of these two nutrients during stem-only harvesting (Morris and Miller, 1994). However, other nutrients, such as phosphorus, potassium and calcium, which are generally deposited in low levels in atmospheric deposition, may be more problematic. The extent to which these and other nutrients become depleted will depend on the initial amounts in the soil and their rate of removal and replenishment (Grigal, 2000).

A considerable amount of work has gone into the status of calcium in forest soils. Much of this work arose following concerns that acidic deposition in some areas, particularly in the southern Appalachian Mountains of the USA, could be leading to significant losses of calcium from forest ecosystems. Calcium losses associated with timber harvesting have also been a concern (Federer *et al.*, 1989). The work in the Appalachians suggested that acidic deposition, a form of air pollution, was destabilizing high-altitude forest ecosystems in the area. While calcium depletion may have been involved in some of the problems that were observed, a variety of other factors were also at play, especially the invasion of the forests by the balsam woolly adelgid (*Adelges piceae*), an exotic pest from Europe, which killed the Fraser fir (*Abies fraseri*) on the mountains (see Figure 5.30).

Full discussions of forest soils and how their nutrition can be affected by forest management practices are provided in Schulte and Ruhiyat (1998) and Binkley and Fisher (2013). A good discussion of how these processes can be affected by external forces, such as atmospheric deposition of pollutants and by climate change, is given by Lukac and Godbold (2011), although this focuses on northern forests, especially those in Europe.

Salination

A special form of soil chemical change is known as salination (also known as salinization), which occurs because of the build-up of salts (Na^+, K^+, Ca^{2+}, Mg^{2+} and Cl^-) in a soil. It can occur through a number of processes, both natural and anthropogenic. Under natural conditions, salt-containing groundwater may be drawn to the surface through capillary action, where it then evaporates, leaving the salts to accumulate. This generally only occurs where the groundwater table is within 2–3 m of the surface. Alternatively, irrigation can lead to salination as the water evaporates, leaving behind salts. Some fertilizers, including potassium, can also result in the accumulation of salts. Salination can cause major losses of ecosystem productivity, with the plants typical of low-salinity soils being replaced by specialized salt-tolerant plants known as halophytes. In Australia, well-known examples of halophytes include saltbushes (*Atriplex* spp.) and samphires (*Halosarcia* spp.). Salination of irrigated croplands is believed to have been one of the causes of the breakdown of early civilizations in Mesopotamia (Jacobsen and Adams, 1958).

While salination occurs naturally, it is most important when associated with irrigation. In Australia, there are an estimated 29 million ha of naturally occurring saline

*Figure 6.16 Left: blue-green saltbush (*Atriplex nummularia*) and right: mound spring samphire (*Halosarcia fontinalis*), two Australian halophytes. Such species occur in both naturally occurring and anthropogenic saline soils.*

*Figure 6.17 River red gums (*Eucalyptus camaldulensis) *at Dunkeld in Victoria, Australia. This species can survive on soils with high salt contents in the surface layers.*

soils, and about 4 million ha (in 2000) of soils salinized by anthropogenic activities (Lambert and Turner, 2000). With its large areas of saline soils, it comes as no surprise that a number of *Eucalyptus* species are capable of growing in them. Twenty-one eucalypt species occur naturally on saline soils. These include *E. robusta* in swampy estuarine sites, *E. camaldulensis* on saline soils of old river valleys, and *E. occidentalis*, *E. sargentii*, *E. halophila* and *E. salicola* on the margins of salt lakes.

Salination of forest soils is relatively rare, except once the tree cover is removed. Some plantations have been established on salinized soils in a deliberate attempt to reduce the extent of salination by lowering the groundwater table. This practice is particularly common in Australia, where the use of trees in the remediation of sali-nized soils is relatively widespread (Lambert and Turner, 2000). In many cases, the trees are planted on soils that formerly held trees but were cleared for agriculture. The choice of tree species depends on the climate and a variety of other factors. For example, in drier areas of the Australian wheatbelt and sheep-grazing areas (400–600 mm annual rainfall), mallee plantations are feasible. In the slightly wetter areas that have mainly been salinized due to irrigation, particularly along the Murray-Darling river system, commercially useful species including river red gum (*Eucalyptus camal-dulensis*), river oak (*Casuarina cunninghamiana*), and possibly Tasmanian blue gum (*E. globulus* ssp. *globulus*) and flooded gum (*E. grandis*) could be used (Marcar *et al*., 1995). The development of such plantations has a number of potential additional effects, including increased carbon storage and increased biodiversity, especially if native species are used.

The salinity problems are closely associated with the groundwater dynamics in the landscape. The management of the problems also requires an understanding of land-scape dynamics. Groundwater is recharged when the precipitation entering the soil exceeds the rate of evapotranspiration. This may occur some distance away from the salinized soils, but the establishment of trees in the recharge area can reduce the rate of groundwater recharge, lowering the water table in the landscape. The groundwater then passes through a transmission zone in which the movement of water is parallel to the surface before reaching a discharge area, where net water movement is out of

Figure 6.18 Examples of salt-tolerant Eucalyptus *species. From left to right: Forest red gum (*E. tereticornis*), salt lake mallee (*E. halophila, also known as* Corymbia halophila*), Salt River gum (*E. sargentii *ssp.* sargentii*) and swamp mahogany (*E. robusta*). Forest red gum shows a particularly high level of intraspecific variation in salt tolerance.*

the groundwater. Trees planted in the discharge area can also help lower water tables, reducing the rate of discharge to the soil surface.

Irrigation is rarely used in forestry except to establish plantations in dryland areas. However, areas of plantations that are currently being irrigated, such as in Australia, China, Egypt, India, Kenya, Pakistan and Sudan, may be affected by this process.

Figure 6.19 Salinized soil on Kangaroo Island, South Australia. The mallee woodland was replaced by a low shrub cover comprising halophytic (salt-tolerant) plants.

Siddiqui (1994), for example, describes problems associated with salinity (and water-logging) in plantations in the Changa Manga area of the Punjab in Pakistan. The salts are carried in water seeping from irrigation canals in the area. In general, however, the small area of irrigated forest land means that this problem is relatively localized.

Mangrove forests represent a special situation. A limited number of species are able to survive an extreme environment that includes daily inundations with saltwater. A number of species of tree, together with a palm and a fern, have developed adaptations to cope with these conditions. A full discussion of mangroves and their ecology is provided by Saenger (2003).

Litter raking

In the past, the litter from coniferous forests in particular was collected and used as bedding for livestock. It has been recognized as a serious problem for some time, and its widespread use until relatively recently in many forested areas means that some forests are still recovering from its effects. For example, in the Lower Spreewald of Brandenburg, Germany, surveys of vegetation change over the past 45 years have revealed that the humus layer is still recovering from past litter raking (Reinecke *et al.*, 2014). The amount of time needed for forest soils to fully recover is uncertain, although one study has suggested that effects may still be evident 130 years after the practice is stopped (Gimmi *et al.*, 2013). In many forests where litter raking has been conducted in the past, there may be significant opportunities for carbon accumulation as the humus layers recover from the past disturbances.

Despite the accumulated evidence of adverse effects of litter raking in European forests, litter raking is still practiced elsewhere. In China, litter raking is an important source of income, but it is still largely done by hand, using rakes. In the southeast USA, litter raking has become increasingly mechanized, and despite the controversy over its effects, its use is expanding (Kelly and Wentworth, 2009). The litter is collected primarily for use in landscaping and comes from several forest types, including loblolly pine (*Pinus taeda*), longleaf pine (*P. palustris*) and slash pine (*P. elliottii*).

Litter raking is significant because of its potential impacts on soil nutrition. However it also affects the diversity of the ground flora and can damage regeneration. While it has been banned in many European countries, there may still be significant removals of coarse and even fine woody debris. As a result, many forests in Europe appear to be 'clean', with almost no woody debris on the soil surface. This has major implications for nutrient cycling and for biodiversity, with the fauna and flora associated with wood decay being adversely affected.

Deliberate modifications

Much of the preceding discussion has been concerned with minimizing the potential damage to soils caused by forestry operations. In most cases, these concerns are primarily related to extensive forestry, where the forest being harvested, and the subsequent regrowth, broadly reflect the natural vegetation of the area. In plantation forests, these concerns are very much less evident. In plantations designed for timber production, the emphasis is on developing a crop as quickly as possible. This means optimizing the seed bed for the trees in question. Such optimization may involve substantial modification of the soil, and is much closer to agricultural practices. It may involve draining a site through deep cultivation and/or removal of the surface organic layers to create an optimal mineral seed bed. The organic layers may be removed by burning, or by physical removal of the organic matter. In such cases, concerns about the soil may be focused on the prevention of erosion, either by surface water or by wind, and a priority is to establish a tree crop as quickly as possible.

A considerable amount of knowledge has been gained about soil preparation techniques through attempts to establish forests on reclaimed mineral workings. There is a long history of such work in Europe, and similar techniques have since been applied throughout the world. Different forms of amelioration may be required. Many such soils are badly compacted, and may require ripping before plants can be established in them. There may be very low nutrients, particularly nitrogen, and may require

*Figure 6.20 Loblolly pine (*Pinus taeda*) plantation in Alabama, USA. The organic matter has been removed prior to the planting, a technique that creates a favourable growing environment for the young trees, and minimizes competition from weeds.*

fertilization or the establishment of nitrogen-fixing plant species, including alder (*Alnus* sp.) and black locust (*Robinia pseudoacacia*). Inoculation of the soil with mycorrhizal fungi may improve initial tree growth. In some cases, the soils may be contaminated with phytotoxic concentrations of elements such as copper, lead and zinc or with pollutants such as oil. Various options exist for such sites, including removal of contaminated soil, or a range of remedial techniques. Full accounts of the use of trees in the reclamation of disturbed land can be found in Moffat and MacNeill (1994) and Montagnini and Finney (2011), as well as a range of other sources.

Nutrient replenishment (fertilization)

In cases where there are insufficient amounts of one or more nutrients critical for tree growth, an option is to add these artificially. This is widely practiced in agriculture, but is less common in forestry, and its effects are quite variable, depending on the nature of the soil, the tree species, the age of the stand and other factors. This topic is dealt with in Chapter 4.

Conclusions

Soils are an essential part of the forest ecosystem. They are dynamic and contain significant amounts of biodiversity and carbon. In many forest ecosystems, there is more biological diversity and carbon in the soil than aboveground. Without the soil, there would be no forest, and the fertility of the soil is an important determinant of the productivity of the forest. Consequently, the maintenance of the soil is an important part of sustainable forest management.

Unfortunately, many forest management operations, if done badly, can seriously damage the soil. On a positive note, a great deal of knowledge exists on how to mitigate this damage. The damage can take many forms, but can be broadly classified into physical, chemical and biological. These are not mutually exclusive, and the soils should be seen as systems: any change in a soil property is likely to have ramifications for many other of the soil properties. There will also be impacts on other parts of the ecosystem, since the soil is the fundamental building block.

Forest managers often take the soil for granted. However, it is a finite resource, and once lost it takes a great deal of time to develop again. The forest ecosystems that were lost in the Mediterranean basin 2,000 years ago and the catastrophic soil loss that followed have yet to recover (see Chapter 1). It is questionable whether they ever will, now that the climate is changing so significantly. In tropical areas, the fast rate of weathering means that soils recover from disturbance more quickly, but they are also more easily lost because of the very high rainfall intensities that characterize most tropical areas. Consequently, forest managers need to pay considerable attention to the maintenance of the soil in the areas that they manage.

Further reading

Binkley, D., Fisher, R. F. 2013. *Ecology and Management of Forest Soils*. 4th edition. New York: John Wiley.

Lukac, M., Godbold, D. L. 2011. *Soil Ecology in Northern Forests: A Belowground View of a Changing World*. Cambridge: Cambridge University Press.

Montagnini, F., Finney, C. (eds.) 2011. *Restoring Degraded Landscapes with Native Species in Latin America*. Hauppauge, NY: Nova Science Publishers.

Osman, K. T. 2013. *Forest Soils: Properties and Management*. Heidelberg: Springer.

Chapter 7

Water and watersheds

R. Dan Moore, Roy C. Sidle, Brett Eaton,
Takashi Gomi, and David Wilford

In the Montreal Process, water and soil are considered together. Here water and soil have been treated separately, as their management represents a crucial part of sustainable forest management. The relevant criterion (Criterion 4: Conservation and maintenance of soil and water resources) has two indicators directly related to water:

4.3.a Proportion of forest management activities that meet best management practices, or other relevant legislation, to protect water-related resources.
4.3.b Area and percent of water bodies, or stream length, in forest areas with significant change in physical, chemical or biological properties from reference conditions.

These two indicators do not really reflect the importance of the relationship between water and forests. Some of the most controversial aspects of forestry, such as its impact on the supply of clean water from forested catchments, are associated with water, and this chapter therefore looks at the relationship in considerable detail.

Hydrological functions of trees and forests

Canopy interception

In forested areas, a portion of rainfall may reach the ground by falling through gaps in the canopy. This component is called direct throughfall. Of the rainfall that is intercepted by the foliage or branches, some will drip off and reach the ground as indirect throughfall or canopy drip. Some intercepted rainfall will be directed along branches to the trunk, where it then reaches the ground as stemflow. The sum of throughfall and stemflow is sometimes called net rainfall or net precipitation. The remaining intercepted rainfall will evaporate back to the atmosphere. This last component represents rainfall that is lost to the terrestrial hydrologic cycle, and is called interception loss. If there is an understory layer of vegetation, throughfall from the overstory will be subject to a second phase of interception processes. In multi-tiered forest canopies, which are common in the tropics and mature temperate forests, rainfall can be intercepted at several levels, creating a vertical cascade of potential losses to the atmosphere.

*Figure 7.1 Red and silver beech in Fiordland National Park, New Zealand. The red beech (*Fuscospora fusca, *formerly* Nothofagus fusca) *is the lighter green canopy growing from the valley floor up the mountain, and the silver beech (*Lophozonia menziesii, *formerly* Nothofagus menziesii) *is the darker green canopy growing above the red beech. Both species are evergreen.*

Snow is also subject to canopy interception, although the processes differ from rainfall interception in that snow must melt before it can drip off the canopy. However, snow can be blown off branches, and clumps of snow can fall off branches, a process called unloading. Unloading of snow from a forest canopy often occurs during warming weather, when branches become more flexible and begin to droop downward in response to the weight of the intercepted snow.

Interception loss can represent a significant fraction of total precipitation. For example, it typically represents about 26% of annual precipitation for coniferous forests and 13% of growing-season precipitation for deciduous forests (Price and Carlyle-Moses, 2003), although the actual amounts will depend on tree species, canopy architecture, and climatic characteristics. Interception loss for conifers is a higher percentage of total precipitation in summer than winter due to the greater availability of energy to evaporate intercepted water. The southern beeches (*Nothofagus* species in South America, *Fuscospora* and *Lophozonia* species in New Zealand) are mostly evergreen broadleaf species, and like conifers the evergreen species can experience substantial interception loss during winter (Figure 7.1). During winter, interception loss from deciduous trees will be diminished by the loss of leaf area.

In addition to its importance in the water balance, canopy interception also affects erosion processes. Canopy drip from relatively short trees and shrubs will reach the ground with a lower velocity than rainfall that has not been intercepted and will thus have lower kinetic energy and less potential to cause surface erosion. However, in taller and uniform canopies, raindrops can sometimes coalesce into larger droplets and actually exert a greater kinetic energy on the soil surface compared to open areas. Soils under large-leaved species, such as teak (*Tectona grandis*) are particularly susceptible, because rainfall intensities where this species grows (the tropics) can be high, water coalesces on the leaves, and the ground cover is often limited by the high level of shading provided by the leaves and the high decomposition rates of any leaf litter.

In coastal and alpine areas that experience heavy fog, fog drip can be an important additional source of water. Fog drip, also referred to as occult precipitation, results from the condensation of atmospheric moisture onto the forest canopy during

Figure 7.2
*Teak (*Tectona grandis*)*
plantation in
Uttarakhand
State, India.
Note the large
leaves and bare
soil, potentially
leading to
erosion problems
during the
intense rains of
the monsoon
season.

Figure 7.3
Fog located at
'World's End',
on the Horton
Plains of Sri
Lanka. The
location marks
the edge of the
Horton Plains,
and the slope
drops 870 m in
a short distance.
Fog frequently
occurs at this
location, and
contributes
significantly
to the total
precipitation at
the site.

fog events. Harr (1982) found that fog drip in the Oregon Coast Range of the USA more than compensated for interception loss, with 1,739 mm of throughfall being recorded below an old-growth forest canopy dominated by Douglas fir over a 40 week period, compared to 1,352 mm of precipitation measured in a nearby clearcut (a 21% increase). However, one of the most detailed studies of fog drip conducted in a tropical rain forest in Taiwan, which had more than 350 foggy days per year, showed that fog

*Figure 7.4
The complex, epiphyte-dominated canopy of a tropical cloud forest, Villa Blanca, Costa Rica.*

drip accounted for only 10% of the annual atmospheric hydrological inputs (Chang *et al.*, 2006).

Fog drip may be particularly important in tropical cloud forests. These forests are generally located at medium to high altitudes (400–2,800 m above sea level [masl]), with an average temperature of 18°C and average precipitation of 2,000 mm yr^{-1}. They are restricted to within 350 km of a coastline and are generally topographically exposed (Jarvis and Mulligan, 2010). Measurement of the water deposited in fog is extremely complex and, as a result, there are considerable variations in estimates (which likely also reflect real differences in the deposition rates). Measurements often refer to horizontal precipitation, which includes all precipitation not captured by a standard rain gauge. Measurements are also complicated by the complex nature of cloud forest canopies, which contain large numbers of epiphytes. These make the calculation of surface areas almost impossible. In addition to the complications associated with the measurement of horizontal precipitation, the reduced evapotranspiration induced by fog also needs to be taken into account when calculating the impacts of fog on the hydrology of cloud forests.

Transpiration

Transpiration is the physiological process by which plants take up soil water via their roots and transport the water through the stem and branches to the leaves, where the water can diffuse through the stomata into the atmosphere. Most tree species have physiological responses to drought conditions that result in closure of the stomata and thus reduce water loss by transpiration, although stomatal response is highly variable among species and depends on a range of environmental factors including sunlight, atmospheric humidity, wind, and soil moisture.

From a water resource perspective, transpiration is a loss of water. Transpiration from a forest stand is typically about three times greater than the evaporation that would occur from an unforested soil surface or grassland experiencing the same weather conditions. The rate of evaporation from a bare soil surface will decline as the surface dries due to the slow rate at which water can migrate upwards from wetter, deeper layers through the overlying drier layer. Trees, on the other hand, extract water from the depth of soil that contains tree roots and thus can access a greater volume of water to sustain evapotranspiration through periods of dry weather.

Influences on snow accumulation and melt

The dynamics of snow accumulation and melt vary depending on regional climate and topography. At inland sites removed from the moderating effect of the oceans, temperatures tend to remain below freezing from late autumn until late winter or spring. There is thus only a minor influence of mid-winter rain and melt events. Snowmelt tends to occur during spring and into early summer. Spring snowmelt generally begins earliest on southerly aspects (in the Northern Hemisphere) and at low elevations.

In temperate coastal mountain regions, snow accumulation and melt dynamics typically vary strongly with elevation. For example, at mid-latitude sites, snowfall occurs at most a few times per year at the lowest elevations (e.g. from 0 to about 300 masl in south coastal British Columbia), and typically melts within one to several days. In the transient snow zone, located between about 300 and 800 masl in south coastal British Columbia, snowfall occurs more frequently than at lower elevations, and a snowpack may persist up to several weeks each season. Above about 800 masl, a continuous snow cover usually forms in late autumn to early winter and lasts until spring or even early summer at higher elevations. Even in this higher zone, frequent warm storms result in rain-on-snow events and mid-winter melt, which can generate some of the highest streamflow events in coastal catchments (Jones, 2000). The boundaries between these snow zones will vary from year to year – typically being higher in warm winters and lower in cool winters – and also with latitude, being lower in the north and higher in the south.

Forest cover has the general effect of reducing snow accumulation through interception loss. Seasonal peak snow accumulation under coniferous forest canopy is typically up to about 40% less than at a nearby open site (Varhola *et al.*, 2010). Forest cover also tends to delay the onset of spring snowmelt and reduce the rate of snowmelt, primarily by shading the snow cover from energy inputs by solar radiation, and also by reducing wind speeds compared to open sites, which results in lower energy inputs by turbulent transfer of heat and water vapour from the atmosphere to the snow surface.

Snow accumulation and melt can be particularly sensitive to forest cover in the transient snow zone. In that zone, snowfall events tend to be accompanied by air temperatures just below the freezing point, and it is not uncommon for periods of above-freezing air temperatures to occur during storms or immediately following periods of snowfall. During these periods of above-freezing air temperature, snow held in the forest canopy will be exposed to warm, humid air and will melt as a result of energy inputs from the atmosphere. In contrast, snow on the ground in an opening will be less exposed to atmospheric energy input than snow in the canopy, and will be less prone to melt away. As a result, snowpacks tend to be deeper and persist longer in openings than under the forest canopy. Furthermore, there is more likely to be a snowpack present at open sites during rain events than under a forest, so that runoff generation at open sites would be augmented by the melting of snow (Leach and Moore, 2014).

Hillslope hydrology in forested landscapes

Hydrologic processes on forested slopes

Soils under forest cover tend to have a thick duff layer and also tend to be laced with large pore spaces caused by the decay of old roots, animal burrows, subsurface erosion channels, and pockets of buried organic material. These macropores function as a secondary porosity in the soil, which may dominate over the porosity of the soil matrix (fine pores) in terms of water transmission. Macropores create preferential flow paths that link up across hillslopes as soils become wetter (Sidle *et al.*, 2001). As a result, forest soils tend to be highly permeable with saturated hydraulic conductivities (K_s) in the upper 10–20 cm (O and A horizons), typically in the range of 10^{-4} to 10^{-3} m s^{-1}. Usually K_s declines exponentially with depth because of increasing clay content, inherent compaction, and lower organic matter concentrations, but this is not always the case, especially when subsoils have high secondary porosity due to features such as cracks.

During rain events, the thick duff layer, if present, protects the soil from the impact of raindrops, which in some soil types can cause 'sealing' and 'crusting' of the soil surface and reduce the soil's infiltration capacity. In tropical forest soils where decomposition rates are high, less litter typically accumulates on the forest floor, and even minor disturbances or heavy rainfall can displace this thin protective layer, resulting in a reduced infiltration capacity. Removal of the surface organic layer by wildfire can also result in soil sealing.

If a forest stand is subject to a hot fire, organic compounds in the upper soil layers will migrate downward as vapours into cooler soil layers below the surface and condense around mineral soil particles. This coating can result in a layer of soil just below the surface that is hydrophobic (water repellent), and which will inhibit deeper percolation of water and thus produce overland flow and sheet erosion of the soil above the hydrophobic layer. This process has commonly been observed in relation to wildfires, but can also occur in association with hot slash burns. Post-fire hydrophobicity appears to be highest in the first year or two following a fire, primarily on soils with a sandy texture. Hydrophobicity may also occur in the natural organic matter of surface soils, but this effect typically diminishes once soils become wet.

The typically high permeability of forest soils promotes the rapid downward percolation of infiltrating rainwater and/or snowmelt. Therefore, rainfall and snowmelt intensities generally will not exceed infiltration capacities, and infiltration-excess overland flow rarely occurs, except at disturbed sites where the surface organic layer has been removed and especially where the soil has been compacted, such as on road surfaces.

In many regions, particularly those influenced by Quaternary glaciation, soils are formed in colluvial and ablation till deposits that lie over relatively impermeable bedrock or compacted glacial till. In these situations, the bedrock or compacted glacial till inhibits the continued downward percolation of water, resulting in the formation of a saturated soil layer with positive pore pressures at the base of the soil profile. The top of this saturated layer is the water table. The water table responds to storms or snowmelt but tends to remain well below the soil surface at upslope areas, especially on steep slopes. However, at downslope sites or in depressions (hollows), especially where the hillslope gradient becomes less steep, the water table can rise to the soil surface.

Where the water table has risen to the surface, the entire soil depth will be saturated, and water that has flowed downslope within the saturated layer from upslope areas can seep through the soil surface and become overland flow. This seepage water is called return flow, because it is rain or snowmelt water that has infiltrated the soil, flowed downslope, and then returned to the surface to reach the stream channel as overland flow. In addition, any rain or meltwater occurring over these saturated zones will not be able to infiltrate the soil surface (because the soil is already saturated and subsurface water is seeping to the surface), and will be deflected downslope with the return flow. The overland flow at locations where the soil is fully saturated is called saturation-excess overland flow, to distinguish it from infiltration-excess overland flow, and is a combination of return flow and rain and snowmelt water onto the saturated soil. Figures 7.5 and 7.6 illustrate typical conditions between and during storm events.

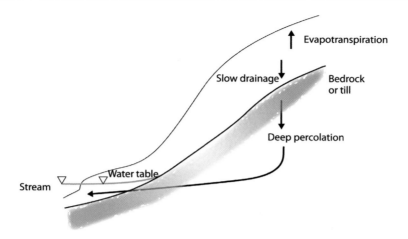

Figure 7.5
Schematic of water fluxes on a slope between storm events.

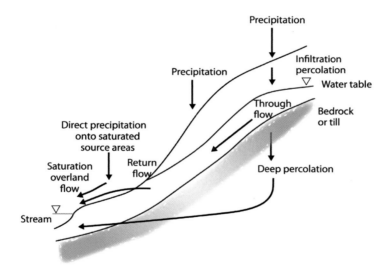

Figure 7.6
Schematic of water fluxes on a slope during a storm event.

Effects of forest management on hillslope hydrology

Infiltration-excess overland flow is rare in undisturbed forest soils due to their generally high permeability. However, as described in the previous chapter, many operations associated with forest management can result in soil compaction and a loss of permeability. On these disturbed surfaces, intensities of rainfall and/or snow melt can exceed the soil's infiltration capacity, resulting in infiltration-excess overland flow. In addition, as mentioned previously, there is the potential for slash burns to result in soil sealing and water repellent soils, significantly reducing the soil's infiltration capacity.

Figure 7.7
Effects of forest
roads and
drainage systems
on downslope
flow of water.

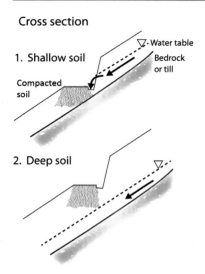

Cross section

1. Shallow soil

Compacted soil

Water table
Bedrock or till

2. Deep soil

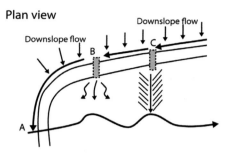

Plan view

Downslope flow

Downslope flow

A

B

C

A. Intercepted water directly flows into a stream via road ditch.

B. Intercepted water flows through culvert and is dispersed and re-infiltrates.

C. Intercepted water flows to a channel via culvert and a gully.

Roads and the associated drainage networks can also influence subsurface flow down hillslopes and water delivery to the stream channel. As shown in Figure 7.7, the road cut can intercept downslope flow and redirect it laterally through a ditch, especially on slopes with shallow soils. If this intercepted water is conveyed directly to a stream channel (as in path A), the net effect will be for the water to reach the stream channel faster than it would have had it followed its original flow path. In pathway B, the water is directed by a drainage relief culvert onto a planar or convex slope, where it re-infiltrates. In this case, the water will have been laterally re-directed but will still dominantly follow subsurface flow paths to the stream. Thus, there may not be a significant change in the timing of water delivery to the stream. In case C, water is conveyed through a culvert into a gully, where it flows as surface flow to the stream channel. As in path A, this pathway involves more rapid conveyance of water to the stream channel. The consequences are not only for faster water delivery to streams but also for the generation of debris flows in gullies or hillslope depressions.

Removal of the forest canopy can also influence hillslope hydrology through its effect on soil moisture content during the transition from a summer dry season to the autumn-winter wet season, which is typical of the hydroclimatic regime in many temperate regions, including western North America. Soils will typically be drier at the end of the summer season under forest cover than at a harvested site due to the differences in transpiration losses. Under dry antecedent soil moisture conditions, much of the rain and/or meltwater inputs may be retained by the soil and thus will not be available to percolate downward through the soil to become groundwater recharge or throughflow. Following harvesting and the associated reduction in transpiration, the reduced soil moisture deficit at the end of summer will allow more rain and/or meltwater to become groundwater recharge or throughflow.

Watershed hydrology in forested landscapes

A watershed is defined in relation to a specific point of interest (POI) along a stream channel, such as the location of a stream crossing, a water intake, or the confluence of the stream of interest and another water body. Specifically, the watershed is defined as the area that contains all locations from which water draining downslope will flow into the stream or one of its tributaries at or upstream of the POI. The watershed boundary is defined by topographic heights of land, or drainage divides.

In North America, the term watershed is synonymous with the terms catchment and drainage basin. However, outside North America, the term watershed is often used to refer to the height of land that forms the watershed boundary.

Runoff generation

Runoff generation refers to the processes and pathways by which water is delivered to the stream channel, as illustrated in Figure 7.7. Different pathways are associated with different velocities of flow. For example, overland flow has much higher velocities than throughflow and will thus reach the stream channel more quickly. Different pathways are also associated with different chemical characteristics. Overland flow, for example, interacts with the upper, organic-rich layers of the soil, and tends to have relatively high concentrations of organic compounds and nitrogen but low concentrations of minerals. Throughflow, on the other hand, tends to have higher concentrations of base cations than overland flow. Deep groundwater, which has longer contact times with the mineral substrate, typically has high concentrations of dissolved silica.

An important concept in understanding runoff generation is connectivity, particularly of overland flow paths and subsurface preferential flow paths. For example, for pathways A and C in Figure 7.7, there is uninterrupted connectivity between the road surface and the stream. As a result, overland flow generated on the road surface, and any sediment eroded from it and the ditch, will reach the stream. For pathway B, the overland flow path lacks connectivity between the road bed and the stream. As a consequence, when the water infiltrates the soil below the culvert, any sediment it carried would be deposited on the soil surface and thus not reach the stream.

Streamflow regimes

The concept of streamflow regime refers to the seasonal distribution of streamflow, which is a function primarily of the interactions between regional climate, vegetation, and topography. In western North America, for example, winters are wet and cool while summers are drier and warmer, and streamflow regimes can be grouped into five primary categories (Trubilowicz *et al.*, 2013). At low elevations in coastal areas, where snowfall represents a small portion of annual precipitation, the pattern of streamflow will follow the seasonal distribution of rainfall, with high flows during winter and low flows in summer and early autumn. These regimes are called pluvial or rain dominated. In the interior of western North America, the majority of winter precipitation falls as snow and melts during spring and early summer. Thus, these nival or snow-dominated regimes exhibit high flows in spring and early summer, with a general decline in flow through summer and into autumn and winter when streamflow is sustained by discharge of deeper groundwater. Catchments with more than about 2% glacier cover exhibit nivo-glacial regimes. Like nival regimes, they have low flows through the winter and high flows in spring and early summer associated with snow melt. However, the period of high flows extends into late summer as a result of continued melting of glaciers after the seasonal snow has entirely melted.

In temperate coastal mountain and coastal-interior transition areas, many catchments have hybrid nivo-pluvial regimes, with rain dominant at the lower elevations and the influence of snow becoming more important at higher elevations. Within these hybrid regimes, some will be snow dominant, with the major period of high flows in spring and summer in association with seasonal snow melt at higher elevations and a secondary period of high flows in late autumn and winter, associated with rainfall onto

Figure 7.8 The presence of snow complicates an understanding of the hydrology of a catchment. While precipitation, in the form of snow, may occur in winter, runoff may be minimal, with peak flows occurring in spring during the melt. Snow may persist into late summer, resulting in a continuous supply of water to streams regardless of the amount of precipitation received during the summer. The picture is a landscape on Vancouver Island, British Columbia, Canada.

the lower-elevation portions of the catchment. Rain-dominant hybrid regimes have the primary period of high flows in late autumn and winter, with a secondary period of high flows in spring and summer.

The definitions of streamflow regimes must be adapted for different regions. For example, in eastern Australia, the seasonal climatic patterns vary from the south to the north. In the south, the general climate pattern is similar to that in western North America, with winters being wetter and cooler than summer. In the tropical climate of the north, however, the seasonal pattern of precipitation is reversed, with summers being wetter than winter, although there is less contrast between summer and winter temperatures than in temperate regions.

The effects of forest harvesting differ among the different hydrologic regimes. For example, the effects of harvesting on hydrological response may be more complex in catchments subject to rain-on-snow events than in rain-dominated or snow-dominated catchments. From the perspective of sustainable forest management, it is appropriate to develop separate watershed assessment and management procedures for the different regime types. For example, in the British Columbia Forest Practices Code, which was in force from 1994 to 2004 and remains the default best practices manual for many forestry practices, separate assessment procedures were developed for coastal and interior watersheds.

Effects of forest management on streamflow

Water yield and low flows

Water yield refers to the total amount of water that leaves a catchment as streamflow over a specific period of time, typically on an annual, seasonal, or monthly basis. For ease of comparing among sites and studies, it is usually expressed as the equivalent depth of water spread over the catchment (i.e. volume divided by drainage area).

Studies globally have found that annual water yield generally increases following harvesting (or decreases following afforestation), consistent with the decrease in both interception loss and transpiration (Bosch and Hewlett, 1982). Differences in annual evapotranspiration, and thus annual water yield, between forested and non-forested catchments increase with mean annual precipitation. That is, the wetter a catchment's climatic regime, the greater the potential for increased water yield following harvesting. This relation between post-harvest water yield increase and annual precipitation also appears to hold for interannual variability at a given catchment. For example, Harr et al. (1975) found that post-harvest water yield increases in western Washington and western Oregon were higher in wetter years. Water yield increases also increased with the percentage of the catchment that was harvested.

The seasonal timing of post-harvest water yield changes depends on the hydroclimatic regime and forest type (coniferous vs. deciduous). In the rain-dominated deciduous forests of the southeastern USA, the greatest increases in water yield occur during the growing season, with the greatest proportional increases in the low-flow months of August to October. Water yield during spring snowmelt seems to be little affected by harvesting in deciduous forests in the northeast USA, likely because leafless deciduous forest has little effect on snow accumulation; the greatest post-harvest water yield increases occurred during the growing season. In the conifer-dominated forests of western North America, most of the increased water yield occurs in the wet autumn and winter months in rain-dominated regimes and during the spring and early summer in snow-dominated regimes, associated with the seasonal melting of the previous winter's snowpack.

Streamflow during extended periods of dry weather is often critical in terms of water supply and aquatic habitat. Under these conditions, streamflow is maintained by discharge of groundwater and the slow drainage of soil moisture. Low flows tend to be less extreme following harvesting, consistent with the reduction in transpiration. However, low flows can become more extreme as the forest regrows (Hicks *et al.*, 1991).

Peak flows

Peak streamflow events are an important component of a channel's streamflow regime, particularly because they are a primary control on sediment transport. In a positive context, peak flows are required for flushing accumulations of fine sediment from the coarse bed sediments used for spawning by fish. The morphology and sediment characteristics of a channel, and thus habitat characteristics, are adapted to and depend on the history of peak flows that it has experienced. The effects of forest harvesting on peak flows have been an ongoing concern because an increase in the magnitude of peak flows has the potential to cause an increase in bed and bank erosion, which can scour salmonid spawning beds and have negative consequences for aquatic species.

Forest operations can influence peak flows by changing the timing and intensity of water inputs. In rain-dominated catchments, input intensities will increase due to the decrease in interception loss following forest harvesting. In nival regimes, forest harvesting typically increases rates of water input by increasing snow accumulation and by increasing snowmelt rates during both spring snowmelt and mid-winter rain-on-snow events. A complicating factor during spring snowmelt is that, in some situations, the acceleration of snowmelt in clearcuts as compared to forests can desynchronize snowmelt over a catchment and reduce peak flows by spreading the snowmelt pulse over a longer period.

Forest operations can also increase the magnitude of peak flows through changes in runoff processes and hillslope flow paths. In areas with drier summers and wetter autumn-winter seasons, soil moisture content during the summer-autumn transition would be higher following forest harvesting, allowing more rain and/or melt to reach the channel rather than being retained as soil moisture. Ground-based

*Figure 7.9
Severe scouring
associated with a
peak flow in the
Bellinger River,
New South
Wales, Australia.
The islands in
the foreground
and on the right
of the photo
were both
overtopped by
the floodwaters.*

*Figure 7.10
Determining
the hydrological
response of
heavily harvested
catchments,
such as this one
near Kelowna in
southern British
Columbia,
Canada, is
complex.*

harvesting operations that result in soil compaction (see Chapter 6) can increase the generation of infiltration-excess overland flow, which has higher downslope velocities than subsurface stormflow. In addition, throughflow can be intercepted at road cuts and then diverted to surface flow paths to the stream channel (see Figure 7.7, flow paths A and C). In all of these cases, the more rapid delivery of water to the stream channel can result in higher peak flows. However, the magnitude of this response depends on the connectivity of overland flow paths (compare flow paths A and B in Figure 7.7).

Hydrological recovery

Following harvesting, establishment and development of vegetation will influence hydrological processes and eventually reduce the magnitude of harvesting-related impacts. The trajectory of recovery will depend on the type of vegetation and its growth rate and on successional processes. Recovery rates for stand-level processes can be quantified using a chronosequence approach. A chronosequence comprises a set of sites chosen to be as similar as possible in terms of soil, climate, and topography, but which have forest stands differing in time since harvest, and typically include relatively recent harvest sites and unharvested reference sites. The process or quantity of interest (e.g. snow accumulation, snow melt rate) is measured at each site, and the percent recovery associated with each site (HR_i) is computed as follows:

$$HR_i = 100\% \frac{x_i - x_0}{x_r - x_0} \qquad\qquad \text{(Eq. 7.1)}$$

where x_i is the measured process or quantity of interest for a given stand, x_0 is the corresponding quantity measured at a recently harvested site, and x_r is the quantity measured at a reference site. The computed values of HR range from 0% for a fresh clearcut up to 100% for a stand that functions like the original mature stand that was harvested. For application, HR can be related to stand age, stand height, leaf area index, or some other measure of stand development.

The ability to quantify rates of recovery is required for assessing cumulative effects of multiple harvesting phases or other land-use activities in a watershed, and is often characterized by calculating the 'equivalent clearcut area' (*ECA*), which can be calculated as

$$ECA = \frac{1}{A_c} \sum_{i=1}^{n} A_i (100 - HR_i) \qquad\qquad \text{(Eq. 7.2)}$$

where A_c is the catchment area (km²), A_i is the area of a given cutblock within the catchment, HR_i is the hydrologic recovery for that cutblock, and n is the number of cutblocks in the catchment. For example, suppose 20% of a catchment was clearcut in the past and the cutblocks have experienced 50% hydrologic recovery towards the pre-harvest condition, and that a further 20% of the catchment has been recently clearcut with no recovery. In this case, the *ECA* for the catchment would be 0.2(100% – 50%) + 0.2(100% – 0%) = 30%.

Application of the ECA concept in forest management faces a number of conceptual and practical challenges. An important consideration is that different processes (e.g. snow accumulation, snow melt, transpiration) may recover at different rates or even follow different trajectories. For example, for mountain ash stands in southeastern Australia, overstory transpiration progressively declines with stand age, understory transpiration initially declines with age for a decade or so and then increases with age, and interception loss increases slightly for a few decades, then declines gradually (Vertessy *et al.*, 2001).

Another challenge to the application of the ECA concept is that the shape of recovery curves for streamflow parameters may not be simple asymptotic forms that vary monotonically from 0% to 100% recovery, as has been assumed for stand-level processes such as snow ablation and accumulation. Evidence from a number of regions suggests that water yield, at least during the growing season, initially increases following harvest, then declines as the juvenile forest stand enters a phase of high water use, and then gradually increases as the stand matures and uses water more efficiently. The magnitudes of the changes and the duration of each recovery phase vary depending on forest type.

*Figure 7.11
Mountain ash
(*Eucalyptus
regnans*) forest,
Tarra Bulga,
Victoria,
Australia.
Mountain ash
is the tallest
flowering plant
in the world,
and there is
evidence that
they can grow
to even greater
heights than the
coastal redwood
(*Sequoia
sempervirens*) of
California, USA.*

*Figure 7.12
It is very unlikely
that a harvested
catchment such
as this one on
Vancouver
Island, British
Columbia,
Canada, will
ever achieve
a 100%
hydrological
recovery to
its initial old
growth state.*

A practical challenge is that much of the existing research on recovery processes has focused on responses to clearcut harvesting. In many jurisdictions, techniques such as variable retention have become more common, and recovery rates for these silvicultural systems have not been well studied, especially over longer (multi-decade) time scales.

The term hydrological recovery implies that a catchment will eventually function as it did prior to harvest. However, this is not necessarily a valid assumption. In the case of conversion of old growth forest to stands that are managed on a sustained-yield basis, the catchment will contain a mosaic of younger stands with reduced biomass compared to the pre-management state. Furthermore, active forest management could involve a shift in species composition relative to pre-management conditions. Finally, climate change over the time period of a typical rotation could modify stand characteristics, forest health, and the hydroclimatic regime of a catchment, and thus its hydrologic functioning.

Effects of forest management on geomorphic processes

Hillslope processes

Hillslope processes include mass movements and surface erosion. Surface erosion is driven by the movement of water over the surface, which can entrain soil material and transport it downslope. Mass movements are driven primarily by the direct action of gravity but can also involve the effects of flowing water, and include landslides, rockfall, soil creep, and dry ravel, and involve the bulk downslope displacement of soil and/ or weathered bedrock via falling, toppling, sliding, spreading, or flowing, or by various combinations. To identify the appropriate management strategies to reduce the impacts of forest practices, clear distinctions must be drawn between these fundamentally different processes. Forest management practices can also influence the runout of mass movements, particularly on alluvial fans.

Surface erosion. Surface erosion involves the detachment, transport, and subsequent deposition of soil by flowing water. Sheet erosion occurs when rainfall intensity exceeds a soil's infiltration capacity, causing overland runoff that is spread relatively uniformly across a hillslope. In most cases, overland flow concentrates into small rivulets as it flows downslope, causing rill erosion. Further concentration of water can deepen rills, causing gully erosion. Gullies can also be formed by collapse of soil pipes, formed by erosion caused by concentrated subsurface flow.

Surface erosion is not common on undisturbed forest hillslopes because the organic horizon of forest soils promotes high infiltration rates and buffers raindrop impact

Figure 7.13 Rills coalescing into gullies. In this case, it appears that the road has triggered the erosion, but increased surface flow may also be associated with the loss of forest cover in the area, Atlas Mountains, Morocco.

Figure 7.14
Extensive
overland flow
and surface
erosion on an
unsurfaced
logging road
in peninsular
Malaysia. (Photo
by Roy C. Sidle.)

(as does lower canopy interception). However, if surface soils are disturbed and compacted, infiltration rates decline and overland runoff may occur during storms, especially in high intensity periods. Infiltration capacities of underlying mineral horizons are much lower than organic horizons; thus, runoff and surface erosion will likely occur if the surface horizons are totally displaced. Tropical soils are particularly susceptible to surface erosion because high decomposition rates usually result in relatively thin organic horizons that can be easily displaced (Sidle *et al.*, 2006).

Several aspects of forest management create opportunities for surface erosion and the potential for sediment to be discharged into streams and rivers, where it can impair water quality and aquatic habitat. Unpaved forest roads, ditch lines, and skid trails are the major sources of surface erosion in actively managed terrain (Figure 7.14). Fire used as a site preparation practice and to clear forests for agriculture is also a cause of surface erosion. However, with proper planning and precautions, problems associated with surface erosion can be substantially reduced.

Road and trail systems in steep terrain are often interconnected, so once runoff initiates on road surfaces, it may travel long distances accumulating erosive energy, creating rills and gullies and transporting large quantities of sediment. Consequently, attention needs to be paid to surfacing heavily used forest roads with rock, implementing adequate road drainage measures (frequent cross drains so runoff velocity does not build up), restricting the use of roads and trails during rainy and wet periods, and maintaining road surfaces and drainage systems. Forest roads need to be planned for long-term management objectives, including future harvesting entries. Road surfaces should be compacted and cut-and-fill slopes should be revegetated to reduce erosion from these disturbed areas. Roads can be in-sloped with an interior ditchline that collects all road runoff and discharges it at frequent intervals through cross drains running under the road bed. These drains should not discharge onto erodible fill material or onto unstable slopes (e.g. hollows). If roads are cut deeply into hillsides (exposing underlying bedrock or compacted till), the road surface will not only collect and transmit rainfall or snowmelt on the road surface, but also subsurface flow that is intercepted by the road cut. This intercepted flow must also be considered in the road drainage design. Skid trails are less frequently used, but are not surfaced and maintained; thus they may be prodigious sediment sources (Sidle *et al.*, 2004). In steep terrain, skid trails should be deactivated and revegetated once harvesting operations are finished. Frequent forest entries (e.g. thinnings) can be problematic because skid trail surfaces are disturbed during each usage. Planning of road and skid trail layout should avoid, as much as possible, the interconnection of these corridors and forest roads in sloping terrain.

Another forest practice that exacerbates surface erosion is the burning of slash in intense fires that may consume the entire organic horizon in severe cases and lead to the creation of a hydrophobic layer just under the soil surface, both of which can result in reduced infiltration capacity and overland flow (Larsen *et al.*, 2009). Some tree species have adapted to regenerate following natural episodic fires that create minor erosion events (e.g. lodgepole pine [*Pinus contorta*], ponderosa pine [*P. ponderosa*], Douglas fir [*Pseudotsuga menziesii*], giant sequoia [*Sequoiadendron giganteum*], and many euca-lypts [*Eucalyptus* spp.]). More severe erosion has occurred when burning has been used as a site preparation measure to retard invasive vegetation growth following harvesting and prior to replanting (see Figure 4.17) as well as to clear tropical forests for agricul-tural uses (e.g. swidden agriculture or slash and burn). Water repellency in surface soils following hot fires promotes overland runoff and surface erosion, especially when most of the organic horizon has been burned. Tropical soils, with their thin organic horizons, are most susceptible. These burning practices have decreased in popularity because of their undesirable environmental impacts; however, fire is still used in some areas (and occurs naturally and due to arson, but also subsistence agriculture). When the organic horizon is consumed by fire, a surficial mass movement process (dry ravel) can be very active on steep slopes. Additionally, these surficial processes (surface erosion and dry ravel) can transport large quantities of sediment into steep headwater channels, often triggering debris flows during a subsequent storm.

Mass movements. Shallow, rapid landslides are the most common mass movements in steep forest terrain and are the geomorphic processes most affected by forest har-vesting due to the shallow soil depths. Shallow landslides are usually triggered by an individual storm or snowmelt event that causes pore water pressure to develop above the failure plane. Deeper-seated mass movements such as rotational slumps, earthflows, and soil creep are less affected by forest harvesting because the depth of movement typically extends below the supporting root systems of trees. However, in tropical and subtropical regions where forests have been converted to pasture, accelerated soil creep may occur along with occasional deeper landslides due to increases in soil water associ-ated with year-round reductions in evapotranspiration from grass cover (Figure 7.15). Deep-seated mass movements usually occur or are reactivated after some accumulation of rainfall or snowmelt. Reactivation may require sustained water inputs over several weeks; thus the movement of these deeper mass movements does not necessarily coin-cide with individual large storms. Soils subject to shallow, rapid landslides typically have low cohesion, while soils in deeper, slow-moving mass movements are typically clay rich, with high cohesion.

The two forest management activities that have the greatest potential to acceler-ate landslide erosion are road construction and timber harvesting. In steep, unstable terrain, clearcut timber harvesting may increase landslide erosion twofold to tenfold compared to similar undisturbed forests, while road construction in steep terrain often accelerates landslide erosion on the affected area by more than a hundredfold (Sidle and Ochiai, 2006) (Figure 7.16).

In temperate forests, the major adverse effect of forest harvesting on slope stability is due to loss of root reinforcement. When trees are cut, their roots begin to decay and lose strength. At the same time, however, root strength recovers as new trees regenerate on the site, although there may be a lag time before the new root systems

Figure 7.15 Deep rapid landslide on a hillside experiencing active soil creep near Manizales, Colombia. The site was originally in evergreen forest cover and was cleared and converted to pasture, thus increasing soil moisture. (Photo by Roy C. Sidle.)

Figure 7.16 Shallow rapid landslides in a steep managed forest (sugi (Cryptomeria japonica)) catchment in southwestern Nara Prefecture, Japan. (Photo by Roy C. Sidle.)

become effective. The period when a site experiences the greatest susceptibility to shallow landslides coincides with the minimum root strength – a function of the decay rate of roots in the original stand and the recovery rate of regenerating trees (Figure 7.17). These rates are affected by tree species, season of harvesting, site fertility, climate, and other environmental factors, but the most vulnerable period for shallow landslides typically occurs 3–20 years after clearcutting. The changes in root strength through time following harvest and the timing of landslide susceptibility have been confirmed through physical testing of roots for different species and ages and through catchment-scale field studies and physically based models. Most studies have focused on clearcuts, but retention of as little as 25% of the forest stand in certain partial cut practices can reduce landslide potential by as much as fivefold compared to clearcut areas (Sidle, 1992). While few studies of the benefits of root reinforcement have been conducted in tropical forests, steep sites where tropical forests have been cleared for agriculture or plantations have often experienced substantial landslide activity.

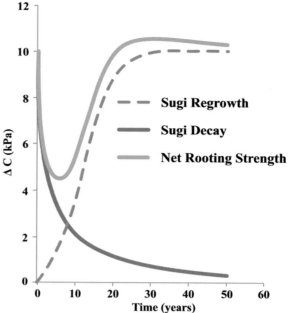

Figure 7.17

Example of tree root strength (ΔC) decay and subsequent regrowth for Sugi *(Japanese cedar,* Cryptomeria japonica); *net root strength is the sum of the decay and regrowth curves. (Modified from DeGraff et al. (2012).)*

Forest roads can destabilize hillslopes in three ways: (1) altering hydrologic pathways and concentrating water onto unstable portions of the hillslope; (2) undercutting unstable slopes, thus removing support; and (3) overloading and oversteepening fillslopes, including the road prism (Sidle and Ochiai, 2006). The relative importance of these destabilizing factors depends on the design and construction standards of the road and associated drainage system, as well as the natural instability of the terrain through which the road is excavated. When drainage systems are poorly designed or non-existent, mountain roads are at high risk for fillslope failures. Roads constructed in mid-slope locations are typically the most unstable because of large quantities of intercepted water and unstable fill and cut materials that need to be disposed of or incorporated into the fillslope. Roads cut through concave mid-slope segments should intercept more subsurface water than those cut through convex or planar slopes; thus, roads along concave slopes should be less stable. Excavating roads near the toe of dormant earthflows can reactivate these failures, as can placing road fill material at the head of dormant rotational slumps. Alteration of hillslope drainage in plateau terrain that is bounded by steeper slopes can concentrate and direct runoff to hillslopes that are not naturally prepared to handle excess flows. Consequently, attention to pre-construction natural drainage patterns is critical.

Sedimentation of fisheries streams in the Pacific Northwest, USA, during the 1970 and 1980s heightened environmental awareness of road-related landslides. Rates

Figure 7.18 Situations such as this represent potential problems. A road has directly crossed a stream channel, and instead of building a bridge or culvert, the water has been directed into a ditch beside the road. Erosion will very likely follow. British Columbia, Canada.

Figure 7.19 Prodigious sediment from road-related landslides, Salween River valley, Yunnan, China. (Photo by Roy C. Sidle.)

of landslide erosion from forest roads in unstable forest terrain within this period averaged about 60 Mg ha^{-1} yr^{-1} (Sidle *et al.*, 1985). Partly because of these road-related landslides, forest harvesting was severely curtailed on public lands in the Pacific Northwest. Application of improved location, construction, and maintenance practices in recent years has reduced landslide erosion along roads in this region, but roads continue to be one of greatest sources of sediment. In Southeast Asia, mountain roads are expanding at a rapid pace due to forest conversion, linking markets in remote villages to cities, hydroelectric power development, and other development activities. Little attention is being paid to road location and construction. Recent field studies in northwest Yunnan Province, China, have shown the highest rates of road-related landslides ever reported – two of the roads surveyed had >33,000 and >48,000 Mg ha^{-1}yr^{-1} (Sidle *et al.*, 2011) (Figure 7.19). These epic rates of landslide erosion emphasize the need for a more sustainable approach to managing these rapidly expanding road systems.

Channel processes and morphology

In landscapes dominated by hills or mountains, small headwater channels tend to be relatively steep, and channel gradient generally decreases in the downstream direction as drainage area, stream discharge, and stream size increase. Steep headwater streams often flow in colluvial sediments eroded from adjacent hillslopes, and the stream bed often includes wood and large boulders that generally cannot be moved by flowing water in the channel. Occasionally, debris flows can scour these channels down to the underlying bedrock, particularly in the uppermost reaches of the headwater drainage system; these channels are believed to undergo cycles of gradual sediment accumulation in the channel followed by episodic evacuation during large debris flow events. Headwater streams are often too steep and/or small to support fish, but can be important sources of nutrients and organic material to downstream reaches.

Downstream from the headwaters, where channel gradients fall below about 0.05 m/m, a transition occurs to an alluvial process regime, where sediment carried by the stream can be deposited to form accumulations such as fans or floodplains. In these stream reaches, the channel bed and banks are commonly composed of alluvial sediment (i.e. sediment that was deposited by and is occasionally moved by the stream). Root networks associated with riparian vegetation can substantially increase the resistance of the banks to erosion by water flowing in the stream channel. In streams where the rooting depth of the dominant riparian tree species is similar to the depth of the channel, riparian forests may prevent such channels from migrating laterally across their floodplain. As a result, forested alluvial streams tend to be narrower than similar, unvegetated channels.

At the reach scale, the morphology of alluvial channels reflects an ongoing response to the following governing conditions: (1) the peak flow regime, which controls the shear stress exerted on the bed and banks and also the capacity for sediment transport; (2) the rate of sediment supply and its size; (3) channel gradient, which determines the supply of potential energy available to perform the work of sediment transport; (4) the cohesion of the stream banks, which controls the resistance to erosion; and (5) inputs of large wood from the riparian forest (e.g. due to windthrow) and landslides. In-stream wood pieces influence channel dynamics by modifying the distribution of water depth and velocity, trapping sediment and forming pools. In larger streams that have the capacity to move large wood pieces, wood can accumulate into log jams that can cause avulsions during flood events (abrupt changes in location of the stream channel).

Forestry operations can influence channel processes and morphology through their influences on the peak flow regime (see the earlier section, 'Peak flows') and on sediment inputs, particularly coarse sediment via mass movements. Increases in peak flow tend to result in erosion of the stream bed (degradation), whereas increases in coarse sediment input tend to result in accumulation of sediment in the reach (aggradation). In cases where riparian forests have been harvested, the decay of streamside tree roots results in a loss of cohesion and a greater likelihood of bank erosion, which can result in widening of the channel, especially when combined with increased peak flows. In addition, riparian harvesting will reduce the supply of in-stream wood to the channel for a period of decades, resulting in a gradual loss of in-stream wood through decay and a loss of its role in shaping channel morphology and aquatic habitat. Riparian forests are of particular importance on alluvial fans where the stream is not topographically confined and relies on the riparian vegetation to prevent avulsions.

*Figure 7.20
A temporary
bridge protected
by riprap,
Chambers Creek,
northern British
Columbia,
Canada.*

*Figure 7.20
A temporary
bridge protected
by riprap,
Chambers Creek,
northern British
Columbia,
Canada.*

These generalizations about headwater versus downstream reaches will not necessarily apply in low-relief landscapes, but the effects of forestry on channel processes and morphology in such regions has been much less well studied and many uncertainties still exist.

Culverts and bridges on forest roads can affect the spatial distribution of sediment storage in channels, such as where culverts are undersized, in-channel bridge supports obstruct flow, or where the cross-sectional profile of the stream is altered. Wide, shallow streams forced to flow through tall, narrow structures can result in sediment deposition upstream and channel scour downstream of the structure. The subsequent downstream deposition of sediment can result in channel avulsions. In areas experiencing high peak flows and/or debris flows, bridges may need to be protected by carefully designed placement of riprap.

Effects of forest management on water quality

Water temperature

Stream temperature influences a range of biological, biogeochemical, and ecological processes, and is therefore a critical aspect of aquatic habitat. For example, increases in summer maximum temperature have the potential for increased mortality or morbidity of cold- and cool-water species such as salmonids. Even where such extreme responses do not occur, relatively small but persistent changes can influence rates of growth and development of fish and the invertebrates they feed on, and thus potentially have profound ecological effects.

Forest harvesting along and near streams reduces shade and thus increases the amount of solar radiation that reaches the stream surface. In an extreme case of clearcut logging followed by a hot broadcast burn of logging debris in the Oregon Coast Range, summer maximum stream temperatures increased by more than 11°C (Harris,

1977). The effect of harvesting is greater in spring and summer than in winter, at least in areas with wetter winters and drier summers. Most research on stream temperature response to forest management has been conducted in areas with temperate climates. It is unclear how tropical streams respond, especially in areas with wet summers.

The magnitude of temperature response varies with the reduction in shade. For example, the retention of forested buffer strips to maintain stream shade reduces post-harvest temperature changes, and buffers that are narrower or only partial-retention provide less protection than wider or full-retention buffers. Stream temperature response also depends on channel morphology. For example, streams flowing in narrow incised channels will be more effectively shaded by the channel banks and will experience less warming than a wider, shallower channel (Gomi *et al.*, 2006).

In addition to the direct effect of reduced shade, forest management can have indirect effects on stream temperature through effects on channel morphology. For example, bank erosion associated with riparian forest harvest or debris flows associated with upslope activities can result in wider, shallower streams that would be more exposed to solar radiation and more sensitive to increased solar radiation due to decreased water depths.

An important consideration for sustainable forest management is the rate at which stream temperatures recover towards their pre-harvest levels. Thermal recovery occurs faster for narrow streams, which can be effectively shaded even by low herbaceous vegetation. For wider streams, taller vegetation is required to provide effective shade. Recovery rates also depend on the species that establish in the riparian zone after harvesting, as well as the site's climate, which influences rates of vegetation growth. For example, shade levels at sites in Oregon that had been clearcut and burned recovered more rapidly in wetter forest types and at lower elevations (Beschta *et al.*, 1987). Results from a number of studies suggest that recovery should occur, or be well underway, in 5–20 years following harvest for streams less than about 10 m wide.

Suspended sediment concentration

Forest management operations can increase soil erosion, often producing higher rates of sediment delivery to the stream channel. High levels of suspended sediment in stream water can damage fish gills, and the increased turbidity (lack of clarity) can interfere with feeding efficiency of fish that use their vision to detect and trap prey. Suspended sediment in the water column can result in turbidities that exceed drinking water standards, and also reduce the effectiveness of chlorination and ultraviolet disinfection systems for eliminating pathogens. Siltation of bed materials reduces permeability and thus the movement of oxygenated stream water into the bed for benthic organisms and fish embryos.

The term sediment yield refers to the total amount of sediment transported out of a catchment by a stream in a specific time interval (typically annually). Forestry-related increases in suspended sediment yield are mostly associated with forest roads and skid trails rather than soil disturbance caused by timber harvesting. Prescribed fire used in conjunction with forest management can also increase suspended sediment yields.

The delivery of eroded sediment to streams depends on the hydrologic connectivity between hillslopes and channels or, for landslides, whether they run out to the stream. Reducing hydrological connectivity from impacted areas, such as roads and skid trails, to streams is important for protection of water quality, in addition to the

Figure 7.21
Different turbidity levels can occur naturally, since the nature of the rock through which a river passes is an important determinant of the sediment load. Here, the confluence of the turbid (whitewater) Rio Solimões and the clear (blackwater) Rio Negro at Manaus in the Brazilian Amazon shows the stark difference in sediment load that can occur.

Figure 7.22
Riparian woodland alongside Turkey Creek, Niceville, Florida, USA. The maintenance of vegetation along streams and rivers is important for maintaining water quality.

measures for minimizing the erosion in the sediment source areas. Riparian buffer zones are often retained along streams to filter sediment and reduce near-channel soil disturbance. Understory vegetation and organic horizons in riparian zones can trap sediment mechanically and increase the opportunity for infiltration. The effectiveness of buffer zones in the context of hydrological processes depends largely on the linkages between the disturbed areas and streams.

Water chemistry

Forest harvesting reduces the uptake of nutrients from the soil by trees, typically resulting in increased streamwater concentrations for mobile nutrients like nitrate-nitrogen (NO_3^-). In addition, post-harvest establishment of nitrogen-fixing species like alder (*Alnus* spp.) can further increase nitrogen pools within the soil. In addition to increases in nitrate concentrations, forestry operations can also result in increased concentrations of base cations such as potassium and calcium, and also dissolved organic carbon. The magnitudes of increased chemical concentrations in streamwater vary substantially from study to study, reflecting the complexity of the processes that control streamwater chemistry (Feller, 2005). Recovery of streamwater concentrations back to or below pre-harvest levels typically occurs within 3 to 10 years.

Changes in streamwater chemistry associated with forestry operations often do not cause concentrations to exceed water quality guidelines (e.g. for drinking water), and the development of best management practices for water quality protection has dominantly focused on reducing changes to water temperature and sedimentation. However, even where changes in water chemistry do not cause obvious water quality problems, they reflect changes in biogeochemical processes and thus changes to the structure and/ or function of both terrestrial and aquatic ecosystems. Unfortunately, it is difficult to interpret the ecological basis or significance of these changes due to the complexity of biogeochemical processes and the interacting effects of changes in light, temperature, and suspended sediment concentrations following harvesting.

Riparian management

The riparian zone is the ecotone between the aquatic and terrestrial ecosystems. Riparian forest provides important roles in the structure and function of the stream channel and the associated ecosystems. The riparian canopy provides shade and thus reduces water temperature, especially during spring and summer. Litterfall from the canopy into the stream provides an important input of organic matter and nutrients that constitute a cross-ecosystem subsidy for stream biota (Richardson *et al.*, 2010). Tree roots provide bank cohesion, which promotes channel stability. The riparian forest is also critical as a source of large in-stream wood pieces, which trap sediment, promote pool formation, provide cover for stream biota to avoid predators (Hassan *et al.*, 2005), and generate log jams that lead to the development of, and moderate flows in, side channels. Retention of riparian buffers protects against soil disturbance adjacent to the stream and can, to varying extents, filter sediment from overland flow.

Harvesting in the riparian zone can have a number of ecological consequences. Increased solar radiation can cause stream warming and also increase rates of primary production. The loss of bank cohesion associated with root decay following riparian harvest can result in bank erosion and a widening of the stream, especially in catchments where peak flows have increased following harvesting. Harvesting of riparian forest also reduces recruitment of in-stream wood via windthrow and a decline in the amount of in-stream wood over a number of decades or even longer. These consequences are most severe in cases where forest land has been converted to pasture or agricultural use.

In response to concerns about the potential impacts of riparian forest harvesting, many jurisdictions have adopted regulations that require retention of a forested buffer strip along streams. Most jurisdictions require the retention of riparian forest cover along larger streams. However, prescriptions vary dramatically among jurisdictions for retention along smaller, usually non-fish-bearing, stream reaches. This is a

cause of debate given that small headwater streams typically constitute up to 80% of the total stream length within a network, thus they are more directly affected by hillslope disturbance than downstream reaches that flow in floodplains, and they are important sources of nutrients and organic matter to downstream fish-bearing reaches.

The ultimate outcome of retaining buffer strips adjacent to water bodies is the prevalence of linear strips of older forest across the landscape and 'doughnuts' around lakes and wetlands. Because this is considered to be 'unnatural' in many settings where riparian forest is subject to periodic disturbances by fire and other agents, a number of forest managers are promoting and adopting emulation of natural disturbance (END) as an emerging paradigm for riparian management. Under the END paradigm, harvesting of riparian forest would be designed to mimic the patterns of riparian stand structure that would result from natural disturbance. However, Moore and Richardson (2012) have cautioned that harvesting does not necessarily mimic the outcomes of natural disturbance. For example, standing dead trees following riparian wildfire continue to provide some shade to the stream and also provide a source of large in-stream wood pieces, functions that are not necessarily mimicked by riparian harvesting.

A challenge to the retention of riparian buffers as a tool to mitigate the impacts of forestry on streams is that they are prone to blowdown. On the one hand, the occurrence of blowdown reduces the effectiveness of the buffer to provide the intended functions. On the other hand, extensive blowdown of a riparian buffer could result in a pulsed input of large wood pieces to a stream, similar to what occurs following a riparian forest fire.

The emphasis placed on the retention of riparian strips of forests is characteristic of extensive forestry. A very different situation may exist in some plantation forests. In the United Kingdom, concerns about the transfer of acidic pollutants to streams resulted in the creation of buffer strips between the trees and streams where no trees were planted. In South Africa and elsewhere, trees have actually been cleared from riparian areas in an attempt to reduce the water consumption associated with afforested catchments.

Figure 7.24 Treefalls and in-stream wood pieces in the Hoh River, Olympic National Park, Washington State, USA. (Photo by David L. Peterson.)

*Figure 7.25 Treeless buffer strip established between stands of flooded gum (*Eucalyptus grandis*) and a stream in the Sabie catchment, KwaZulu-Natal, South Africa.*

Conclusions

Over the last century, research and operational trials have led to an extensive body of knowledge about watershed processes and values, and their interactions with forestry operations, which can help to guide the sustainable management of forests and watersheds. Much of this knowledge has been synthesized in 'state of knowledge' reports (e.g. Pike *et al.*, 2010) and formalized as best management practices and procedures for watershed assessment and analysis (e.g. Sidle *et al.*, 2006). In addition to the existence of gaps in our understanding, a challenge in applying this accumulated knowledge lies in the need to account for spatial variations in watershed processes and forest dynamics, which limit the transferability of research results among regions or, indeed, even among watersheds within a region.

Further challenges are introduced by the changing environmental, socio-economic, cultural and political contexts within which forests and watersheds are managed. Climate change has had and will continue to have pervasive influences on forest ecology and dynamics, as well as hydrological processes, such that knowledge based on historical research may not remain valid into the future. For example, changes in forest health and species suitability associated with changing moisture availability will influence rates and trajectories of stand development, with fundamental implications for characterizing hydrologic recovery. In addition, increasing air temperature can cause shifts in hydrologic regimes – e.g. from snow dominated to hybrid rain/snow, or hybrid to rain dominated. Partly in response to changing social values, forest practices have been evolving, with a move away from traditional clearcut harvesting to variable retention systems in some jurisdictions. A related change is the increasing dependence on second-growth stands as primary forest continues to be converted to managed stands or removed from the working land base as parks or other forms of protected areas. In these cases, the baseline for assessing effects of forestry is now the second-growth forest, not the primary forest condition. Another changing context is the shift to alternative forest products, such as biofuel production. One aspect of this complex and evolving paradigm is the increased pressures on timber resources in developing countries, where best management practices are either not in place or not carefully enforced. As a consequence of these evolving contexts, there will be an ongoing need to continue research and adaptive management trials to evaluate the effects of contemporary forest practices.

Increasingly, forestry may be just one land use within a catchment among many others, such as recreation and the associated infrastructure (including ski areas and other large-scale developments), mining, and oil and gas development. In many cases, the watersheds may have an existing historical legacy of past forestry or mining activity. Therefore, it is important to assess the cumulative effects of forestry in light of past, current and future activities.

Further reading

Bonell, M., Bruijnzeel, L. A. 2005. *Forests, Water and People in the Humid Tropics: Past, Present and Future Hydrological Research for Integrated Land and Water Management*. Cambridge: Cambridge University Press.

Bren, L. 2014. *Forest Hydrology and Catchment Management: An Australian Perspective*. Berlin: Springer.

Chang, M. 2013. *Forest Hydrology: An Introduction to Water and Forests*. 3rd edition. Boca Raton: CRC Press.

Furniss, M. J., Staab, B. P., Hazelhurst, S., Clifton, C. F., Roby, K. B., Ilhadrt, B. L., Larry, E. B., Todd, A. H., Reid, L. M., Hines, S. J., Bennett, K. A., Luce, C. H., Edwards, P. J. 2010. *Water, Climate Change, and Forests: Watershed Stewardship for a Changing Climate*. General Technical Report PNW-GTR-812. Portland, OR: US Department of Agriculture, Forest Service, Pacific Northwest Research Station.

Sidle, R. C., Ochiai, H. 2006. *Landslides: Processes, Prediction, and Land Use*. Water Resources Monograph No. 18. Washington, DC: American Geophysical Union.

Chapter 8
Forests and carbon

John L. Innes

Wood contains almost 50% carbon, and so it is not surprising that the forests of the world represent a carbon store of global significance. The total biomass of forests is believed to be about 677 petagrams (Pg; 1 PgC = 10^{15} gC), with trees constituting 80% of the world's biomass (Kindermann *et al.*, 2008). About half of the terrestrial carbon sink is in forests, with carbon being stored in biomass both aboveground and belowground. Managing this resource well requires an understanding of the ways that carbon is accumulated and stored in forests, and also in the way that it is released back into the atmosphere.

The management of global carbon has become so important that managing forests specifically to accumulate and store carbon has become important. Like all special uses of forests, there is a choice of concentrating on the maximization of carbon, or accumulating and storing carbon while also generating some of the other benefits that well-managed forests provide. The Montreal Process indicators for the criterion 'Maintenance of Forest Contribution to Global Carbon Cycles' are: 5.a Total forest ecosystem carbon pools and fluxes; 5.b Total forest product carbon pools and fluxes; 5.c Avoided fossil fuel carbon emissions by using forest biomass for energy.

The global carbon cycle

Carbon is a key element for all life on Earth. Most carbon is stored in carbonate sediments below the Earth's surface, with only about 0.2% of the total amount being available to organisms in the form of carbon dioxide (CO_2) or decaying biomass. Carbon can take a number of forms: it can be a free gas in the atmosphere or a dissolved gas in fresh and saltwater; it can exist as carbohydrate molecules in organic matter, as hydrocarbon compounds in rock, and as mineral carbonate compounds. The atmospheric pool of CO_2 is derived from the respiration of plants and animals, both on land and in the oceans. There is also some outgassing of CO_2 and carbon monoxide from volcanoes and geothermal areas. In recent centuries, emissions of CO_2 from the burning of fossil fuels has been a major source of atmospheric CO_2.

Ciais *et al.* (2013) divide the global carbon cycle into two domains, a fast one and a slow one. The fast domain consists of relatively large fluxes and rapid turnover of carbon reservoirs. These carbon reservoirs consist of atmospheric carbon, oceanic carbon, carbon in surface ocean sediments, and carbon stored on land in vegetation, soils and freshwaters. The slow domain consists of the carbon stored in rock and sediments.

There is some transfer between the two: carbon in the slow domain can be transferred to the fast domain through processes such as volcanic outgassing, chemical weathering of rocks and erosion. The burning of fossil fuels represents a major change in the balance between domains, as it has greatly increased the transfer of carbon from the slow to the fast domain.

The most important carbon-bearing trace gas in the atmosphere is carbon dioxide (CO_2). The atmospheric concentration in 2011 was approximately 390.5 ppm, and in 2013, measurements of over 400 ppm were recorded for the first time since measurements began. This corresponds to a mass of 828 PgC. Other important carbon-bearing trace gases include methane (with an atmospheric mass of about 3.7 PgC) and carbon monoxide (0.2 PgC) (Ciais *et al.*, 2013).

A significant carbon flux occurs between the atmosphere and oceans. The dominant form of carbon in the ocean is dissolved inorganic carbon (DIC, about 38,000 PgC), although there is also a substantial pool of dissolved organic carbon (DOC, about 700 PgC). Carbon is taken up by phytoplankton in oceans and deposited on the seafloor as calcium carbonate when they die. However, at any given moment, the amount of carbon in phytoplankton is small (3 PgC).

Terrestrial systems contain between 450 and 650 PgC in vegetation, and between 1,500 and 2,400 PgC in dead organic matter in litter and soils (Ciais *et al.*, 2013). There is a further 300 to 700 PgC in wetland soils, and about 1,700 PgC in permafrost soils. In all cases, there is considerable uncertainty, reflecting the difficulties of measurement. Overall, terrestrial ecosystems represent a net sink for carbon, although there are major regional variations. Terrestrial systems take up about 123 PgC annually (Beer *et al.*, 2010), although about half of this is returned to the atmosphere through respiration, leaving about 57 PgC in the terrestrial sink. The land area located in the tropics has been assumed to be a net source of carbon because of deforestation and forest degradation, but even today the very large uptake of carbon in tropical forests means that the losses may be offset by gains (Ramankutty *et al.*, 2007), emphasizing the potential of planned management of tropical forests to increase carbon sinks. The temperate biome in the Northern Hemisphere is generally considered to be a sink, as is the boreal biome, although the large-scale disturbances that typify boreal forests means that, in certain regions, the forest can switch from being a sink to a source and back again.

The biggest source of anthropogenic CO_2 emissions is the burning of fossil fuels. In 2010 and 2011, this contributed 9.2 ± 0.8 and 9.5 ± 0.8 PgC yr^{-1}, respectively, compared with the 1.1 ± 0.8 PgC yr^{-1} arising from land-use change (Houghton *et al.*, 2012; Peters *et al.*, 2013). Total cumulative emissions from 1750 to 2011 have been 375 ± 30 PgC, including 8 PgC from the production of cement (Ciais *et al.*, 2013). The emissions from land-use change have varied over time, but by far the most important emissions have occurred since the onset of the industrial era (1750). This is reflected in the concentrations of CO_2 recorded in ice cores. These indicate that 7,000 years ago, atmospheric CO_2 concentrations were around 260 ppm. From 7,000 years ago to 260 years ago, average atmospheric concentrations of CO_2 rose by about 20 ppm, despite major land-use changes in Europe and North America that are estimated to have released 27 PgC between the years 800 and 1750 (Pongratz *et al.*, 2009). Between 1750 and 2011, atmospheric concentrations of CO_2 rose by 110 ppm, representing an increase in atmospheric carbon of about 240 PgC. There is not a direct link between the amount of CO_2 emitted and the amount in the atmosphere, as the atmospheric concentrations represent the net effect after absorption of CO_2 by oceans and terrestrial sinks. Land-use changes are estimated to have released 180 ± 80 PgC into the atmosphere between 1750 and 2011; terrestrial ecosystems unaffected by land-use change have accumulated 160 ± 90 PgC of anthropogenic carbon in the same period (Ciais *et al.*,

Figure 8.1 The 120 MW Van Eck coal-fired power station near Windhoek in Namibia. Anthropogenic emissions of CO_2 are far more important than those associated with deforestation, which in 2014 accounted for only 11% of total carbon emissions worldwide.

2013). Emissions from the burning of fossil fuels have been rising rapidly since the start of the industrial era. While some of the emissions associated with anthropogenic activities such as fossil fuel burning and land-use change have been absorbed by oceanic and terrestrial sinks, the rate of emission increase has surpassed the capacity of these sinks, and they have in addition been compromised by the impacts of global warming (Fung *et al.*, 2005). However, there has been an increase in the carbon stored in terrestrial ecosystems, which has been explained by a combination of enhanced photosynthesis at higher CO_2 concentrations and nitrogen deposition, as well as changes in climate that have resulted in longer growing seasons at mid to high latitudes (Ciais *et al.*, 2013).

The primary concern over CO_2 is the role that it plays as a greenhouse gas (GHG). It is not the only greenhouse gas: methane (CH_4) and nitrous oxide (N_2O) are also important. However, from the perspective of forest management, it is CO_2 that is by far the most important, as forests represent a significant terrestrial pool of carbon.

Primary production

Much carbon management is about net primary production (NPP). This is the result of two major processes: the accumulation of biomass and the loss of biomass through organic detritus production, exudation, volatilization, leaching and herbivory. Traditionally, forestry has concentrated on aboveground NPP, yet belowground NPP may be just as or even more important, especially in some tropical forests (Priess *et al.*, 1999). Ensuring that this belowground production is as carefully managed as the aboveground production is an important aspect of carbon forestry.

The aboveground capture of carbon begins when carbon dioxide (CO_2) is assimilated by plants through photosynthesis into reduced sugars. About half the gross photosynthetic products (GPP) are released by plants in the form of autotrophic respiration (R_a), which occurs as plants need to synthesize and maintain living cells. As a result,

CO_2 is released back into the atmosphere. The remaining amount (GPP – R_a) goes into net primary production (NPP), consisting of the foliage, branches, stems, roots and reproductive organs of plants. As these plant parts die, or the whole plant dies, the dead organic matter forms the substrate for animals and microbes. These also release CO_2 back into the atmosphere through their heterotrophic respiration (R_h). Over the course of a year, undisturbed forest ecosystems usually show a small net gain in the amount of carbon in the ecosystems, known as the net ecosystem production (NEP). Any significant disturbance, such as a storm or harvesting operation, can result in the export of carbon from the system, either by physical removal of the carbon or by a spike in R_h. Soil organic matter (described in detail later in this chapter) represents an important storage site for forest ecosystem carbon, and in many forest types it may be the most important carbon store.

The exchange of carbon into ecosystems is often considered in terms of fluxes. The exchange of carbon into the ecosystem through photosynthesis is considered to be a positive flux, whereas its loss through respiration is considered a negative flux. The net ecosystem exchange (NEE) differs between day and night:

$$\text{Day NEE} = P_g - R_p - R_m - R_s - R_h \qquad\qquad \text{(Eq. 8.1)}$$

$$\text{Night NEE} = -R_m - R_s - R_h = -R_c \qquad\qquad \text{(Eq. 8.2)}$$

where P_g is the gross photosynthesis (the same as GPP), R_p is photorespiration, R_m is maintenance respiration, R_s is synthesis or growth respiration and R_c is total ecosystem respiration. Gross ecosystem production (GEP) includes photorespiration (R_p), but because this is usually small, GEP and P_g are generally assumed to be about the same.

A major component of belowground production is root production. This can be broadly divided into fine roots (<2 mm diameter) and structural roots (>2 mm diameter). In addition to root production, there are a number of other sources of production, including cyanobacteria in the upper few millimetres of the soil and mycorrhizae. Since production by free-living cyanobacteria is dependent on ambient light, this form of production, of limited significance (<5% of total NPP) even in unforested areas, is not likely to be important under a forest cover. However, cyanobacteria are also present as symbiotic partners in several epiphytic lichen species, and production by these may be important in some environments.

In contrast, production through the symbiotic associations between fungi and trees (arbuscular mycorrhizae and ectomycorrhizae) can be significant. While the mycorrhizae are an important source of nutrients for trees, the primary material supplied by trees to the symbiotic partnership is carbon. As a result, the soil and its associated flora represent an important carbon sink for forests.

Secondary production

Secondary production involves the creation of new body tissues from primary production. If this comes from living plant sources, the grazing food chain is involved. If the sources are dead plant material, the detrital food chain is involved. In both cases, dissipation of energy occurs, and the organisms involved are known as heterotrophs.

There is a huge variety of heterotrophs, and they range in size from bacteria to various burrowing mammals, although the biggest organisms may actually be genetically identical fungal bodies. Heterotrophs ingest organic carbon and associated nutrients, and are responsible for converting these into carbohydrates, lipids and proteins. As with all organic growth, the conversion is not perfectly efficient, and a proportion (40% or more) of the chemical energy is dissipated as metabolic heat and evolved carbon dioxide (Coleman and Crossley, 1996). Decomposition, strictly speaking, is attributable to microbial activities, since very few animals are capable of digesting plant litter. However, the soil fauna, which is usually distinguished from the soil microbes, is responsible for the comminution of plant material, increasing the surface area available for microbes.

Soil organic matter

The organic matter content of soil (which has a carbon content of about 45%) ranges from almost 0% in very young soil to over 80% in some organic soils. The majority of forest soils have about 0.5%–20% organic matter in the surface 20 cm of mineral soil, although some montane and boreal forests may have more, and some dry tropical forests have less. Perry (1994) divides soil organic matter into six types, and summarizes them as follows.

Living tissues of plants and their microbial symbionts. These include roots, mycorrhizae and mycorrhizal hyphae.

Soil biomass. This includes all living organisms with the exception of plants and symbiotic fungi. It includes a huge diversity of species, including microbes, protozoa and invertebrates, and may constitute a significant amount of the carbon in the soil. The microbial biomass constitutes about 1%–5% of the total soil organic matter, but still outweighs the total animal biomass in forests.

Exudates and leachates. Exudates are derived from roots, mycorrhizal hyphae and microbes, whereas leachates are derived from living and dead plant tissues. They include sugars, amino acids, nucleic acid derivatives, vitamins, enzymes, waxes and a range of other compounds (Richards, 1987). They represent a small proportion of the total organic matter, but are critical for the functioning of ecosystems.

Litter. Litter comprises fresh or slightly decomposed plant materials and dead biomass. As with coarse woody debris, the amount on the forest floor can vary substantially between different types of forest.

Coarse woody debris. Coarse woody debris, together with dead wood greater than 2.5 cm in diameter, can be an important component of the soil organic fraction. However, it varies greatly between ecosystems, both naturally and as a result of human interventions (particularly firewood collection).

Humus. This comprises a range of organic compounds that are either synthesized and released by microbes and invertebrates as litter decays or that originate directly from plant and animal material without being transformed (such as lignin and some lipids). With a few exceptions, humus accounts for 80%–90% of the soil organic matter. It consists of aromatic compounds, carbohydrates, amino acids and chains of methylenes (Newman and Tate, 1984).

Carbon allocation

After accounting for autotrophic respiration ($R_m + R_s$), the remaining carbon is allocated to new tissues, storage reserves and protective compounds. There are major differences between ecosystems, and between the plant species within ecosystems, in how carbon is located. There are also seasonal shifts, especially among perennial plants. No carbon storage will occur until all the basal metabolic requirements are met, including storage reserves to cover periods when photosynthesis is not occurring (Waring and Running, 2007). The actual process of carbon allocation once basal requirements are met is poorly understood, and while there are numerous theories, each has its limitations. This likely reflects the diversity of environmental conditions experienced by trees and the diversity of tree forms (evergreen vs. deciduous, coniferous vs. broadleaved).

Modelling carbon fluxes and storage

Most attempts to determine carbon fluxes and storage rely heavily on modelling. This is a problem, as models rarely, if ever, give 'correct' answers. Instead, they give an approximation of what might be happening given a number of assumptions. The nature of these assumptions is absolutely critical, and as science has progressed it has become evident that many early assumptions were unjustified. As a result, our knowledge in this area is being continuously and rapidly updated. In addition, different models generate different results, and the differences can be quite substantial.

Global carbon budgets

Global carbon budget models can be broadly classified into top-down and bottom-up approaches (Spalding *et al.*, 2012). Top-down approaches are based on measurements of atmospheric CO_2, changes in the concentrations and models estimating the fluxes between the atmosphere and land and oceans (e.g. Dolman *et al.*, 2010). Top-down approaches generally consist of either measurements of atmospheric CO_2 concentrations or inverse modelling. The measurement approach relies on partitioning carbon sinks on land and in oceans through measurements of changes in atmospheric O_2/N_2 concentrations and measurements of $^{13}C/^{12}C$ ratios (Keeling *et al.*, 1996). The inverse modelling approach also examines atmospheric concentrations of CO_2, but looks at regional distributions across space and time and then uses transport models to derive estimates of sources and sinks.

Bottom-up approaches involve measurements at the Earth's surface that are used to infer changes in atmospheric concentrations. They include inventories, bookkeeping approaches and process-based modelling (Spalding *et al.*, 2012). Bottom-up approaches are the most likely to be adopted in regional or local carbon budget models, and so are described here in more detail than top-down approaches.

Inventory models are based on data on forest area, timber stocks and forest growth. These data are then converted to biomass estimates to estimate how much carbon is in the vegetation. The accuracy of the original data is obviously paramount to obtaining good estimates of carbon stocks and fluxes, as is the reliability of the biomass estimate conversions. Spalding *et al.* (2012) detail a number of sources of error, including inconsistent definitions of forest cover, incomplete or incorrect data, inconsistent treatment of areas with low tree cover, inconsistent treatment of areas of land that have temporarily changed (such as harvested areas), and inconsistencies and inaccuracies in conversion factors, which are influenced by variables such as the age class distribution and species composition of the forest. Generally, better information is available for temperate and boreal forests than for tropical forests as they have been studied longer and are usually simpler in structure and form. Other problems with the inventory approach include a lack of information on all the different components of a forest ecosystem, especially the belowground carbon.

Bookkeeping models are based on estimating changes in carbon stocks by recording changes in land use and inferring the subsequent changes in carbon stocks. As they are based on land-use data, they generally miss changes in carbon stocks caused by natural disturbances, unless the disturbances result in a change in land use. This can result in the inclusion of substantial errors, especially at national, regional and local scales (Kurz and Apps, 1999).

Process-based models purport to generate estimates of carbon stores and fluxes through an understanding of the ecosystem processes that determine global carbon budgets (Nilsson *et al.*, 2007). The models used are generally terrestrial biogeochemical models (TBMs) or dynamic global vegetation models (DVGMs). The models are extremely complex and rely on large amounts of data for their parameterization.

As indicated in the previous paragraphs, the different models produce very different results. In addition, the error bars surrounding estimates can be substantial. Spalding *et al.* (2012) provide a comparison of some of the estimates of annual carbon fluxes in different biomes. A top-down inverse modelling approach suggests that there may be an annual carbon sink of 2.1 (± 0.8) PgC yr^{-1} in northern mid-latitudes, whereas a bottom-up land-use change approach provides a figure of 0.03 (± 0.5) PgC yr^{-1}. There is some evidence that better models are beginning to result in the convergence of the results from top-down and bottom-up approaches (Malhi *et al.*, 2008), but there is clearly a long way to go before we fully understand carbon fluxes and stores at a global scale.

Estimates of carbon emissions attributable to land-use change in the 2000s are provided in Table 8.1. There are substantial differences in the estimates produced in

Table 8.1 Estimates of net land to atmosphere flux from land-use change (PgC yr^{-1}) for the 2000s by region. Positive values indicate net CO_2 losses from land ecosystems affected by land-use change to the atmosphere. Uncertainties are reported as 90% confidence interval.

	Land Cover Data	Central and South Americas	Africa	Tropical Asia	North America	Eurasia	East Asia	Oceania
van der Werf *et al.*, 2010[a, b]	GFED	0.33	0.15	0.35				
DeFries and Rosenzweig, 2010[c]	MODIS	0.46	0.08	0.36				
Houghton *et al.*, 2012	FAO-2010	0.48	0.31	0.25	0.01	−0.07	0.01	
van Minnen *et al.*, 2009[a]	HYDE	0.45	0.21	0.20	0.09	0.08	0.10	0.03
Stocker *et al.*, 2011[a]	HYDE	0.19	0.18	0.21	0.019	−0.067	0.12	0.011
Yang *et al.*, 2010[a]	HYDE	0.14	0.03	0.25	0.25	0.39	0.12	0.02
Poulter *et al.*, 2010[a]	HYDE	0.09	0.13	0.14	0.01	0.03	0.05	0.00
Kato *et al.*, 2013[a]	HYDE	0.36	−0.09	0.23	−0.05	−0.04	0.10	0.00
Average		0.31 ± 0.25	0.13 ± 0.20	0.25 ± 0.12	0.05 ± 0.17	0.12 ± 0.31	0.08 ± 0.07	0.01 ± 0.02

Notes:

[a] Method as described in the reference but updated to 2010 using the HYDE land cover change data.

[b] The 1997–2006 average is based on estimates of CO_2 emissions from deforestation and degradation fires, including peat carbon emissions. Estimates were doubled to account for emissions other than fire, including respiration of leftover plant materials and soil carbon following deforestation (Olivier *et al.*, 2005). Estimates include peat fires and peat soil oxidation. If peat fires are excluded, estimate in tropical Asia is 0.23 and pan-tropical total is 0.71.

[c] CO_2 estimates were summed for dry and humid tropical forests, converted to C and normalized to annual values. Estimates are based on satellite-derived deforestation area, and assume 0.6 fraction of biomass emitted with deforestation. Estimates do not include carbon uptake by regrowth or legacy fluxes from historical deforestation. Estimates cover emissions from 2000 to 2005.

FAO = Food and Agriculture Organization (UN); GFED = Global Fire Emissions Database; HYDE = History Database of the Global Environment; MODIS = Moderate Resolution Imaging Spectroradiometer.

From Ciais *et al.* (2013).

different studies. This reflects the many uncertainties associated with the models used to calculate the fluxes. Terrestrial uptake of carbon is estimated to have been 1.5 ± 1.1, 2.6 ± 1.2 and 2.6 ± 1.2 PgC yr^{-1} in the 1980s, 1990s and 2000s, respectively (Ciais *et al.*, 2013), excluding the effects of land-use change. After taking into account emissions associated with land-use change, there is still a net landwards sink. This sink has intensified, being 0.1 ± 0.8 PgC yr^{-1} in the 1980s, 1.1 ± 0.9 PgC yr^{-1} in the 1990s and 1.5 ± 0.9 PgC yr^{-1} in the 2000s (Ciais *et al.*, 2013). The sink was originally identified as being in the northern extratropics, but the tropics may also represent an important net sink (Stephens *et al.*, 2007). The extent to which there is an important net sink in the tropics remains controversial, primarily due to the lack of reliable data, whereas the presence of a substantial net sink in the northern extratropics has been confirmed in a number of studies (Ciais *et al.*, 2010).

Limitations of terrestrial carbon cycle models

While the results of modelling studies often receive very high profile attention, great care is needed in their interpretation, as they contain numerous flaws and shortcomings. Ciais *et al.* (2013) acknowledge this, pointing out that some models miss important processes such as various forms of disturbance and ecosystem dynamics, including migration, fire, harvesting, insect outbreaks and drought. Many processes related to the decomposition of carbon are also missing, especially those related to permafrost and wetland areas. Nutrient dynamics are also largely missing from many models, as are the effects of air pollutants such as ozone, sulphur dioxide and the oxides of nitrogen. Human management effects, including tillage, fertilization and irrigation, are largely missing. Essentially, as pointed out by Ciais *et al.* (2013), carbon cycle models are still at a very early stage of dealing with land use, land-use change and forestry.

Management actions to increase the flux of carbon to forest lands

There are many steps that a forest manager can take to increase the amount of carbon sequestered by forests. These naturally work best when done in conjunction with natural processes. However, of the range of techniques available, not all are compatible with each other. A basic choice lies between maximizing carbon uptake and maximizing carbon storage, as younger forests generally take up more carbon than older forests, whereas older forests store a lot more carbon than younger forests.

Carbon dioxide fertilization

Modellers are convinced that an important sink for carbon has grown as a result of the fertilizing effects of CO_2. It is well-known that as CO_2 rises, growth increases and the water-use efficiency of trees increases. This is supported by numerous experimental studies, but some of these experiments have suggested that the effect might not be maintained with continuous increases in CO_2. There is a fundamental problem here related to experimental design. In nature, atmospheric CO_2 concentrations have been increasing, although with marked seasonal variations in concentration. Experimental studies tend to adopt a different approach, exposing young trees to steady levels of a higher concentration (for example, 450 ppm CO_2). This problem has been partly circumvented by the use of Free-Air CO_2 Enrichment (FACE) experiments, where plants are grown in the open air and specific amounts of CO_2 are added. However, the majority of these continue to expose plants to a fixed amount of CO_2. The majority of longer-term FACE studies are located in temperate regions, and many of these have begun to show a fairly consistent pattern of elevated carbon accumulation being maintained over time. However, some show a lack of CO_2 fertilization effect for some plant species, so great care is needed in making inferences about future trends.

One possible cause for the lack of response or the decline in response after an initial effect is nutrient limitation. According to Ciais *et al.* (2013), nitrogen and phosphorus deficiencies are very likely to be involved in these nutritional limitations, with nitrogen being the primary limitation in temperate and boreal forests and phosphorous being the primary limitation in the tropics. However, as shown in Chapter 4, other nutrients can also be important, especially in the tropics. Monitoring the nutritional status of forests over time is clearly important if any benefits from the fertilization effects of increased CO_2 are to be maintained.

In established forests, there is evidence for increasing sequestration, both in temperate and boreal forests (e.g. Bellassen *et al.*, 2011) and in the tropics (e.g. Lewis *et al.*, 2009). The extent and cause of this increase is still controversial.

Afforestation

One of the most frequently cited means of increasing both carbon sequestration and carbon storage on bare land is to plant trees. In many cases, trees are being planted on land that once held trees but which were cleared for agricultural purposes. Conversion of these lands to forests generates a significant increase in the carbon stored onsite. Similarly, conversion of natural grasslands to forest increases the carbon stored onsite, but may conflict with biodiversity objectives if the area of remaining grassland is limited. For example, the remaining areas of tallgrass prairie in North America are much too ecologically valuable to be converted to forest. However, other grasslands, such as the Patagonian pasturelands (which were originally shrub-grass steppe) provide a good opportunity for afforestation, and both natural regeneration with Chilean cedar (*Austrocedrus chilensis*) and afforestation with ponderosa pine (*Pinus ponderosa*) have resulted in increased soil carbon, as well as major increases in aboveground biomass (Laclau, 2003). The purpose of most afforestation is to increase wood supply, but significant areas are also planted with trees for environmental remediation, and today trees in some areas are planted specifically as a means to sequester carbon.

A major potential issue with afforestation is the loss of soil carbon that may accompany soil preparation techniques. In Scotland, Scandinavia and the southern USA, the drainage of peatland for afforestation causes significant losses of carbon as the peat dries and oxidizes. These carbon losses can exceed any gains derived from reduced emissions of methane form the peat, and may even exceed the carbon stored in aboveground biomass (Cannell *et al.*, 1993). Carbon losses can be reduced by maintaining a high water table, suggesting that deep drainage should not be undertaken on such sites (Carroll *et al.*, 2012). Elsewhere, the removal of any organic matter in order to create a mineral soil seed bed can also result in significant carbon losses.

Similar losses can also occur in tropical and sub-tropical areas. For example, losses of soil carbon have been associated with afforestation of wet grasslands in Ecuador, primarily due to the soils drying out and oxidizing (Farley *et al.*, 2004). As with other cases, careful management of the water table is required, and afforestation may be better directed towards less sensitive sites.

While afforestation is seen as being a major potential sink for carbon, reforestation is also important, given the amount of land that has been deforested in the last 50 years. Concerns exist over the replacement of tropical rain forest with monospecific plantations, but reforestation can take many forms (Elliott *et al.*, 2013). While

*Figure 8.2
Former farmland
afforested
specifically
for carbon
near Wenzhou
in Zhejiang
Province, China.*

*Figure 8.3
A small area of
forest replanting,
using a variety
of tree species
found in
tropical rain
forest, Atherton
Tablelands,
northern
Queensland,
Australia.*

the diversity of a tropical rain forest will not be recreated through plantings, there is an increasing tendency for small patches of land to be planted with a diversity of rain forest trees as part of landscape restoration projects. Individually these are insignificant, but collectively and together with patches of second-growth forest, they have the potential to form a very large carbon sink while at the same time creating forested areas that have many of the values of a semi-natural rain forest (Chazdon, 2014).

Increased rotation lengths

In tropical forest management, the most frequently prescribed cutting cycle is 30 years (Sist *et al.*, 2003). Moving to a longer cycle would not only result in more sustainable timber yields, but would also significantly increase the amount of carbon stored in the aboveground forest biomass. Much the same applies for temperate and boreal forests, although the timescales involved are generally longer (Foley *et al.*, 2009). However, any such increase in rotation length must take into account a number of factors, including potential lost income if trees are harvested later than the optimal rotation time, and the increased risks of disturbance as a stand ages (especially increased risk of windthrow).

Soil management

Most soil management associated with forest operations involves preparation of the ground for the next generation of trees following a harvest operation. Techniques that remove the surface organic matter, including scarification and burning, will result in increased losses of carbon, but may be a prerequisite for the successful regeneration of some forest types. For example, prescribed fire followed by aerial seeding is a common way of regenerating eucalypt stands in parts of Australia. Establishing the amount of carbon sequestered in such situations will involve careful measurement, particularly of carbon incorporated into the soil profile.

Increasing offsite carbon stocks in wood products

The woody products taken from a forest represent a net loss of carbon from the forest ecosystem. However, the carbon is not necessarily immediately emitted into the atmosphere. Although the majority of wood that is taken from the world's forests is used as firewood, and therefore results in carbon emissions to the atmosphere, a significant proportion goes into longer storage in the form of buildings, furniture and other products. At the end of the product's life, some of this may be recycled, further delaying the emission into the atmosphere. There are major opportunities to reduce the rate of carbon emissions in this area: much paper is now recycled. The wood stored in buildings is generally held for longer, although in North America, where most family homes have a significant wood component, buildings are generally poorly constructed and replaced a lot more quickly than in other parts of the world. Very little of the wood from these buildings is recycled.

Enhancing product and fuel substitution

Although much of the wood taken from forests is burned, this firewood substitutes for other forms of energy. In most developing countries, efforts are being made to convert such use to more efficient forms of energy, although this often involves the use of fossil fuels. Ideally, firewood should be replaced by other forms of renewable energy, or at least produced on a sustainable basis. There has also been a major push to use wood as a substitute for fossil fuels in developing countries, and this has created new markets, such as that for wood pellets. Wood pellets have generally been produced from waste streams within the forest products sector, but forests are now being grown in some parts of the world specifically for pellet production. The management of forests for bioenergy production is examined in detail by Kellomäki *et al.* (2013).

Density management

From the perspective of carbon management, the greater the biomass density on a site, the more carbon will be stored. However, increased density comes with a number of potential risks, such as increased risks of fire in some forests. As with some of the other approaches described here, careful management will be required to balance any potential gains in carbon storage with any losses of other values associated with the forest.

Diversifying the forest

While the number of studies supporting the idea that a more economically diverse forest would hold more carbon is growing, Del Cid-Liccardi *et al.* (2012) provide an interesting example from Sri Lanka. It is based on a study by Ashton *et al.* (2001) of the financial implications of silvicultural activities in Sri Lanka. They argued that following harvesting, planting of a number of economically important species would not only provide revenue between timber harvests but could also increase the carbon stored in the forest. Activities included planting economically useful shrubs such as cardamom (*Elettaria cardamomum* var. *major*) around advanced regeneration, planting medicinal vines (e.g. *Coscinium fenestratum*) and rattans (e.g. *Calamus zeylanicus*) and planting understory palms such as fishtail palm (*Caryota urens*) that could later be used for tapping sugar.

This form of enrichment planting has a number of side benefits, including making the forest more valuable to local people and therefore reducing the risk of it being converted to other forms of land use with consequent releases of carbon.

Considerable benefits may also be gained from making plantation forests more diverse. Wei and Blanco (2014) found that diversifying subtropical plantation forests in China could result in significant increases in the amount of carbon stored in the forest. For example, a mixed Chinese fir (*Cunninghamia lanceolata*) and *nanmu* (*Phoebe bournei*) forest would have 67.5% more carbon than a pure Chinese fir forest over a 100-year simulation period when managed using more sustainable methods (40-year

*Figure 8.5 Species being used in the example of enhanced silviculture in Sri Lanka. On the left, cardamom (*Elettaria cardamomum var. major), *and the right, fishtail palm (*Caryota urens).

rotation, stem-only harvesting) than are currently practiced in Chinese fir silviculture (25-year rotation, stem and slash removal, site preparation by understory burning).

Similarly, in temperate forests, strong arguments have been made in favour of diversifying at least some planted forests. In Europe, beech (*Fagus sylvatica*) and Norway spruce (*Picea abies*) were traditionally grown in single species stands. Mixing the two species in the same stand not only results in improved biodiversity but likely makes the stand more resilient to climate change. It may also improve the financial viability of the stand, especially if the costs of disturbances and log price variations are included in the calculation (Griess and Knoke, 2013).

Better management

One factor that should not be forgotten is that many carbon releases from forests attributable to anthropogenic activities are the consequence of deforestation and forest degradation. There are documented examples in which managed forests may be less susceptible to deforestation than strictly protected areas, primarily because local people realize that there are economic benefits to be gained from the managed forest, whereas reserves that are not supported locally may be subject to conversion to other forms of land use (Durán-Medina *et al.*, 2005).

However, in many countries, particularly in the tropics, there are a number of forces that mitigate against the implementation of sound management techniques. These include lack of clarity over forest tenure, a lack of enforced environmental legislation, limited access to markets, lack of expertise and a lack of financial incentives to implement good practices (Del Cid-Liccardi *et al.*, 2012).

One possibility in tropical forests is provided by the implementation of Reduced Impact Logging (RIL) (Pinard and Putz, 1996). The implementation of RIL could have important implications for carbon management as it is designed to reduce the damage and mortality associated with logging operations. This is particularly important in species-diverse forests, where only a small number of trees per hectare may be of commercial interest. Minimizing the damage to the remaining trees would reduce carbon losses associated with the loss of damaged trees. Similarly, RIL results in more attention being given to reducing soil disturbance, and therefore will result in fewer emissions of carbon from the soil in comparison to more conventional logging techniques. RIL also involves making detailed inventories of the forest stock prior to making any interventions. Del Cid-Liccardi *et al.* (2012) point out that this would enable a manager to determine the frequencies of different species in the landscape, their contribution to the overall carbon stock of the management unit and the potential impacts on the carbon stock of the selective removal of certain species. This is important given the interspecific variations in carbon storage in tropical forests (Kirby and Potvin, 2007).

Del Cid-Liccardi *et al.* (2012) argue that tropical forest management needs to move away from broad prescriptions, such as diameter cutting limits, towards stand-level prescriptions within a broader landscape context. They argue that this will enable carbon and other attributes to be better managed, rather than focusing exclusively on the extraction of timber. This will require the development of long-term, strategic forest management plans – which today are very rarely used in tropical forests. They also emphasize the need to understand two important stratification processes when dealing with species mixtures in tropical forests. Late-successional species that occupy particular vertical strata in the mature forest canopy are referred to as 'static' stratification. Species that move through different vertical strata as they move upwards towards the canopy are referred to as 'dynamic' stratification. Understanding this relationship is important for optimizing the amount of carbon stored in a tropical forest.

In temperate and boreal forests, a variety of steps can be taken to reduce carbon losses associated with poor management. During thinning operations, care is needed to ensure that the remaining trees are not damaged, since damage could result in subsequent mortality and decay onsite. There is considerable debate over resiliency treatments, especially in the USA, where fuel reduction programs (either thinning or prescribed fire) can result in short-term losses of carbon but prevent the sudden release of large amounts of carbon in stand-replacing disturbances (Carroll *et al.*, 2012). Treated stands carry lower biomass (and therefore carbon) than untreated stands, but the benefits appear to outweigh the reduced carbon stocks.

Harvesting wood from a site results in an immediate loss of carbon from the ecosystem. Further losses occur through increased litter and soil respiration. Harvested sites may remain a net source of carbon for 10–30 years following harvest, but then become carbon sinks as trees regrow (Carroll *et al.*, 2012). The extent of carbon losses following harvesting is quite controversial, with conflicting results from different ecosystems. However, evidence seems to be amassing that the losses are quite limited, and can be further reduced by careful management of the slash left onsite and by choosing the most appropriate silvicultural treatment for the site.

Maximum storage of carbon can be obtained by maintaining old forests where they exist. When these are replaced by secondary forests there is always a significant loss in the amount of carbon stored onsite. Such forests are sometimes thought to be carbon neutral, but there is increasing evidence that they continue to sequester carbon over time. In a few cases, extremely old forests may start to deteriorate – this is often because they do not represent the true climax for the site, and they are in the process of being replaced by another form of forest.

Forest restoration

There is increasing interest in the restoration of degraded forests. Some of this is related to carbon sequestration and storage, and involves some of the techniques described in the previous section. However, as should now be evident, forests provide a myriad of services, and so the restoration of degraded forests and deforested lands has become a priority. Estimates of the amount of land available for restoration vary widely, from 350 million to 2 billion ha, although in reality a variety of legal, social and cultural factors may mean that the actual area that could be restored is much less. The differences in the estimates of the land available for restoration arise from the confusion over what constitutes 'degraded' (Lamb, 2014). There is no globally accepted definition of this, despite the importance of REDD+, the acronym given to countries' efforts to reduce emissions from deforestation and forest degradation, and to foster conservation, sustainable management of forests, and the enhancement of forest carbon stocks. In particular, some would consider any harvesting activities undertaken in a forest as degradation, whereas most in the forestry sector would disagree strongly. Some conservation efforts have attempted to restore forests to some hypothetical state from the past (Burton and Macdonald, 2011), but this presupposes that the nature of these past forests is well understood, and that those forests would be capable of surviving not only today's environmental conditions but also those of the next 100 years. Such an approach also needs to determine what constitutes a natural forest and whether pre-industrial human influences should be reproduced or not. This issue is particularly apparent in Australia, where Aboriginal peoples have modified forests since their arrival on the continent more than 40,000 years ago, but is also apparent in more recent modifications. For example, the structure and form of ancient woodlands in England was a direct result of medieval management practices (Rackham, 2006). Contemporary restoration techniques therefore focus on the functional diversity of tree species assemblages (Aerts and Honnay, 2011) and the re-establishment of a flow of goods and services from forested lands (Lamb *et al.*, 2012).

There is considerable political interest in forest restoration. The 2011 Bonn Challenge called for the restoration of 150 million ha of deforested and degraded forest lands by 2020; if this is achieved, it is estimated that it would sequester an additional 1 Gt CO_2e per year. The 2014 New York Declaration on Forests maintained this target, but added that the rate of global restoration should be increased thereafter, resulting in the restoration of an additional 200 million ha by 2030.

Figure 8.6 Plantations of exotic conifers on a hillside adjacent to Loch Fyne, Scotland. The impact on the scenery is clear, and scenes like this resulted in significant opposition to afforestation projects in Scotland.

While forest restoration has been undertaken for a long time, it has not always met with public approval. For example, the re-establishment of forests in the Scottish Highlands, using exotic conifers (mainly from the Pacific Northwest) rather than native trees, resulted in considerable public opposition as people had become used to the entirely anthropogenic, treeless landscape. With reduced emphasis on timber production, and more on public amenity and nature conservation, some of these forests have been restructured, although this action in itself could be considered a form of restoration. The problem, it seems, was with the idea that the restoration efforts were entirely aimed at providing a source of timber, and therefore focused on a narrow range of exotic species. Similar concerns arose in central Europe, particularly in Germany over the preponderance of mono-specific stands of Norway spruce (*Picea abies*), giving rise to the development of 'near-to-nature' forestry techniques.

Although some restoration schemes have focused on the development of timber resources, others have focused on environmental protection, with some of the major afforestation schemes in China (including the Grain to Green Project, the Three Norths Shelterbelt Program and the Natural Forests Conservation Program) being good examples. Others have focused on the conservation of biodiversity: the Atlantic Forest Restoration Pact in Brazil aims to restore 15 million ha of tropical rainforest by 2050 (Pinto *et al.*, 2014), although even with this program the impact on livelihoods is being considered very carefully as it involves the conversion of degraded pasturelands and abandoned agricultural lands to forest (Melo *et al.*, 2013). Many of the individual projects are quite small: in 2015, BNDES (the Brazilian development bank) provided US$ 12 million to 14 projects involving the restoration of about 3,000 ha of Atlantic Forest. More than half of the total area involved the restoration of forests within existing protected areas.

The move away from an emphasis on timber production in forest restoration has seen the emergence of forest landscape restoration. This attempts to balance ecological integrity with people's livelihoods. This is very different to more traditional approaches to forestry, where the emphasis has been on the production of timber, and the focus has been on stands and forests, excluding considerations of the surrounding lands. Contemporary forest restoration focuses instead on the complete bundle of goods and services provided for forests, with the emphasis being placed on forest function (Stanturf *et al.*, 2014). The Global Partnership on Forest Landscape Restoration identifies four main principles of forest landscape restoration:

- The restoration of a balanced and agreed package of forest functions;
- The active engagement, collaboration and negotiation among a mix of stakeholders and other parties;
- The adoption of a landscape approach;
- Adaptive management.

The importance of a landscape-scale approach to forest restoration is emphasized. Deforestation is a problem primarily associated with agriculture, and unless the people involved can be successfully involved in restoration efforts, further forest damage and destruction is likely. The adoption of a landscape approach reflects current thinking that this is the most appropriate scale to deal with many of the broader issues facing forests today (Sayer and Maginnis, 2005b). The emphasis on goods and services also recognizes that timber is only one forest product, and that there are numerous others, many of which are critical to the livelihoods of local people. As a result, forest landscape restoration places more emphasis on social science than has traditionally been done in forestry (Stanturf *et al.*, 2012), although the preponderance of natural science in forestry education has been diminishing for some time (Innes, 2009).

The Global Partnership on Forest Landscape Restoration points to some impressive examples. The Republic of Korea estimates that it has received a fiftyfold return on investment on the work it has done to improve and restore forests: between 1953 and 2007, the forest cover increased from 35% to 64% of the country's area. In Costa Rica, the forest area increased from less than 30% in 1987 to 50% in 2010, enabling the country to develop a very successful tourism industry. In Tanzania, the people of Shinyanga in the north of the country restored 2 million ha of forest and agricultural land, and in doing so they doubled household income.

Stanturf *et al.* (2014) recognize four overarching strategies to restoration: rehabilitation, reconstruction, reclamation and replacement. All of these are relevant to the management of carbon in forests, but they also reflect the broader need to ensure that all forest functions are addressed. Lamb (2014) argues that ecosystem functioning and biodiversity are closely related, drawing on studies such as that of Cardinale *et al.* (2012), which demonstrate that there is a considerable body of scientific evidence demonstrating links between biodiversity and ecosystem functioning. The available evidence indicates that loss of biodiversity within an ecosystem reduces the efficiency with which resources such as nutrients, water and light are captured, reduces the production of biomass, and reduces decomposition and recycling rates. Biodiversity increases the stability of ecosystem functions over time, increasing ecosystem resilience. When biodiversity changes, there are non-linear impacts on ecosystem processes. In addition, loss of diversity across trophic levels probably has greater effects on ecosystem functions than loss of diversity within any one trophic level. These results have been used to promote the idea that plantation monocultures will be less useful in providing ecosystem services than more diverse systems (Gamfeldt *et al.*, 2013; Lamb, 2014).

The restoration of functional landscapes in Europe has been linked to a remarkable resurgence in many wildlife species (Deinet *et al.*, 2013). This has occurred as a result of both species-specific conservation and the development of a number of habitat initiatives that helped restore some of the functions of forest landscapes.

*Figure 8.8
The European
subspecies of
the grey wolf
(Canis lupus ssp.
lupus) has seen
a remarkable
recovery in the
last 30 years in
some parts of
Europe. Grey
wolves are
an extremely
successful and
adaptable
species, which
has undoubtedly
helped their
recovery.*

Table 8.2 Advantages and disadvantages of different methods of initiating forest restoration.

	Advantages	Disadvantages
Natural regrowth	• Low cost • Can treat large areas • Little ecological knowledge needed	• Prone to further disturbance as site may not be recognized as undergoing restoration • Difficult to restore all flora • May result in sparse or patchy distribution of trees • Rate of regrowth may be slow
Direct seeding	• Relatively low cost • Seed can be dispersed from the air	• Not always effective • Needs large seed quantities due to low establishment rates • Seed of some species may be unavailable • Best used on cleared sites so that competition from weeds is minimized
Planting seedlings	• More reliable • Efficient use of scarce seed resources • Ensures preferred species are established • Can plant at required locations and densities • Weed control easier	• More costly because of nursery and planting costs • May be difficult to raise seedlings of all species in species-rich forests

From Lamb (2015).

Lamb (2015) describes a number of ways in which forest restoration can be initiated. These include natural regrowth, direct seeding and planting seedlings, each of which has its advantages and disadvantages (Table 8.2). The choice of method will depend on the circumstances and on which of the strategies described in the table is being followed.

Rehabilitation

Rehabilitation involves the restoration of desired species composition, structure or processes to a forest ecosystem that still exists but has become degraded. Management actions are aimed at altering the system in such a way that natural processes lead to the desired function. This obviously requires very clear objectives from the outset, a feature of all successful contemporary restoration activities. While many rehabilitation projects are aimed at restoring late seral conditions, others may be aimed at other seral stages, particularly if the aim is to provide habitat for certain endangered species. For example, the near-threatened Kirtland's warbler (*Setophaga kirtlandii*) requires young (4- to 20-year-old, 2–4 m tall) jack pines (*Pinus banksiana*), and management for the recovery of this species has involved prescribed burns and clear-felling. This has been controversial, however, and there are suggestions that other species may be suffering because of the focus on the requirements of a single species (Corace *et al.*, 2010). Similarly, in southeast Australia, attempts to rehabilitate the river red gum (*Eucalyptus camaldulensis*) forests along the Murray River have been exceptionally controversial, partly because the history of these forests is complex (Colloff, 2014), and understanding of their dynamics is limited. These examples illustrate the dilemmas facing the management of restoration efforts.

Stanturf *et al.* (2014) describe two approaches to rehabilitation, namely conversion and transformation. Conversion involves the removal of an existing overstory and replacement by one or more other species, whereas transformation occurs more gradually, involving partial removals and species replacement. Both approaches may be aided by natural processes such as wildfires, windstorms and pest outbreaks. An example of such management is provided by the restoration of longleaf pine (*Pinus palustris*) ecosystems in the southeast USA. In stands with high concentrations of loblolly pine (*P. taeda*), restoration not only involves the removal of mature loblolly pine, but also requires the use of prescribed fire to prevent the rapid regrowth of the loblolly pine (Knapp *et al.*, 2011).

*Figure 8.9 River red gums (*Eucalyptus camaldulensis*) on the Murray River at Echuca, Victoria, Australia. The condition of this ecosystem is disputed, as is the need for restoration.*

*Figure 8.10 Forest restoration in the form of rehabilitation at Glenmore Forest, Scotland. In this case, exotic Sitka spruce (*Picea sitchensis*) has been removed, leaving native Scots pine (*Pinus sylvestris*) to provide seed for natural regeneration. These old pines are colloquially known as 'granny pines'.*

Reconstruction

Reconstruction is the process of restoring native plant communities on land previously used for other purposes, such as agriculture. It involves a lot more work than rehabilitation, and Stanturf *et al.* (2014) argue that it may involve soil treatments such as increasing organic matter content, decreasing bulk density, reducing the weed seedbank, outplanting seedlings and direct seeding. It can also involve passive approaches – allowing natural regeneration to take effect, and the two can be combined. For example, the restoration of koa (*Acacia koa*) forest at Hakalau Forest National Wildlife Refuge in Hawai'i has involved a combination of planting (active management) and suckering (passive management) in an attempt to convert grassland into koa forest (Scowcroft and Yeh, 2013).

Reconstruction of tropical forests is particularly difficult, but does occur. The primary means of seed dispersal for tropical forest trees is by animals (particularly birds), which are generally reluctant to move into open areas. Even if they reach the desired site, there are many limiting factors, including aggressive pasture grasses, adverse microclimatic conditions, limited soil nutrients, and high seed predation and seedling herbivory (Holl, 2013). Despite these problems, there are very real possibilities to regenerate degraded land in the tropics, and many examples of how this might be done (Chazdon, 2014).

Reclamation

Reclamation occurs on land that has been severely degraded. In many cases, the soil may be badly damaged, either through erosion or though chemical pollution. Considerable experience has developed in this area, mainly as a result of attempts to restore vegetation around severely polluted sites, such as mine spoil and industrial deserts (see Chapter 5 for more information on the impacts of pollution and the restoration of industrial deserts). Work often involves soil amelioration, seeding or outplanting

Figure 8.11 Mature koa (Acacia koa) forest on Hawai'i, the largest of the Hawaiian Islands. Considerable efforts are being made to try to restore deforested areas to this type of forest. These forests have a single canopy species, reflecting the isolation of the Hawaiian Islands, although another species, 'ōhi'a lehua (Metrosideros polymorpha) may develop in the shade created by the koa trees.

seedlings, irrigation, and weed control. Sometimes non-native plants are used in the early stages, either for bioremediation or as nurse plants for native species (Parrotta *et al.*, 1997).

Replacement

The final approach described by Stanturf *et al.* (2014) involves the replacement of existing native vegetation with other species, and is particularly aimed at the adaptation of forests to climate change. Recent severe pest outbreaks, such as the emerald ash borer (*Agrilus planipennis*) in eastern North America, have also prompted the need to replace susceptible species in some forests, and restoration objectives need to consider both the susceptibility of a species to pest attack and its ability to survive under current and future climatic conditions (e.g. Dalgleish *et al.*, 2016). In the context of climate change, the replacement of an existing species or genotype by one from a different climate is referred to as assisted migration (Williams and Dumroese, 2014). This is also a controversial approach, and concerns have arisen over the potential ecological implications of assisting the movement of species. Such concerns arise because of the risk of unintended consequences – there are so many examples of well-meaning introductions that have gone wrong that many are now very concerned about any attempt to move species artificially.

Conclusions

When the Montreal Process was developed, the primary concern was that forests be allowed to maintain their contribution to global carbon cycles. A forest that has been degraded or converted to another form of land use cannot achieve this. However, in the intervening period, there has been increasing interest in managing forests specifically as carbon stores, a form of management sometimes referred to as carbon forestry. This involves a number of techniques aimed at increasing the carbon stored on a site, and potentially involves some forgone revenue. For example, managing a plantation estate to maximize financial returns is not consistent with managing a plantation estate to maximize carbon onsite. However, if the carbon stored offsite in wood products is factored into the equation, then a different picture emerges. To date, most schemes aiming to increase carbon sequestration have failed to take wood products into account, so management has concentrated more on maximizing the storage onsite.

Managing forests for carbon is likely to become increasingly important as the political community begins to take the threat of climate change seriously. Most countries have committed to reducing their carbon emissions, and in Paris in 2015 many political leaders stressed the importance of forests in achieving national and global targets. This will require careful implementation. It is insufficient to lock up carbon in forest reserves: these forests change over time, and as discussed in Chapter 5, natural dynamics can play a big role in destabilizing forests. The argument is sometimes made that natural forests will look after themselves, but it is increasingly difficult to find forests growing under completely natural conditions. The rise in atmospheric CO_2 concentrations is ubiquitous, and forests everywhere are responding to this anthropogenic change. Global warming will also have an effect, and there is already evidence of the destabilization of some forests growing under already marginal conditions. Consequently, forest managers will need to play a major role in ensuring that forests globally not only actively sequester carbon but that the major stores of carbon in forests are maintained.

Much future management will take place under the auspices of REDD+, especially now that the role of forests in climate change is receiving so much attention. A major obstacle however remains the funding of REDD+: while promises have been made, the amount of money that has actually been made available falls far short of the need. An example of what is possible is provided by the Amazon Fund, which in 2015 was supporting 77 projects with a total of US$ 552 million. The fund has received US$ 917 million in donations: US$ 882 million from the Norwegian government, US$ 28 million from the German government and US$ 7 million from Petrobras, Brazil's oil and gas company. Both Norway and Germany have indicated that they will contribute more to the fund.

The interest in carbon sequestration has also created major opportunities for forest restoration. This could take a number of forms, but the basic principle is to increase the forest area of the world through a thoughtful process that ensures that the goods and services provided by the forests are adequately restored. This is most likely to occur through the restoration of forest ecosystem function, but also requires the involvement of local people. If the goods and services do not meet the needs of local people, then there is a strong possibility that the restoration efforts may be in vain. Restoration goes beyond REDD+, which after all is about reducing emissions. Restoration is, among other things, about increasing sequestration and storage of carbon while providing other benefits. It involves a skill set that many current foresters do not have and, as a result, will be a major but absolutely critical challenge for foresters.

Further reading

Angelsen, A. (ed.) 2009. *Realising REDD+: National Strategy and Policy Options*. Bogor: Centre for International Forestry Research.

Ashton, M. S., Tyrrell, M. L., Spalding, D., Gentry, B. (eds.) 2012. *Managing Forest Carbon in a Changing Climate*. Dordrecht: Springer.

Chazdon, R. L. 2014. *Second Growth: The Promise of Tropical Forest Regeneration in an Age of Deforestation*. Chicago, London: University of Chicago Press.

Elliott, S., Blakesley, D., Hardwick, K. 2013. *Restoring Tropical Forests: A Practical Guide*. Kew, UK: Royal Botanic Gardens.

Kellomäki, S., Kilpeläinen, A., Alam, A. (eds.) 2013. *Forest Bioenergy Production: Management, Carbon Sequestration and Adaptation*. New York: Springer.

Lamb, D. 2014. *Large-Scale Forest Restoration*. London: Earthscan from Routledge.

Lorenz, K., Lal, R. 2010. *Carbon Sequestration in Forest Ecosystems*. Dordrecht: Springer. 277 pp.

Ravindranath, N. H., Ostwald, M. 2008. *Carbon Inventory Methods: Handbook for Greenhouse Gas Inventory, Carbon Mitigation and Roundwood Production Projects*. Dordrecht: Springer.

Rietbergen-McCracken, J., Maginnis, S., Sarre, A. 2007. *The Forest Landscape Restoration Handbook*. London: Earthscan.

Chapter 9

The changing socio-economic contributions of forestry

Harry Nelson, Ngaio Hotte, and Robert Kozak

The Montreal Process focuses on the socio-economic contributions of forests in Criterion 6: maintenance and enhancement of long-term multiple socio-economic benefits to meet the needs of societies. This is a vast topic, arguably not particularly well handled by the criterion and, in this book, we look at the socio-economic contribution in several chapters. This chapter focuses primarily on the direct socio-economic contributions from natural and semi-natural forests and how those contributions have changed in recent decades. Chapter 10 deals with social contributions (sub-criterion 6.4), and Chapter 11 looks at public participation (covered by both Criteria 6 and 7). Chapter 12 deals with international policies surrounding sustainable forest management, which is mainly covered by Criterion 7, although there is some overlap with Criterion 6.

Criterion 6 is divided into a number of sub-criteria and indicators, as follows.

6.1 Production and Consumption. Indicators: 6.1.a Value and volume of wood and wood products production, including primary and secondary processing; 6.1.b Value of non-wood forest products produced or collected; 6.1.c Revenue from forest based ecosystem services; 6.1.d Total and per capita consumption of wood and wood products in round wood equivalents; 6.1.e Total and per capita consumption of non-wood forest products; 6.1.f Value and volume in round wood equivalents of exports and imports of wood products; 6.1.g Value of exports and imports of non-wood forest products; 6.1.h Exports as a share of wood and wood products production, and imports as a share of wood and wood products consumption; 6.1.i Recovery or recycling of forest products as a percent of total forest products consumption.

6.2 Investment in the Forest Sector. Indicators: 6.2.a Value of capital investment and annual expenditure in forest management, wood and non-wood forest product industries, forest-based environmental services, recreation, and tourism; 6.2.b Annual investment and expenditure in forest-related research, extension and development, and education.

6.3 Employment and Community Needs. Indicators: 6.3.a Employment in the forest sector; 6.3.b Average wage rates, annual average income, and annual injury rates in major forest employment categories; 6.3.c Resilience of forest-dependent communities; 6.3.d Area and percent of forests used for subsistence purposes; 6.3.e Distribution of revenues derived from forest management.

6.4 Recreation and Tourism. Indicators: 6.4.a Area and percent of forests available and/or managed for public recreation and tourism; 6.4.b Number, type, and geographic distribution of visits attributed to recreation and tourism and related to facilities available.

6.5 Cultural, Social, and Spiritual Needs and Values. Indicators: 6.5.a Area and percent of forests managed primarily to protect the range of cultural, social, and spiritual needs and values; 6.5.b The importance of forests to people.

As discussed in Chapter 1, forests have been important to society for millennia for shelter, sustenance, fuel, and forest products for use and exchange. In addition to these consumptive uses, forest ecosystem services, including cultural and spiritual values, have become increasingly recognized and progressively important in how forests are managed.

Recognition of the importance of the socio-economic impacts of forest management has long been a part of the discussion around forest resource use and perhaps, more importantly, forest misuse. In a 1960 address to the World Forestry Congress, the director of the Forestry and Forest Products Division of the Food and Agriculture Organization of the United Nations (FAO) stated that 'the realization is spreading that the diminution of forest areas may have far-reaching consequences because forests provide not just wood but a host of other benefits or human utilities' (Glesinger, 1960). While the discussion has evolved since 1960, the essence of Egon Glesinger's argument remains the cornerstone of the importance of recognizing those values in forest management: forests are much more than sources of timber, and the benefits that people derive from forests are far-reaching and inextricably dependent on sustainable management of forest resources.

Over the past few decades, the concept of sustainable forest management (SFM) has taken root, reflecting a shift from managing forests for timber harvests and associated economic benefits such as jobs and income to more inclusive consideration of the values that forests can provide. In many regions of the world, the contribution of forests to people's livelihoods is greater than their formally recognized economic contribution, particularly when one considers the massive informal economy. There can be even more important social benefits that people and communities obtain from access to forests. However, the economic contribution of forests remains an important topic because of the scale of employment opportunities and income that they generate and the value of those economic activities to forest-dependent communities, rural regions, and national economies. These economic activities and returns (both public and private) also continue to exert a strong influence on forest policy and management, as well as the ability to implement and achieve SFM objectives. Beyond that, the economic benefits also contribute to how forests are valued at the broader national and international scales, where society faces the challenge of how best to meet the needs of a growing population and expanding economies, while still sustaining the global ecosystem upon which we rely.

Thus, while recognition has grown of how forests can contribute to these other values, especially to people's livelihoods (FAO, 2014), concerns remain that forests are not being sustainably managed: the world's forest area continues to decline (especially in natural forests), with over 240 million ha lost between 2000 and 2015 (FAO, 2015), primarily in tropical countries, and accompanied by a wider degradation and loss of ecological integrity in the remaining forests.

It would be impossible to cover all aspects of the socio-economic contributions of forests in this chapter. Many different indicators have been proposed, and this is the criterion that has created the greatest difficulty for the Montreal Process. In part, this reflects the diversity of forests and the range of benefits that they provide. It also reflects the equally diverse social settings and institutional arrangements that govern forest access and use, which can shape particular linkages and create unique paths within different regions and countries. Despite this complexity, there are similar forces and processes at play, common to all countries, that influence how forests and forest lands are managed and utilized. Here, we first identify how widespread changes in technology, markets, and trade have affected the traditional socio-economic contributions generated from natural forests at the global level and now increasingly from semi-natural forests (those forests managed mainly for timber production, including plantations). We then examine new, alternative, and non-timber products emerging from natural forests, where many of the socio-economic contributions still remain poorly recognized. Finally, we review the global forces shaping land use and economic development and the implications for forest transitions and changes in the area and stock of future forests.

The socio-economic contribution of forests

Forests make up 31% of the world's landmass, slightly less than the amount of agricultural land (at 33%) (FAO, 2015). While forests are ubiquitous, their official contribution to global economies has historically been small, reflecting that much of the forest was either inaccessible or had little or no commercial value. The monetary contribution of forests to the global economy is estimated at US\$ 600 billion or 0.9% of annual gross domestic product (GDP) in 2009 (FAO, 2014). Agriculture, by comparison, accounts for 3% of annual global GDP.

These monetary values do not capture the full socio-economic contribution of forests: it is estimated that between 1 billion and 1.6 billion people rely on forests for a significant portion of their household income or livelihoods and that more people are employed in the informal economy than the formal economy (Agrawal *et al.*, 2013). An estimated 13.2 million people are employed in the formal forest sector, and over 41 million are employed in the informal sector (FAO, 2014). The benefits are even greater when consumption of forest products for fuel and shelter is included, but it is difficult to quantify those benefits as non-industrial data is lacking (FAO, 2014). These contributions to society's well-being become even greater when other forest values are brought into the picture, including the contributions of forests to biodiversity, climate regulation, and non-marketed goods and services such as cultural and spiritual values.

Wiersum (1995) provides an historical overview of how the idea of sustainability has been approached in forestry. While there is general agreement on the norms involved, there are challenges in operationalizing the concept. Of the four norms identified by Wiersum (1995), three are centred on benefits to society: the maintenance of yields of forest products for human needs, including consumption; the sustenance of human institutions that are forest dependent; and the sustenance of institutions to protect forests, of which a subset is ensuring maintenance of socio-economic conditions for populations living near forests. The fourth is the maintenance of forest ecological characteristics.

Box 9.1   Measuring socio-economic benefits.

There are multiple challenges in assessing the socio-economic benefits of forestry. These include a lack of commonly agreed definitions, and the problems associated with the need to include both clear economic benefits and more difficult-to-quantify benefits such as social justice, cultural and spiritual values, the preservation of harmony, and security (FAO, 2014). Hickey and Innes (2008) evaluated criteria and indicators for possible use in sustainable forest management in British Columbia, Canada, where socio-economic contributions can be found in two criteria: socio-economic benefits and society's responsibility. Hickey and Innes developed a set of 47 questions by which to evaluate indicators under all six key criteria; to illustrate the diversity of those contributions, more than half (24) of the questions address just those two criteria. Examples of indicators of economic benefits are the contribution to the regional economy and employment, including not only the level but also the kind of employment opportunities. Indicators of society's responsibilities involve the well-being of Aboriginal and forest communities, as well as their participation not only in forestry-based economic activities and the forest sector, but also in decision-making.

Paralleling the variety of indicators are two other challenges. First, while there are direct links between some indicators such as employment and income and their impact on individual well-being, in other cases those links may be more tenuous or indirect, especially where they may reflect more fundamental structural issues in society rather than simply outcomes from forest management (such as improved access or clarity of rights). Second, while relatively good data exist for activities associated with the formal forest sector, these statistics do not cover the full range of economic activities associated with forests. Many forestry-based activities are not captured or recognized in national accounts, either because they are not identified explicitly (such as the production and marketing of non-wood forest products) or because they are part of the informal economy and are not measured (such as for the provision of fuel wood). Criteria are evolving to consider not only these contributions, but also other elements of well-being, including the provision of shelter, health, and security, and contributions to improving equity (Agrawal *et al.*, 2013).

The economic contributions of forests can be captured in several ways, including employment, wage income, government taxes and revenues, returns to firms, and contributions to the economy (GDP). Publicly reported data exists at a global level for three of these indicators – employment, economic activity, and trade – and there are consistent time series for these socio-economic contributions and the ways in which they are changing over time. Employment is a key indicator, as it is the primary means by which people earn income and can support their livelihoods. Value-added statistics capture not only the economic contribution of the forest sector to national economies but also the income available to all aspects of production within the forest sector. Export statistics show changes over the time period in both the level and kind of products traded. One advantage of these statistics is that they are also available at a more disaggregated level, where they can show differences in the types of forestry activities taking place. The four subsectors for which information is available include: forestry and logging; wood processing; pulp and paper processing; and furniture construction (which includes both wooden furniture and other types).

Changes in industrial harvest levels

Forests are not distributed equally at a global scale: a few countries account for most of the forestry-related economic activity. Table 9.1 illustrates the 10 countries with the highest industrial roundwood harvest levels in 2013. African countries, such as the Democratic Republic of Congo, are conspicuously absent from this list, despite possessing considerable forest resources. This is because they are not engaged in corresponding levels of industrial economic activity. If the total volume of roundwood production is considered, including industrial and non-industrial harvests, a very different

Figure 9.1
Employment is
a particularly
valuable
indicator of the
socio-economic
contribution
of forests. The
picture here is
of a log sort
in Limpopo
Province, South
Africa.

Table 9.1 Top 10 forestry countries (by harvest level in 2013).

Countries	Roundwood production (million m³)		Forest resources (million ha)	Type of forests	Employ* (thousands of persons)	Trade (exports)*	
	1990	*2013*				*1990*	*2011*
USA	509	294	304	Temperate/ plantations	1,172	20,657	37,062
Russian Federation	292	180	809	Temperate	752	6,909	10,206
China	90	169	205	Tropical/ plantations	4,833	1,862	43,881
Brazil	145	149	520	Tropical/ plantations	1,036	3,905	9,693
Canada	162	147	310	Temperate	299	31,965	28,245
Sweden	53	63	28	Temperate	114	13,013	20,903
Indonesia	38	63	115	Tropical/ plantations	637	6,384	10,107
India		50		Sub-tropical/ tropical/ plantations			
Finland	43	49	22	Temperate/ boreal	82	10,605	15,635
Germany	85	42	11	Temperate	444	16,261	42,381
World	1,954	1,737	4,033		17,843	208,824	421,160

Note: * Forest products + wooden furniture. Trade in constant US$ millions (2011).

Source: FAOStat (2014, 2015).

Table 9.2 Top 10 producers of roundwood worldwide in 2013.

Country	Total roundwood production (million m³)
India	357
China	347
USA	334
Brazil	269
Russian Federation	194
Canada	148
Indonesia	118
Ethiopia	108
Democratic Republic of Congo	85
Nigeria	74
World	3,591

Source: FAOStat (2104, 2015).

picture emerges (Table 9.2). The difference between Tables 9.1 and 9.2 reflects the amount of wood that is harvested that does not enter the industrial roundwood sector. Instead, almost half of all roundwood harvested is used as fuelwood. Much of this fuelwood is harvested within the so-called informal sector.

Data extending back two and a half decades show that global industrial roundwood harvest levels have remained relatively steady, decreasing slightly from 1,954 million m³ in 1990 to 1,737 million m³ in 2013.

FAO statistics do not distinguish between roundwood harvested from plantations and roundwood from other sources, such as natural forests. As such, this data masks a shift between the two sources, and it is estimated that about two-thirds of the timber supply now comes from planted forests (FAO, 2010). The following discussion about employment levels, economic activity, and trade aggregates both sources, even though it would be useful to make the distinction, as done by some countries in their State of the Forest reports. One approach, as a first approximation, would be to ascribe the statistics on a proportionate basis by country. At the global level, this approach would show a diminished economic contribution from natural forests, reflecting the increased timber supply harvested from plantations.

Changes in employment levels

While global harvest levels have remained relatively unchanged in recent decades, forest sector employment decreased from just over 19.9 million to 17.8 million between 1990 and 2011 (FAOStat, 2014). Two of those sectors, logging and related services and wood processing, involve activities closer to the forest resource because of the nature of production activities, and hence can be indicative of what is happening at sub-national levels in forest-dependent regions. Trade in industrial roundwood has been increasing, but is still small relative to overall harvest levels. In the pulp and paper sector, increases in the use of recovered fibre and trade in pulp and raw material have attenuated what were historically close links between where fibre was being sourced and where it was subsequently processed. When broken down by subsector, employment in logging and related services has fallen by 84%, employment in the furniture sector has risen, and employment in the other two areas has remained approximately constant (Figure 9.2).

Figure 9.2
Total
employment in
forestry by sub-
sector. (Source:
FAOStat (2014).)

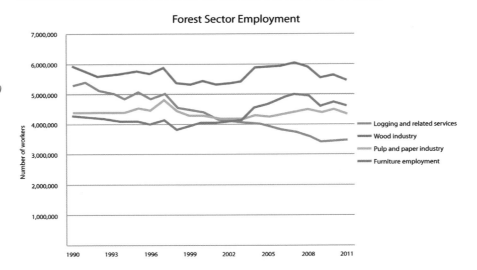

Figure 9.2
Total
employment in
forestry by sub-
sector. (Source:
FAOStat (2014).)

Figure 9.3
Self-loading
logging truck.
Advances in
equipment in
the logging and
related sectors
have resulted
in reductions
in the number
of people
employed.

Regional patterns reveal that employment levels have fallen by 41% in Europe, 31% in North and Central America, and 14% in Oceania between 1990 and 2011, while increasing slightly in South America, Asia, and Africa (25%, 18%, and 18%, respectively) with respect to the 1990 values.

Technology

Examples of labour-saving technologies include the mechanization of what were previously manual jobs in areas such as felling and the introduction of automation and control equipment into manufacturing, reducing the need for workers (Blombäck *et al.*,

2003). The impacts of labour-saving technologies have been most evident in logging and related forest services.

Considerable care should be taken when considering the effects of technological change on employment numbers. In Figure 9.4, two workers are shown grading lumber. The installation of sensors that perform this task would not only have replaced them, but also the graders working on other shifts. For a mill running three shifts and also operating on weekends, this could mean the loss of six full-time jobs and six part-time jobs. In areas where labour is cheap, there may be a reluctance to invest in labour-saving technology, not only for economic reasons, but also because providing employment may be a deliberate strategy (especially in government-owned enterprises).

Figure 9.4 Technology has also resulted in significant reductions in the number of employees needed in the processing sector. The manual grading being performed in this photo has now been replaced by sensors in modern sawmills that grade the lumber automatically.

Figure 9.5 Moving a log manually at a log sort in Hebei, China. This photo, taken in 2015, shows that technological solutions are not always adopted. In some cases, this is because the technology is too expensive; in others, it is because of a desire to provide employment opportunities.

Planted forests

Another significant development in forestry has been the emergence of planted forests as an increasingly important source of harvested timber. There is a long history of intensively managed and planted forests in some regions, particularly in Europe (see Chapter 1), and their subsequent expansion has been pursued as a strategy to promote economic development in many regions of the world. The financial returns from planted forests often exceed those of natural forests, particularly in more northerly temperate regions (Sedjo, 1999). These returns have been increasingly recognized and embraced by the private sector, with planted forests now considered an attractive investment class in their own right. Supported by both public and private investments, the area of planted forests has grown rapidly in recent years, especially in the Southern Hemisphere and within tropical regions. Planted forests now cover 290 million ha, accounting for 7% of the world's forest area, and their area increased by 110 million ha between 1990 and 2015 (FAO, 2015).

The expansion of planted forests offers economic benefits, as yields are higher under planted forests compared to natural forests. However, planted forests can negatively impact forest communities in a number of ways, including displacing local people who might have been dependent on lands classified by forest authorities as degraded and, therefore, available for conversion to planted forests. Planted forests can also involve the conversion of natural forest, a problem that is particularly apparent in Indonesia and other tropical countries. The development of planted forests can also reduce access to forest resources and economic opportunities and can trigger changes in land ownership that have negative socio-economic consequences (Charnley, 2005). Bull *et al.* (2006), in their review of global subsidies for planted forests, note the potential for such subsidies to lead to the conversion of natural forests, with subsequent negative impacts. They also highlight the important effect that increased supply coming from planted forests can have on forest product prices.

Planted forests have created employment opportunities, both directly in the activities associated with the planted forests and in downstream activities (prominent examples

Figure 9.6 Farmhouse abandoned in favour of the development of a planted forest, Fujian, China.

are the development and rapid expansion of the pulp and paper industries in Brazil and Chile). This explains why the Americas was the only region where employment grew, and why employment has remained relatively steady in tropical countries relative to temperate countries (Figure 9.2). However, just as planted forests can contribute to regional and local economies, they can also have an impact on other forest economies through interlinked markets and trade. For example, increased supply coming from planted forests can affect product prices. Bull *et al.* (2006) found that prices for pulpwood from natural forests in countries with expanding areas of planted forests fell to match the prices of pulpwood from planted forests. At the global level, increased supply of products from planted forests through trade can also affect global forest product markets, decreasing prices for forest products and hence for logs (Figure 9.7), holding all other factors constant. Most timber supply studies investigating the expansion of planted forests either look at global changes in harvest levels and sources of timber supply, or more localized regional impacts. Li *et al.* (2008) considered the impact of eliminating illegal logging (calculated at 6%–12% of industrial roundwood supply in 2004), and estimated that by eliminating that additional supply, world roundwood prices would rise by 1.5%–3%, there would be increased production in temperate regions (which have more efficient industries), and value-added activities in those regions would rise by 2%–4%.

The emergence of plantations can also affect the nature of production processes, with potential socio-economic consequences for forest-dependent peoples and communities. Planted forests are capital intensive, which can put smaller landowners at a disadvantage. Charnley (2005) found that, in the US South and elsewhere, the expansion of planted forests and the forest industry led to land consolidation, as smaller landowners were unable to compete. Reyes and Nelson (2014) noted similar negative impacts from plantations in Chile, where there were also reduced employment opportunities for locals, as firms brought in contractors from outside the region because of lower costs. However, Charnley (2005) argued that in Australia, planted forests have had more positive effects, increasing income for landowners.

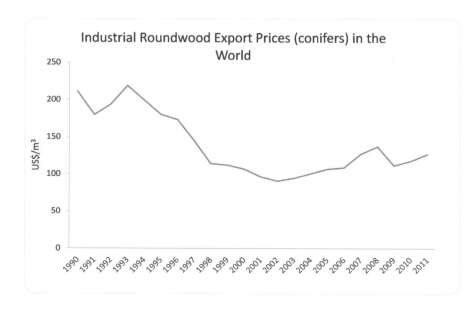

Figure 9.7 Real roundwood conifer prices, 1990–2011. Note: prices estimated from FAO Stats and deflated by using world GDP deflators (World Bank Database, 2011 = 100). (Source: FAOStat (2014).)

Changes in value of the forestry sector to economies

Another important indicator related to socio-economic contributions is the value added by the forest sector to national economies. Value-added statistics, in addition to showing the importance of forestry to national economies, can also shed light on the income being generated within the different sectors. Value added, as used in national economic accounts, is measured as the difference between revenues and the costs of all inputs purchased from other sectors. That value added is then income available to be paid to the various factors of production, including wages paid to employees, returns to capital, any returns to land or forest charges, and profits. How that income is then distributed depends on how the various factors are utilized and the costs of those factors of production.

Between 1990 and 2011, global GDP, which can be measured through value added through various parts of the economy, grew by 76%, but total value added in the forest sector remained relatively constant, growing only slightly (Figure 9.8). As a consequence, forestry's share of world GDP has decreased since it did not grow as much as value added in other sectors in the global economy.

There has also been a shift in that the location of forestry activity, driven mainly by increased exports coming from China (see Table 9.2). This has influenced the steady growth of Asia's share of international forest product sales since 2000 (see Table 9.3).

Figure 9.8 Total value added in the forest sector. (Source: FAOStat (2014).)

Table 9.3 Changes in value added by region during the period 1990–2011 (US$ millions, 2011).

Region	Logging and related services	Wood industry	Pulp and paper industry	Furniture industry	All sub-sectors
Africa	4,129	228	–1,130	–552	2,675
Asia	27,312	29,871	52,666	12,600	122,449
Oceania	1,170	299	–870	–1,138	–539
Europe	–2,694	–16,966	–19,683	–23,478	–62,821
North and Central America	4,468	–9,167	–16,489	–5,453	–26,641
South America	–1,291	1,924	4,563	1,217	6,411
World	33,094	6,189	19,056	–16,803	41,535

Source: FAOStat (2014).

Value added in forestry is now lower in North America, Europe, and Oceania than it was in 1990, while it has increased in all other regions, most notably in Asia. Table 9.3 shows changes in forest-based value added by sector and by region. In Europe, it has decreased across all sectors, while it fell in all of the manufacturing sectors in North America, with the exception of logging and forestry-related services.

Trade

A third indicator of socio-economic contributions is trade. Trade statistics are important for two reasons. First, national governments commonly promote trade because of the associated economic benefits, including export earnings which are seen as providing positive net benefits to the national economy. Second, changing trade patterns are also indicative of more structural changes in the world economy, with globalization now having integrated local and national markets with international markets. This has affected the competitiveness of the forest sector in many countries, where countries that were once traditional manufacturers of forest products have seen their market share shrink, as either new products or new suppliers emerge, and production levels have subsequently fallen along with a corresponding decrease in economic activity. This can be seen in changes in trade balances in the developing countries in Figure 9.9, which have historically been exporters of forest products.

Figure 9.9 shows that, while trade balances have remained positive in the developing world, they have been declining in Europe, and recent rebounds in North and Central America have offset some of the previous declines that took place in those regions. In the developing world, Africa's trade balance has been deteriorating, while Asia shows a large trade imbalance, principally due to large volumes of raw materials imported into China.

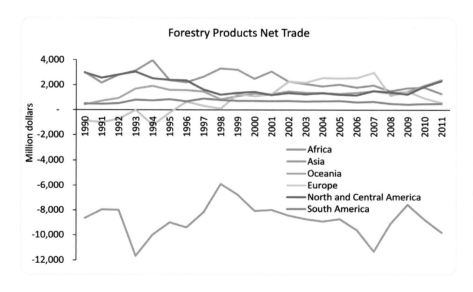

Figure 9.9 Forestry products net trade (exports minus imports) by region (in constant US$ 2011). (Source: FAOStat (2014).)

*Figure 9.10
Logs stacked for
export to China
at the port of
Astoria, Oregon,
USA. The export
of logs is a direct
consequence of
removing trade
barriers, but may
cause problems
locally if mills
are unable to
get sufficient
numbers of logs.*

*Figure 9.10
Logs stacked for export to China at the port of Astoria, Oregon, USA. The export of logs is a direct consequence of removing trade barriers, but may cause problems locally if mills are unable to get sufficient numbers of logs.*

Economic and social impacts

In summary, there have been several significant changes affecting the forest sectors of all countries, and hence, their socio-economic contributions in terms of employment and income. Technology, where it has been adopted, has increased labour productivity, reducing the need for workers across all the different sectors of forestry, while increasing globalization and modernization are changing patterns of forest use and trade (Nikolakis and Innes, 2014). The emergence of plantations has reduced pressures on natural forests elsewhere by intensifying production on a smaller land base, but has also changed the competitive landscape in the forest sector. These shifts in production patterns, and increased reliance on fast-grown timbers, have also had other consequences, namely economies of scale favouring large-scale production, whether in growing timber or in manufacturing forest products. Where forest plantations are being established or expanded and where there are existing forest-dependent communities, they can potentially trigger changes in land ownership, industry structure, and local markets.

Overall, globalization has meant that forest products prices have either fallen or grown at less than the rate of economic growth, reducing relative returns. This has led to lower economic margins that then lower potential returns to forest lands, particularly as firms face increased capital requirements to remain competitive. Reduced margins also mean that wage levels have not kept pace, making it difficult to attract workers in developed countries, a concern expressed in both Europe and North America (FAO, 2003). With the growing importance of plantations, the net economic benefits from natural forests are even smaller today than they were two and a half decades ago. Indeed, this combination of factors, including external competition, has challenged traditional forestry business models in the developed world based on the harvesting of natural forests. The net effect has been a redistribution of economic activity, shifting from developed countries that have historically relied on natural forests to those countries, both developed and developing, utilizing natural forests. The result of these changes associated with increased demand for other values (which we discuss next) on the traditional or industrial forest industry reliant on natural forests is that the higher cost of extraction, combined with changed management emphases, has made the supply and potential contribution from the natural forests less predictable. These changes have even greater consequences for local forest communities, since there are fewer jobs and less income now generated from those forests. In Europe, Blombäck *et al.* (2003) note that the combination of increasing productivity, coupled with stable or steady harvest levels, are likely to result in decreased employment opportunities in rural areas, where forestry has been historically important.

While economic contributions from the formal forest sector have diminished in relative importance in many regions, this has been offset by growing awareness and recognition of the importance of the benefits from the informal sector. This includes non-timber goods and services for consumption, income generation, and employment. While data regarding these non-timber goods and services is more sparse than timber harvest data, a recent report estimated that incomes in 2011 (for both forest owners and workers) from non-wood forest products (NWFPs) was: US$ 88 billion; US$ 2.5 billion in payments for ecosystem services; and US$ 33 billion for wood fuel and construction materials (FAO, 2014). NWFPs are discussed in more detail in the following section, in particular emphasizing their importance in the developing world, where their contribution far exceeds that from the formal forest sector within those countries.

Value of non-timber forest products

For western forest economists steeped in the industrial model of forestry, non-timber forest products (NTFPs) include fuelwood, smallwood, and other NWFPs that are harvested and used by people in all countries around the world both as a food source and to generate income, meet cultural needs, and diversify income sources (Shackleton *et al.*, 2011). The FAO has a much more appropriate definition, stating that NWFPs are 'goods of biological origin other than wood derived from forests, other wooded land and trees outside forests' (FAO non-wood forest products website: http://www.fao.org/forestry/nwfp/6388/en/). According to this definition:

> NWFP may be gathered from the wild, or produced in forest plantations, agroforestry schemes and from trees outside forests. Examples of NWFP include products used as food and food additives (edible nuts, mushrooms, fruits, herbs, spices and condiments, aromatic plants, game), fibres (used in construction, furniture, clothing or utensils), resins, gums, and plant and animal products used for medicinal, cosmetic or cultural purposes.

As shown in Table 9.4, many forest-based plant products are processed and used as inputs for manufacturing pharmaceuticals, cosmetics, food additives, industrial enzymes, bio-pesticides, and personal care products. An estimated 80% of the population in most developing countries is believed to rely on traditional, plant-derived medicines (Rao *et al.*, 2004) (Box 9.2). According to the IUCN Plants for People website (http://iucn.org/about/work/programmes/species/our_work/plants/plants_projects_initiatives/plants_for_people_/), the value of the global trade in medicinal plants is estimated at US$ 60 billion annually. In 2005, the total reported value of removals of all NTFPs was US$ 4.7 billion (Agustino *et al.*, 2011). Plant products (US$ 3 billion), including those for food (US$ 1.3 billion), accounted for the bulk of this value. Asia (US$ 1.7 billion) and Europe (US$ 1.8 billion) reported the highest market values; however, lack of data in other regions is likely to produce underestimates of total value generated. This compares with a total value of international trade in NTFPs estimated at US$ 11 billion.

Table 9.4 Examples from around the world of the extent, size, and value of the trade in different NTFPs.

Product	Location and economic contribution
Medicinal plants	Bangladesh – some 12,000 tons of dried medicinal plants worth around US$ 4.5 million are sold annually from rural areas (SEDF/IC, 2003). Southern Africa – the trade in medicinal plants is valued at US$ 75–150 million per annum with some 35,000–70,000 tons of plant material traded each year (Mander and Le Breton, 2006).
Baskets	Botswana – commercial buying started in early 1970 in Ngamiland District. In that first year, US$ 500 worth of baskets was bought from a handful of women; by 1990, this increased to US$ 115,000 per year to more than 2,000 women. By 2000, the value of the trade was some US$ 350,000 per year.
Gums and resins	Ethiopia – the value of gum and resin exports from 2001 to 2003 amounted to US$ 2.8 million, US$ 3.3 million, and US$ 4.1 million, respectively. Natural gum tapping and collection activities create seasonal employment opportunities for 20,000–30,000 people (Roukens et al., 2005).
Woodcarvings	Kenya – the woodcarving industry is worth over US$ 20 million annually in export products and employs some 60,000–80,000 carvers supporting over 400,000 dependents (Choge, 2004).
Honey	Zambia and Tanzania are two dry forest countries exporting the largest volumes of honey. In Zambia in 2005, 219 tons of honey were exported with a value of US$ 491,000, while Tanzania exported 466 tons with a value of US$ 674,000. Volumes exported have risen by 20%–30% since 2001 (ITC, 2006).
Oils – shea butter	Burkina Faso – shea butter provides income to about 300,000–400,000 women (Schreckenberg, 2004). Imports of shea butter to Europe from Sahelian countries were estimated at US$ 13 million in 1999 (Schreckenberg, 2004).
Insects	Botswana – the trade in mopane worms was valued at UK£ 4.42 million in 1995 and employed as many as 10,000 local people (Styles, 1995)
Wood and charcoal	Tanzania – in 2002 some 21.2 million cubic metres of wood, equivalent to 625,000 ha of woodland, were used for 43.7 million bags of charcoal with a net annual value of US$ 4.8 million (Scurrah-Ehrhart and Blomley, 2006).

Adapted from Shackleton *et al.* (2011).

The dearth of aggregated data on NTFPs and the lack of information systems that integrate this data into regional, national, or international decision-making have resulted in chronic undervaluing of their economic contributions compared to formally recognized products, such as timber. While national governments maintain data about the value of some individual NTFPs that are traded through formal markets (e.g. Christmas trees, maple syrup), exchanges of many NTFPs are either not tracked or are traded for subsistence (i.e. in non-cash markets) (Agrawal *et al.*, 2013).

Governance systems routinely focus management activities on those products with the highest reported market values (i.e. timber resources) and, consequently, systematically neglect to maximize the potential monetary and non-monetary values derived from NTFPs (Agrawal *et al.*, 2013). For accounting purposes, it is possible to categorize NTFPs as 'tradable' and 'non-tradable'. Tradables can be exchanged through markets beyond the immediate vicinity of their collection; non-tradables cannot. Since tradables can be accessed by a much larger market, they are subject to greater fluctuations in demand based on consumer preferences or the shifting availability

*Figure 9.11
A typical
non-timber
forest product:
beeswax (in
this case from
Lesotho). Apiary
products are
better quantified
than most non-
timber forest
products as they
are often sold in
formal markets.*

of substitutes (Agrawal *et al.*, 2013). Non-tradables are somewhat protected from these fluctuations, but may be supplanted if new foods or materials become available. The United Nations Environment Programme (UNEP, 2013, p. 146) reports that global market demand for tradable NTFPs continues to increase, spurred by growing demand from consumers in developed countries and among Asia's middle classes for natural and organic products in sectors such as cosmetics, pharmaceuticals, and foodstuffs. Trade in goods via the Internet is reported to have increased consumer access to NTFPs from around the world. Dietary supplements, particularly those promoted to combat aging, are favoured by the baby boomer generation, while 'superfoods' (e.g. those containing antioxidants) are preferred by the younger generation (UNEP, 2013).

In addition to their market value, non-tradable NTFPs make an important contribution to food security in developing countries across Asia, Africa, and Latin America. Paumgarten and Shackleton (2011) found that women and children are the main consumers of wild foods, and Shackleton *et al.* (2011) reported that women, the elderly, and the less educated benefit most from the use and sale of NTFPs.

While subsistence and commercial harvesting of NTFPs has the potential to support sustainable livelihoods and even create incentives for the conservation of valuable species and their habitats, the unfortunate reality is that, once a significant market for an NTFP is discovered or developed, the pressure to overharvest intensifies (Shackleton *et al.*, 2011). Compounding factors include poverty, food insecurity, and environmental stressors. Two multi-case study research projects from the Centre for International Forestry (CIFOR) and the UNEP showed that, in almost all cases examined, non-cultivated NTFPs declined following commercialization (Belcher *et al.*, 2005). However, hope remains that, by complementing wood production, the production of NTFPs can motivate a shift towards more sustainable management of global forests (Agustino *et al.*, 2011).

Several organizations have established product certification schemes as a way to encourage and promote sustainable production of NTFPs. Four categories of certification schemes – forest management, social, organic, and product quality – have been developed to signal the sustainability of NTFPs to consumers (Walter *et al.*, 2003). In particular, the FairWild Foundation, established in 2008, offers its Fair-Wild Standard and certification system for wild, foraged products to protect vulnerable plant species and local ecosystems and safeguard the livelihoods of collectors (see http://www.fairwild.org/). The standard was developed in accordance with the International Convention on Biodiversity and the Non-Detriment Findings Process of CITES (the Convention on International Trade in Endangered Species of Wild Fauna and Flora).

Box 9.2 Medicinal and aromatic plants (MAPs) in Nepal.

UNEP (2012) identifies Nepal as a country with a natural wealth of medicinal and aromatic plants (MAPs) that could support economic development. Approximately 700 species of medicinal trees, plants, and fungi grow in Nepal, of which 250 are endemic (UNEP, 2012). These plants include kutki (*Picrorhiza kurroa*), chiraito (*Swertia* spp.), lauth salla (*Taxus* spp.), yarcha gumba (*Ophiocordyceps sinensis*), panch aunle (*Dactylorhiza hatagirea*), pakhanved (*Saxifraga ligulata*), harro (*Terminalia chebula*), barro (*Terminalia bellirica*), amala (*Phyllanthus emblica*), and neem (*Azadirachta indica*) (UNEP, 2012), although not all of these are forest products. For example, yarcha gumba is restricted to the alpine zone, well above the treeline. Of the nine species of *Swertia* that are harvested, overharvesting of *S. chirayita* has already led to its severe depletion. Only 10% of manufacturing occurs in Nepal; the majority of value is added to MAPs in India. Market prices of the highest value plants have risen in recent years, leading to overexploitation. In 2009, the estimated value of MAP exports was US$ 9.8 million (UNEP, 2012).

Singapore accounts for the largest proportion of exports, by value, largely because it imports very high-value products. India is the second largest consumer of MAP exports from Nepal. Nepal continues to suffer from widespread unemployment, particularly among disadvantaged populations and in rural areas – the same areas in which many valuable MAPs are harvested (UNEP, 2012). UNEP has proposed several measures that may help these populations to strengthen access to markets, including: raising awareness about sustainable harvesting practices; conducting regular inventories of resources; conducting research to identify varieties for cultivation; linking research, development agencies and businesses; supporting adoption of international sanitary and phytosanitary standards; improving producers' access to credit; developing transparent, fair marketing channels; and ensuring royalties paid to collectors align with market prices.

Production and trade of MAPs is widely believed to have high potential to reduce poverty, particularly among women and disadvantaged people in rural areas of developing countries (UNEP, 2012). Unfortunately, harvesting of MAPs in many countries remains unregulated, and unsustainable practices risk degradation of critical ecosystems that produce MAPs, particularly those that fetch a high price in the market.

Figure 9.12 Insect larvae are an important part of the diet of many forest-dwelling people. The ones shown here, from an unidentified species, are from the forests of the Congo basin. It will likely be some time before these become a major export product to western countries.

A recent FAO report estimated that, in 2011, total income from NTFPs, including animal-based, medicinal plants, and plant-based NTFPs – excluding insects, which contribute to the traditional diet of at least 2 billion people (van Huis *et al.*, 2013) – was US$ 88 billion (in 2011 prices). Income from the informal production of wood fuel and for home construction in 2011 was US$ 33 billion (2011 prices) (FAO, 2014). The report also estimated the benefits of consumption for these different forest goods, reporting consumption, percentage contribution to food supply, the amount of energy from wood fuel, and people using wood fuel (FAO, 2014). These further expand our knowledge of the range of benefits, as well as the people, benefiting from forests.

Small-scale forest enterprises

Historically, dialogue on the socio-economic contributions that the forest sector makes has been dominated by the interests of large export- and commodity-oriented corporations. This is not surprising given the immediate and measurable impacts that 'big business' has, and it goes a long way in explaining why much of the global discourse, lobbies, and even research efforts have focused on business models that perpetuate the larger-scale production of forest goods (and increasingly services). While directing efforts here makes some conceptual sense, oftentimes lost in this forestry landscape are the massively important socio-economic contributions made by smaller forest-based enterprises. It is the recognition of these broader benefits that has reengaged decision-makers' interest in finding ways to facilitate these kinds of enterprises.

Small-scale forest enterprises, sometimes known as small and medium forest enterprises (SMFEs), are small and medium-sized enterprises (SMEs) that occur within the specific context of the forest sector. According to most global definitions, SMEs have fewer than 250 employees, but definitions in the forest sector have evolved to incorporate more explicit metrics of enterprises employing fewer than 100 workers, consuming no more than 20,000 m³ of roundwood annually, having fixed capital investments of less than US$ 500,000, and annual profits of no more than US$ 30 million, but typically much less (Mayers, 2006). Most small-scale forest enterprises are, in fact, very small, and definitions can also include microenterprises, where: activities occur largely at the household or individual levels; employment comes from family members, friends, and neighbours; and salaries are generally inconsequential. In line with this, Mayers (2006) estimates that approximately 80% of the funding for small-scale forest enterprises comes from the owners of the enterprises, or their friends and families.

Small-scale forest enterprises can take on a variety of forms – from privately held enterprises and partnerships to community-owned enterprises and producer associations. Typically serving local and domestic markets by providing much-needed forest-based goods, they produce a wide range of forest-based products and services, spanning the spectrum of timber-based products (such as commodities, value-added goods, and bioenergy), NTFPs (such as foods and medicines), and increasingly ecosystem services (such as carbon and ecotourism). Like most forest sector businesses, employees of these enterprises generally rely on forest-based activities as their primary sources of income. However, one important feature that distinguishes small-scale forest enterprises is that they tend to be vested in the communities within which they operate and provide a practical alternative to the large-scale Fordist business model that focuses on global exports of commodity wood products like lumber and pulp and paper (Donovan *et al.*, 2006). From a social point of view, this means that small-scale forestry has the potential not only to provide forest goods and services sustainably, but to do so in a manner that can 'generate wealth that stays within the communities, provide local employment and improve the livelihoods of the rural poor, and directly address questions related to the restitution of customary and indigenous tenure and access rights' (Kozak, 2007, p. 3). As such, small-scale forest enterprises are increasingly being upheld as a viable and tangible means of concurrently addressing tenure issues, poverty, and unemployment for impoverished forest-dependent communities around the world.

For these reasons, small-scale forest enterprises are being endorsed – by civil society and policy-makers alike – as a potential means of alleviating poverty and providing meaningful employment in developing economies, but their socio-economic contributions in the developed world should not go unnoticed. Small-scale forest enterprises contribute over one-third of all forest-based employment in the USA, while it has been estimated that 90% of the wood processing firms in the European Union employ 20 or fewer workers (Hazely, 2000). In fact, it can be argued that the promotion of small-scale forest enterprises in these regions is fast becoming a pressing issue and a means of mitigating against declining employment levels, as the forces of globalization and an increasingly competitive global business climate alter the dynamics of the traditional forest sector.

There is a distinct lack of economic information about the contributions that small-scale forest enterprises can and do make, similar to that observed for other forest-based activities as discussed earlier. In part, this is due to government agencies not taking an active role in data collection, but is exacerbated by many small-scale forest enterprises operating within the informal sector, especially in developing economies. The informal sector refers broadly to the production of market-based goods and services, where incomes go unreported to authorities in order to evade taxation and/or costly and time-consuming registration protocols. More specifically, the informal sector comprises a range of activities, from manufacturing and selling value-added wood products in local wood markets to illegally harvesting fuelwood for subsistence. The most recent estimates put the number of people in the world collecting fuelwood/charcoal at approximately 840 million (FAO, 2014). The size of the informal sector itself is estimated to be 41 million people, although some estimates have placed it as high as

Figure 9.13
A small-scale
sawmill near
Tsukuba in
Japan, employing
fewer than five
people (although
the owner-
operator's family
includes the local
forester who
supplies logs to
the mill, and a
local builder,
who purchases
the lumber).
Small-scale mills
such as this are
very common,
yet have received
much less than
attention than
large-scale
producers of
dimensional
lumber.

Figure 9.14
Bags of charcoal
stacked by
the roadside,
Samburu, Kenya.
The development
of a new road
from Nairobi has
enabled this area
to be opened
up for charcoal
production by
local people, but
current levels of
production are
not sustainable.

140 million individuals. In either case, the informal sector in forestry – much of which consists of small-scale enterprises – is substantial. This has two important implications. First, the exact socio-economic contributions that small-scale forest enterprises make are unknown, as this information is rarely captured in national-level statistics. Second, a good deal of thinking needs to take place on whether or not it makes sense – from a socio-economic point of view, at least – to legitimize and legalize this sector (Kozak, 2009).

Even excluding the informal sector, there is much debate on the precise economic contributions made by small-scale forest enterprises. Some country-level data exists – for sub-Saharan Africa, Southeast Asia, the Amazon region of South America, and Central America – but it is scant and, in many cases, dated. Globally, it has been estimated that some 20 million individuals are formally employed in small-scale forestry operations, a sizable portion of whom work in small wood processing facilities. In many countries around the world, small-scale forest enterprises account for more than 80% of all forest-based business activities, and gross value-added contributions by this sector are estimated to exceed US$ 130 billion (Mayers, 2006). Importantly, the SME industrial segment – which includes small-scale forestry enterprises – has been characterized as robust and 'one of the faster-growing industrial sectors in the world' (Canby, 2006, p. 4).

Despite the considerable socio-economic contributions that small-scale forest enterprises make, there remain issues with respect to their ongoing success as a viable business model within the forest sector. In particular, three recurring themes are cited as common constraints to success: the smaller scale of operations, scarce access to raw materials, and a lack of business capacity and capital. Notably, many of these barriers are issues for small-scale forest enterprises in both developed and developing forest economies.

The small scale of SMFEs makes it difficult, if not impossible, for entrepreneurs engaged in these activities to compete for market share against large-scale business interests, multinational corporations, and concessionaires that dominate the global forest sector, largely with the sale of low-value, commodity forest products to export markets. Current thinking suggests that this is in fact a fool's errand, and that a more prudent strategy would be to concentrate efforts by manufacturing and selling higher-value forest products and services to increasingly lucrative local and domestic markets (Kozak, 2009). The argument here is that SMFEs can work much more closely with local customers, tailoring products to their needs. Another possible means for small-scale enterprises to scale up revolves around the creation of formal business associations, informal business networks, and clusters. Such aggregations of entrepreneurs have been shown – through articulating common goals and achieving collective efficiencies – to help small-scale forest producers lessen the burden of scale by providing, for example, access to markets that would otherwise be unattainable, improved institutional support, opportunities for cost-sharing and information exchange, environments that are conducive to building meaningful supply chain relationships, increased access to capital, and lowered risks (Biggs and Shaw, 2006). Moreover, these sorts of associations and networks can play an important role in policy reform by acting as advocates and the voice of small-scale forest entrepreneurs.

The issue of scale also directly relates to a scarcity of raw material supplies for small-scale forest enterprises. One of the most significant business barriers for small-scale forest enterprises is the inability to secure raw materials, particularly for those engaged in woodworking activities. Some of the major reasons cited include: an inability to pay the high and oftentimes fluctuating prices for logs; inaccessibility due to poor infrastructural conditions; restricted access to forests and low availability of 'in-demand' logs as a result of large-scale forest producers being allocated the highest quality logs and species; limited access to funds for maintaining sufficient inventories to propel businesses forward; and more broadly, ill-conceived forest policies and trade practices which favour large corporations (Arnold *et al.*, 1994). One possible solution is to facilitate and foster long-term alliances between small-scale forest enterprises and larger forest products companies, the idea being that the small businesses could conceivably utilize lesser-known species (which large companies typically do not want) to manufacture high-value wood products (Kozak, 2009).

Last, but not least, many small-scale forest enterprises around the world lack two critical elements required to ensure business success: business skills capacity and working capital. Small-scale forestry operations in developing economies generally lack capital, assets, and secure access to financing, instead relying on friends and family for funds and assistance in daily operations. There also seems to be a general deficiency in managerial expertise, entrepreneurial skills, and time, meaning that marketing plans, promotional strategies, and even simple bookkeeping practices tend to be ignored. This situation is by no means different for small-scale forestry enterprises in developed economies. For example, small and medium-sized enterprises in the Canadian value-added wood products industry face similar barriers related to financing for business expansion, market research, increasing market reach, and upgrading labour skills for employees, while opportunities abound for the provision of managerial training. Several strategies for overcoming some of these barriers – generally within the context of developing regions – have been put forward, including, for example, catalyzing government agencies to implement business capacity and support programs, enabling the provision microcredit and other financing schemes vis-à-vis group lending programs through business associations, promoting the formalization of the small-scale forestry sector, creating stable forest policy environments, and advocating the tenets of sustainable forest management (Tomaselli *et al.*, 2012).

Valuing forest ecosystem services and non-consumptive uses

The previous sections illustrate the diversity of goods that forests can provide for human consumption and the associated employment opportunities and income that they can generate. Sustainable forest management recognizes that forests provide a wide range of values to human society, and there has been growing interest in determining the total economic non-physical value of ecosystems. Estimates of the value of the other goods and services that forests provide beyond timber, paper, and fibre are in the trillions of dollars (UNEP, 2011).

The economic framework within which these contributions are assessed and evaluated is based on a utilitarian approach in which monetary values are measured as a way to create a common metric across different ecosystem goods and services, not all of which have a market value. In the classification of these ecosystem services, some forest services may provide indirect roles, such as regulating climate, reducing erosion, or providing flood control that benefit humans, while in other cases, humans may derive aesthetic, spiritual, or cultural benefits or even the maintenance of forest biodiversity without directly consuming those goods or services. The total economic value (TEV) framework provides a way of organizing these uses into different types, including use (consumption) and non-use (non-consumptive), and direct and indirect uses and associated values. However, not all of the components of the TEV can be directly added together, as some uses are exclusive.

There are several purposes to this TEV quantification. Beyond assessing the contribution of ecosystems to well-being, it is also meant to assist decision-makers with better information on the choices and consequences of different decisions by helping them understand how and why different actors use ecosystems in the way they do (and why it can lead to degradation). In the context of forestry, this economic analysis can inform the selection of forest management policies by illustrating their impacts on environmental values, or it can examine the impact of other policies around land use, agriculture, or renewable energy on forests. Generating such information is also useful in explaining the difference between individual incentives and socially preferred outcomes, where the distinction between a financial and an economic analysis, and at what level that analysis is performed, becomes important. A financial analysis examines the financial or cash return to whoever is undertaking the activity. In the context of forestry, this could be a forest owner, forest worker, or firm. It will use the prices and costs they face (even where there may be evidence that such prices or costs are not reflective of the true value), and focuses on the returns they receive from engaging in different activities, ignoring any costs that they do not bear. By contrast, economic analyses (such as cost-benefit analyses) adopt a broader approach. They not only consider these financial returns, but also consider whether or not any prices may be distorted. They extend the scope of analysis to consider the broader set of values, including the other benefits provided through such services as the provision of clean water, biodiversity, recreation, and climate regulation, to highlight just a few. They do so by constructing the total economic value of these goods and services. Total economic value encompasses all uses, including direct, indirect, consumptive, and non-consumptive uses. It is in this context that efforts have been made to quantify the value of forest goods and services, especially in tropical countries.

Analyses that include the broader picture often reveal that actions that are financially profitable are actually harmful to society. Conversely, actions that benefit society may not be in the best interest of private firms. Indeed, this distinction lies at one of the fundamental problems around sustainable forest management, namely that while there are positive returns from sustainable timber management, shorter-term timber harvesting is generally more profitable (Pearce *et al.*, 2003), especially in tropical countries. These types of analyses also reveal that different perspectives may yield different socially optimal outcomes. How the benefits and costs of different forest management strategies or exploitation are viewed at the national versus international level is one of the fundamental issues underlying efforts to develop a coherent international forest policy. Understanding these differences as a way of developing policies or mechanism to address these issues underlies much of the work in this area, especially around ecosystem valuation, including international efforts such as The Economics of Ecosystems and Biodiversity (TEEB), which is identifying the economic benefits of maintaining biodiversity as a way of reversing its decline (see http://www.teebweb.org).

Different techniques can be used to elicit these values. Some techniques depend on estimates of the replacement cost if those services had to be provided through built infrastructure (i.e. flood control). Others are designed to extract the value of certain environmental attributes from market transactions or proxies reflecting actual payments. The most widely used approach has been willingness-to-pay (WTP) surveys, especially for some non-use values, such as the desire to maintain species diversity in an undisturbed forest or maintain the option to visit such places, where values can only be derived through such surveys. However, there are a variety of techniques that can be used, evolving from what were originally straightforward contingent valuation surveys (asking people what they would be willing to pay for that good or service) to more complex survey instruments, such as choice experiments that ask people to choose between alternatives (where the attributes of those alternative are varied in a systematic way) as a way of assessing underlying values. There are also methods such as benefit transfer, where measurements from one area are used to assess a similar area. One consequence of such varied techniques, including differences in what is being evaluated (preservation versus restoration, whether it is for species or landscape protection, or even the size of the area under consideration), who the potential beneficiaries are, and what kind of change is being evaluated, is that there can be a wide range in estimates.

The challenge has been that, while such studies have proliferated, they appear to have had little effect in changing either policies or decision-making at the international level. While efforts continue to generate more information about these values, increasing attention is being paid to see whether or not these benefits could be turned into economic opportunities that would then allow owners or governments to either enhance or better capture these socio-economic benefits of forests while also support-ing SFM objectives. Different approaches towards recognizing those contributions are examined in the next three sections. The first two involve creating economic value out of goods and services that have not traditionally been part of the formal economy or previously recognized. The first of these is payment for ecosystem services (or PES), whereby either new markets (such as forest carbon) are created or the scheme links beneficiaries with the providers of such services. The second is the encourage-ment of non-consumptive uses of the forest through recreation-based businesses. The third approach involves changes in forest governance, including modifying rights, as a way of better aligning individual and community that rely on those forests with a more direct role in management. These each illustrate a different approach and, it should be emphasized, are not necessarily exclusive – indeed, they can be mutually supporting.

Ecosystem services

The use of market-based instruments to address environmental degradation and maintain and restore ecosystems and their functions has received considerable atten-tion in the past decade. PES in particular has gained momentum, providing financial incentives for the provision of ecosystem services. Such schemes are also receiving attention for the socio-economic benefits that they can provide, especially for marginal-ized or disadvantaged groups that rely on the ecosystems that provide those services. Wunder (2008) argues that well-targeted PES programs can deliver welfare gains to disadvantaged participants (although they are seldom very large) who exercise control over lands of strategic environmental value. However, to deliver those gains there are four influencing factors that reflect, in part, governance arrangements. These are *eligi-bility*, or the recognized right to land and the ability of participants to exclude others from areas; a *desire* to participate in PES (which factors in opportunity costs, trust, transaction costs, and perceived risks); an *ability* to participate in PES schemes, includ-ing skills, capacity, and agency; and finally *competitiveness*. These all affect the ability of disadvantaged groups to be efficient providers of ecosystem services.

Wunder (2008) identifies five criteria that must be satisfied for a PES scheme. It must (1) involve a voluntary transaction, (2) be for a well-defined ecosystem service, (3) be purchased by an ecosystem service buyer, (4) be from an ecosystem service provider, and (5) be from a provider who has secured the ecosystem service (conditionality). A number of schemes exist, some long-standing and others are fairly new. Examples of PES in forestry range from schemes involving payments to forest landowners (either by government or from third parties) to the provision of clean water, carbon sequestration, and maintenance of biodiversity, or several of these services at the same time. In 2011, total income from PES to forest landowners was about US$ 2.5 billion (in 2011 prices) (FAO, 2014). Yet, while there is considerable experience with PES activity around the globe, generally the preconditions to successfully implement PES programs are not well understood, progress has been slow, and PES has been mainly applied on private lands, with far fewer examples on collectively owned or government-owned lands.

Forest ecotourism

Tourism provides an avenue to monetize the amenity value of maintaining and enhanc-ing natural forests while generating employment and economic activity. In particular, ecotourism is seen as one way in which the values expressed by the developed world can be expressed through economic opportunities around forestry in both developing countries and developed countries.

It can be difficult to distinguish 'ecotourism' from 'tourism' because many urban and non-urban tourism industries depend on the quality of the natural environment to attract visitors. Ecotourism is a smaller segment of what may be called 'nature-based tourism': travel to relatively undisturbed or uncontaminated natural destinations (Gössling, 1999). Ecotourism is distinguished by being compatible with conservation goals, even supporting conservation through income and education, while minimizing threats to local cultures. Ecotourism typically offers a focus on local culture and wilderness and involves travel to destinations where flora, fauna, and cultural heritage are the primary attractions.

In the case of forest-based ecotourism, partnerships are developed between forest owners and tourism operators to provide goods and services such as supplies, equipment, accommodation, guides, and access. Of all forest-based industries, the ecotourism business model may make the strongest case for biodiversity conservation because revenues are biodiversity-dependent. Because it involves non-consumptive use of forest resources, ecotourism typically has low negative impact on the environment and is well-suited to achieving conservation objectives (Agustino *et al.*, 2011).

In places such as Kenya and Madagascar, ecotourism accounts for a large proportion of GDP. If supported by proper management and planning, ecotourism has the potential to alleviate poverty. Certification schemes can help ecotourism businesses demonstrate commitment to biodiversity conservation and meet growing demand for services in both developing and developed countries that has been driven by growing consumer awareness about threats to biodiversity (Ten Brink *et al.*, 2012). Certification is also an important tool to ensure that some of the benefits flow to local people.

Ecotourism can be a valuable source of foreign exchange, particularly for developing countries, and one of the most commonly cited reasons for establishing protected areas is to profit from ecotourism. Key benefits include the development

Figure 9.15 Lake Moeraki Wilderness Lodge at Arthur's Pass in New Zealand. There are many such lodges specializing in ecotourism, generally offered at a premium to more traditional tourist locations. While a niche market, operators have realized that such lodges attract a particular (and growing) group of higher-income customers.

and operation of lodgings and generation of government tax revenues, creation of new educational and training opportunities through interactions that expand social networks, and establishment of incentives for nature conservation through collection of user fees, development of natural products, and establishment of privately managed reserves on the periphery of publicly owned parks and protected areas. With an increasing number of tourists seeking out ecotourism opportunities in protected areas worldwide, the industry has attracted investments from private companies, governments, non-governmental organizations (NGOs), and the international community.

The global ecotourism industry is not without its problems. The term ecotourism has been appropriated by some traditional operators that do not operate in accordance with conservation, sustainability, or cultural objectives. Some ecotourism operations fail to become economically self-sustaining or to generate the expected conservation or cultural benefits, and some operations even generate negative impacts on wildlife and local communities, particularly where the number of tourists exceeds the environmental capacity of the location to accommodate them. Conflict has arisen over land use at the interface between agricultural and wild areas, where the opportunity cost of conserving forest land is measured against the value of cultivation (Miller, 2012). Subsistence farmers in developing countries, like Costa Rica, have complained of displacement in circumstances where national governments have forcibly seized land to set aside for parks. Definitions of ecotourism also typically do not address environmental or resource impacts of travel itself, such as the associated solid waste production, non-renewable resource use, and greenhouse gas emissions, which can produce negative environmental impacts in other regions even while supporting local conservation objectives. Despite these issues, the global ecotourism model has demonstrated tangible successes. Global estimates suggest that the total flow of revenues from developed to developing countries is in the range of US$ 28.8 billion per year (Kirkby *et al.*, 2011).

Figure 9.16 An ecotourism group at Lake Clark National Preserve, Alaska, USA. Such groups are typically small and aim to cause minimal impact.

Forest governance and rights

Another approach towards enhancing sustainable forest management involves changing underlying governance arrangements. Unfortunately, forests are not always managed sustainably, particularly where forests are treated as a 'common-pool resource' that can be accessed by any user. The phenomenon whereby the environment (and its associated ecosystem services) deteriorates through the individual actions of multiple resource users is known as the tragedy of the commons (Hardin, 1968). Resource depletion occurs when each individual resource user is motivated by the direct benefit to them of extracting the resource and the cost of collective resource degradation is delayed. Competition by resource users for desirable NTFPs, combined with an inability to exclude others from using the resource, can create incentives for rapid, unsustainable extraction in the pursuit of short-term benefits.

Given the importance of forests to local communities and the influence that rights and access can have on those communities, there have been sub-national efforts to formalize or strengthen those rights and the ability to participate. Perhaps the best known of these efforts are the widespread initiatives in many countries to promote community forestry.

With the emergence of decentralization as a major theme in forest governance and a strong push for empowering forest-dependent people, the establishment of community forests in many parts of the world has been widely suggested as a means of improving the socio-economic realities of forest-dependent communities. The participatory framework in which forest management decisions are made has the potential to address the issues communities deem most important. Greater access and control over forest resources allows forest users to draw greater benefits from the resources and make strides towards empowerment.

In an attempt to identify and use criteria and indicators for successful community forest enterprises, Macqueen (2012) studied a number of community forests in the developing world. He found that successful community forests need three enabling conditions to achieve success: accessible commercial forest rights, processes of enterprise-oriented social organization, and infusion of competitive business skills. While community forests were by no means universally successful in achieving greater socio-economic outcomes in their communities, successful cases met these conditions.

While there are a number of success stories in the push for increased community involvement in forest management, community forestry initiatives, especially in regions that have well-established large-scale forestry operations, are often plagued by top-down management where the interests of outsiders are placed above those of the community. Additionally, community forestry models are often based on large-scale logging practices, resulting in little to no consideration for local preferences or management goals. Elsewhere there may be capture by local elites that can worsen rather than improve inequality.

The broader forces affecting forests

This chapter has looked at how technology, globalization, and modernization have affected the economics of the forest sector and the subsequent socio-economic implications. These forces are also at work at a higher level, influencing how forests are valued and utilized. A widely held idea is that there are forest transitions, whereby countries pass through predictable stages in which forests are first exploited, degraded, and then recover as demand for the other values they provide increase as countries become wealthier (Mather, 1992). Such observations reflect the historical development of many developed countries with abundant natural forests, where they served first as an engine for economic development and provided the basis for the development of a large, established forest sector. More recently, Lambin and Meyfroidt (2010) have suggested that other factors, including the impacts of modernization, globalization, and trade, land tenure, and national policies, can also influence these outcomes, and Martin (2015) argues that it would be foolish to place too much reliance on the theory in the hope that deforestation will decrease over time.

Conclusion

The economic returns from the formal forestry sector have historically provided the basis for policies and institutions and have financed investment in both the resource base and in capital (both physical and human). These returns then generated the economic benefits that justified those investments, but they have been diminishing in importance in recent years. Despite this, the importance of the larger value of forests to society is increasingly recognized, including the socio-economic benefits from forests not traditionally considered in forest policy-making (FAO, 2014). The Rio+20 Outcome Document emphasizes the important contributions of forest ecosystems to sustainable development by providing goods and services that are environmentally sound, enhance food security and the livelihoods of the poor, and invigorate production and sustained economic growth. Supporting these is now a long-standing discussion on how to balance economic development with those environmental goals, including international processes articulating sustainable forest management, such as the Montreal Process.

While these discussions take place, globalization and modernization have had significant impacts on forests, how those forests are used, and the benefits that society derives from them, and continue to affect them. While these changes are not uniform, trade has linked markets and increased competitive pressures, technology has reduced the need for workers, and plantations are increasingly meeting the increased demand for forest products. One consequence is the diminishment of economic benefits captured from natural forests and where these impacts occur at the local scale. Evidence is also mixed as to whether or not increasing economic growth and wealth will necessarily result in more or fewer forests, or what the quality of those forests will be.

In the developed world, management paradigms have been shifting to favour sustainable forest management, but the economics have not. Reduced returns mean less incentive to invest in forestry, whether by the private sector or public sector, and forestry's contribution to rural economies, which historically justified policy attention and political consideration, has been diminishing. In the developing world, falling values for natural forests mean less income and greater likelihood of either degradation or conversion, either because of weak governance or national priorities placed on economic growth. At the same time, increased public demand has affected the availability and cost of harvesting wood from natural forests by increasing the amount of environmental regulation and scrutiny that wood products produced from natural forests receive. The result of these management changes associated with natural forests, and the markets for their products, is that they will make the supply and potential contribution from these traditional sources of timber less predictable.

Finding ways to finance sustainable forest management and increase the socio-economic contributions from natural forests beyond just the formal forest sector that has traditionally provided these benefits is important, since increasing the value of standing forests will help offset these pressures. But as highlighted in this chapter, poor statistics and a paucity of data limit our understanding of the full range of benefits and how to encourage more sustainable activities on both an individual and collective basis.

Further reading

Bauhus, J., vander Meer, P., Kanninen, M. (eds.) 2010. *Ecosystem Goods and Services from Plantation Forests*. London: Earthscan.

Heal, G. 2000. *Nature and the Marketplace: Capturing the Value of Ecosystem Services*. Washington, DC: Island Press.

Hyde, W. F. 2012. *The Global Economics of Forestry*. Washington, DC: RFF Press.

Innes, J. L., Hoen, H. F., Hickey, G. (eds.) 2005. *Forestry and Environmental Change: Socioeconomic and Political Dimensions*. Wallingford, UK: CABI Publishing.

Nikolakis, W., Innes, J. L. (eds.) 2014. *Forests and Globalization: Challenges and Opportunities for Sustainable Development*. London: Routledge.

Panwar, R., Kozak, R., Hansen, E. (eds.) 2016. *Forests, Business and Sustainability*. London, Routledge.

Pokorny, B. 2013. *Smallholders, Forest Management and Rural Development in the Amazon*. London: Routledge.

Sabogal, C., Guariguata, M. R., Broadhead, J., Lescuyer, G., Savilaakso, S., Essoungou, J. N., Sist, P. 2013. *Multiple-Use Forest Management in the Humid Tropics: Opportunities and Challenges for Sustainable Forest Management*. FAO Forestry Paper 173. Rome: Food and Agriculture Organization of the United Nations.

Wagner, J. E. 2011. *Forestry Economics: A Managerial Approach*. London: Routledge.

Zhang, D., Pearse, P. 2012. *Forest Economics*. Vancouver: UBC Press.

Chapter 10

Social, cultural and spiritual (SCS) needs and values

Janette Bulkan

Criterion 6.5 of the Montreal Process addresses social, cultural and spiritual (SCS) needs and values, and provides two specific indicators against which member countries can report. The first (Indicator 6.5.a) is quantitative, and focuses on 'area and percent of forests managed primarily to protect the range of cultural, social and spiritual needs and values'. The second (Indicator 6.5.b) is qualitative, seeking a description of 'the importance of forests to people'. These two indicators provide an acknowledgement of SCS values but, as is shown in this chapter, they fall far short of what needs to be understood, assessed and reported.

SCS needs and values in relation to the natural world have been recorded and reported for all societies in historical time. While social or material needs may appear to be plainly utilitarian, they are generally grounded in the worldviews and values of specific peoples. In other words, how a forest is perceived, what foods are deemed edible, what other products are extracted from it, how much is extracted and during which season, and whether parts or all of that forest are valued in a sacralised sense tend to be culturally determined. Values can be defined as 'the preferences, principles and virtues that we (up)hold as individuals or groups' (Chan *et al.*, 2012, p. 10). In turn, those values are assumed to influence decisions on environmental uses (including needs) (Dietz *et al.*, 2005) (Chapter 11 also overviews values as applied to public participation).

Understanding in a broad sense the past and current SCS needs and values of both majority and minority population groups globally, and possible drivers of change, are critical in a world of more than 7 billion people who are consuming ecological services 1.5 times faster than can be renewed by the planet's self-regulatory systems (Global Footprint Network, 2010).

This chapter surveys the SCS needs and values of Indigenous Peoples and Local Communities (IPLCs) and of majority societies, and examines why they are important. The impact of the capitalist mode of production on SCS values in general is considered. The safeguards for SCS needs and values, and the metrics used, in some intergovernmental and international processes are summarized. In a final section, the successes and challenges of protecting or sustainably managing the remaining natural landscapes infused with SCS needs and values are examined.

SCS needs and values of Indigenous and forest-dependent peoples

Martinez Cobo's working definition of Indigenous peoples for the United Nations (UN) in 1981 captures the importance accorded by them to SCS needs and values, a point that has often been less prominent in subsequent UN formulations:

> Indigenous communities, peoples and nations are those which, having a historical continuity with pre-invasion and pre-colonial societies that developed on their territories, consider themselves distinct from other sectors of the societies now prevailing in those territories, or parts of them. They form at present non-dominant sectors of the society and are determined to preserve, develop and transmit to future generations their ancestral territories, and their ethnic identity, as the basis of their continued existence as peoples, in accordance with their own cultural patterns, social institutions and legal systems.
>
> (Cobo, 1981, para. 362)

Indigenous peoples are estimated to number about half a billion, and traditional or forest-dwelling peoples an additional 0.8 billion, within the global total (Chao, 2012). Indigenous identity is rooted in 'ancestral territories', and mediated by 'cultural patterns' that have an ancient provenance. In other words, SCS needs and values are central to indigeneity. The decision on who is recognized as Indigenous varies among countries (USA and Canada included). Legal status as a member of a named Indigenous group can be determined by percentage of blood type. At the other end of the spectrum, self-ascription is sufficient in some countries. The Canadian Constitution includes the Métis people in the category of 'Aboriginal peoples' along with First Nations and Inuit peoples. The Métis, who maintain a distinctive ethnic identity, trace their ancestry to the marriages between European fur trappers and Canadian First Nations and Inuit women in the 18th century.

Traditional peoples are social groups or peoples who do not self-identify as Indigenous and who affirm rights to their lands, forests and other resources based on long-established custom or traditional occupation and use.

> [They] may be described as peoples who live in and have customary rights to their forests, and have developed ways of life and traditional knowledge that are attuned to their forest environments. Forest peoples depend primarily and directly on the forest both for subsistence and trade . . . In South Asia, Southeast Asia and Africa, for example, agriculturalists [who] have a long history of using forest produce and of regulating access to forest resources may not see themselves as different from the national population yet they claim rights in forests based on custom.
>
> (Chao, 2012, p. 7)

The principal distinctions between Indigenous and traditional peoples, then, are the shorter length of historical attachment to a particular territory, and part- or non-Indigenous ethnicity. Traditional peoples include the *caboclo* communities of the Amazon and the forest-dependent African populations of Colombia, Ecuador and Peru. Traditional peoples are social and political communities, with a stated attachment to particular territories, on which the survival of their specific cultures is dependent. Race and ethnicity are not the determining factors. As a leading Canadian Aboriginal lawyer has explained:

> In its recent decision on Métis Aboriginal rights recognized in section 35 [of the Indian Act], the SCC [Supreme Court of Canada] views Métis communities

as descendant communities of historical groups that are, as such, social and political communities and not 'racial' groups united merely by birth. The judicial test for proof of Aboriginal rights in Canada is based on the idea that those rights are vested in historical communities that are social and political in nature. The judicial conception of Aboriginal rights is predicated on the social and constitutional value of culture: not biological cultures evoked by the term 'racial groups' but human communities formed by social and political relationships.

(Chartrand, 2010, p. 133)

Not all IPLCs practice their customary or traditional lifestyles on a full-time basis. However, they maintain that their survival, continuity and inter-generational transmission of their Indigenous knowledges are dependent on the integrity of their traditional lands, and on recognition of their customary land rights by the nation-states in which they are generally minority and marginalized populations. Afro-Colombianos of Colombia, for example, assert that their culture will disappear if the mangrove forests become sites of industrial shrimp aquaculture; the transhumance practiced by the Saami of Scandinavia is similarly dependent on recognition by landowners of Saami customary rights for winter grazing of their reindeer herds on the lichens found on trees in old-growth forests.

SCS values are generally encoded in traditional ecological knowledge (TEK) systems that are passed on in stories, songs, ceremonies, everyday practices and other culturally specific forms. Key attributes of TEK systems are that they 'tend to be holistic, recognizing the connections and interdependence of everything' (Turner and Spalding, 2013, p. 29); and kincentric, or

the understanding that all of the other life-forms on Earth, from bears to salmon, from cedar trees to berry bushes, as well as even rivers, mountains and other geographic features, are sentient beings who are relatives, or kin, of humans, related to us both practically and spiritually.

(Turner and Spalding, 2013, p. 29)

These beliefs in turn are expressed in reciprocity towards other life forms, as 'in the "contingent proprietorship" model, in which rights of individuals and leaders to lands and resources is balanced by responsibility for sustaining resources and communities within their influence for future generations' (Turner and Spalding, 2013, p. 29). Unsurprisingly, for most of human history, the SCS needs and values of majority societies shared much with those of IPLCs – more ecocentric than homocentric, and ownership or access more communal than individual. The divergence from this pattern would follow from the privatization of land and the means of production, urbanization and the separation of the vast majority of society from a lived experience with the natural world.

SCS needs and values of majority societies in the historical record

The earliest expressions of SCS are on record from the early State-level societies of Mesopotamia, Greece, India and China/Japan. In many places rules were formalised from ancient times. There was an historical appreciation of natural forests, woodlands and trees as providers of goods and services – goods in the form of timber, poles, fuel, fodder, fruits and nuts, and extractives (resins, gums, latex); services such as shelter from weather and human enemies, perennial flows of clean water, and protection from avalanches, storms and floods. There was early understanding of the negative effects of overharvesting and forest degradation – for example, literature from ancient Greece, China and India about denuded hills losing topsoil after forest degradation and defor-estation (Grove *et al.*, 1998).

The valuing of forested landscapes as the source of cultural/religious inspiration and location for rituals, sacred landscapes, trees and groves have been reported from every continent: sacred groves in India and Africa; home of the masters of animal spirits for Indigenous peoples globally; in China, *feng shui* or sacred forests that can be traced back to the Song dynasty (960–1279) and which are still protected by local communities; and sacred forests in Europe as described, for example, in *The Golden Bough* (Frazer, 1958). Once viewed as relics of the past, sacred forests are increasingly recognized as protected areas in the present and important as part of a mosaic of man-aged and protected natural areas.

In Europe, before the Industrial Revolution that began in the mid-18th century, the dependence on forest lands was more direct for the majority of people. In England, for example, soon after King John signed the Magna Carta in 1215, groups began to assert their traditional rights to forests based on centuries of customary use (see Chapter 1). The Charter of the Forest did not create new rights but gave courts a reference point in hearing oral arguments about competing claims, as written laws gradually replaced a variety of local practices and unpredictably variable interpretations (Osmaston, 1968). This provided an early demonstration of what are now termed multi-stakeholder nego-tiations for recognition of respective rights and responsibilities. In Europe, there were legal controls over forests for some 500 years, from the 13th to 17th centuries, in parallel with mining controls. Those controls preceded the institution of forest man-agement principles overseen by Colbert in France from the 1660s, Cotta in Germany in the 18th century and Knuchel in Switzerland. The social needs and values of forests in Europe can be deduced from the gradual adaptations to cope with irregular seeding and intermittent or insufficient regeneration, from about 1550 onwards, including yield control by area in lowlands and by tree numbers in mountain selection forests.

In Europe as well as globally, where communal or customary rights to forests were respected and sometimes recorded in writing, there was a greater willingness to set communal benefits above individual gains when failure to agree and enforce forest pro-tection rules risked disaster to the whole community (Ostrom, 1990). One well-known and surviving example is provided by the *Bannwald* of Andermatt in Switzerland, which has been protected by village agreement since 1397 as a defence against winter avalanches. A wedge of forest on the steep slopes above the village splits down-coming avalanches and blocks or diverts the snow slides away from the village (Figure 10.1). A study in 2009 tested the effectiveness of this protection by modelling and confirmed its utility as well as the importance of maintaining the forest in good condition (Johann *et al.*, 2012). The Andermatt communal agreement became the basis for local laws and regulations in many mountainous countries. The safe situation of the village encour-aged winter tourism over hundreds of years.

During much of human history, access to forest lands was open to all social classes although usually under strict conditions. In State-level societies, there was a broad understanding that forests can usually provide multiple simultaneous benefits – royal hunting, commercial timber, local underwood and household non-timber forest prod-ucts (NTFPs). When different strands of the 'bundle of rights' are apportioned to distinct categories of persons, then those rights coexist and interact, rather than over-lap (Albion, 1926). Along with common features and differences, there was also the recognition of both the power and fragility of Nature, hence the need for management controls. There was an appreciation that some forests can be cropped rotationally like agricultural fields, as farm fallow cropping systems and woodland coppice systems are much alike. One rule was to impose different lengths of felling cycles for products of

Figure 10.1
The village of
Andermatt,
Switzerland (in
the foreground),
with the Bannwald
on the left.

Figure 10.2
A traditional
fence in Dorset,
UK. The wood
came from short-
rotation coppice.

different sizes: annual lopping for winter fodder, two to four years for osier beds for basketry (and similar cutting cycles of reed beds for roof thatch), three to five years for hazel coppice for sheep hurdles, seven years for poles for buildings, and three or more decades for standard trees for timber (Matthews, 1989).

Over time human ingenuity led to the overcoming of individual muscular weakness by organized application of energy to transform natural systems. The increased demand for wood fuel and charcoal, added to steadily rising populations from the Neolithic period (ca. 10,000–3000 BC), drove the deforestation and forest degradation first recorded in the Middle East and Western Europe and, later, in every part of the globe (see also Chapter 1). The expansion of monotheistic religions, and the use of selective passages from the Christian Bible by Western Europeans, provided ideological justification for the domination or the subduing of Nature to the needs and desires of colonial rulers. By the end of the 16th century, the European States were establishing colonies globally, justified by narratives of progress, 'improvement' and divine will (Weaver, 2006). Communal access rights to land were also weakened following the growth and expansion of the capitalist system from the 19th century, which was premised on constant expansion in the search for new sources of raw materials and consumers, and in which ownership of the means of production was privatized.

Gradually, as States formed (characterized by growing urban centres, with greater differentiations among distinct social classes, professions [guilds] and wealth), the supply chains linking people to forests have grown longer and more complex. Furthermore, the capitalist economic system that is premised on growth and consumption has become firmly established globally, despite half a century of warnings about the limits to growth in a world of finite resources (Meadows *et al.*, 1972). Social stratification, power and privilege facilitated the cordoning off of forested areas in both the West and the colonies for hunting or recreation. Hunting in the game parks that were established in East African colonies, for example, was reserved for Europeans. Native Africans hunting for bushmeat for subsistence, to control predators of their cattle and in fulfilment of a culturally important practice were categorized as poachers and criminalized (Adams, 2003).

SCS values as expressed in national parks and wilderness

In the USA from the 1870s onwards, social and cultural values in relation to forest lands found expression in the creation of national parks, generally established on the traditional lands of the Indigenous peoples who had been driven off or exterminated beforehand or often excluded thereafter (Colchester, 2004). The movement to protect the remaining American wilderness coalesced around the personalities of John Muir and Aldo Leopold. There is a vast literature by environmental ethicists and activists in the global North on the various philosophical frameworks used to analyse the dominant Western values towards the natural world. These include strong and weak anthropocentrism that place humankind's needs at the centre (Callicott, 2005), and biocentrism and the deep ecology movement that argue in favour of the intrinsic value of all forms of life. It became harder for societies in the global North to make the link between their modern lifestyles embedded in industrial societies and deforestation and forest degradation happening far away.

Criticisms of those dominant social and cultural value systems have come from environmental historians, ecological economists, Marxists and others in both the global North and South. William Cronon, in an influential and controversial book and in his articles, argued that wilderness was 'entirely a cultural invention . . . [or] cultural construction' of the USA. According to Cronon, 'wilderness posed a serious threat to responsible environmentalism at the end of the twentieth century', as it lulled people into imagining that they could continue with their environmentally destructive lifestyles while being able to get away periodically to pristine landscapes. Cronon further charged that 'celebrating wilderness' was 'an activity mainly for well-to-do city folks' (Cronon, 1996, p. 15).

From the mid-20th century, as forests grew back on abandoned farmland in the eastern USA and Western Europe, a dominant view was that environmental degradation was chiefly a problem in the developing world, generated by poverty and overpopulation. In response to news stories of faraway eco-disasters, a large section of the Western media and academia subscribed to neo-Malthusian arguments and tended to ascribe blame for the assorted ills of environmental degradation to over-population in the global South and a commensurate lack of environmental values. The argument was that the direct dependence of the Third World poor on forest access meant that they could not afford the luxury of protecting the environment. This idea was sometimes expressed as an 'environmental Kuznets curve', which proposes that 'environmental degradation displays an inverted-U shaped pattern over time. It is low prior to economic development, increases in the course of economic development, and then decreases when income (GDP) reaches a certain level' (Sunderlin *et al.*, 2005, p. 1389).

Marxists geographers in the global North, including Noel Castree, David Harvey (1996) and Neil Smith (2008), rejected that line of argument as narrow and self-serving. They suggested that greater negative externalities on a per capita basis were generated under the Western industrial system than was the case in the global South (Castree, 2000). Further, they asserted that Nature was not outside of Western society, but harnessed to the capitalist mode of development, which in turn was not designed to serve the interests of all classes in any nation state. In their analyses, the critical problems were not that the poor in or outside of the West lacked social or environmental values, but rather that the asocial capitalist system and its associated 'uneven development' favoured individuals or firms that controlled the means of production (land, labour, capital), producing goods in response to market signals, generally without bearing the environmental costs – that is, the negative externalities induced by the activities of private enterprises whose costs must then be carried by society at large.

'The environmentalism of the poor'

Prominent critics of the dominant view of a progressive Western environmentalism include the Indian historian Ramachandra Guha and the Spanish ecological economist Joan Martinez-Alier. Guha's main criticisms of 'wilderness thinking' were fourfold: first, that the anthropocentric/biocentric distinction was of little help in understanding the dynamics of ecological degradation; second, that 'deep ecology' ignored the most serious problems with environmental implications, which were over-consumption and militarism; third, that deep ecology was in essence an elaboration of the American wilderness movement; and fourth, that in other cultures 'radical' environmentalism expressed itself very differently (Guha, 2006). As Cronon had noted a decade earlier, Guha pointed out that the social values embodied in deep ecology ran 'parallel to the consumer society without seriously questioning its ecological and socio-political bias' (Guha, 1993, p. 99). The consequence of setting aside wilderness areas in India had resulted in a direct transfer of resources from the poor to the elite castes that controlled non-governmental organizations (NGOs) like Project Tiger. Guha celebrated those he saw as true environmentalists – embodied, for example, in the Chipko movement that erupted in Uttarakhand in 1973, led by village women who protected what were sacred groves in their eyes by forming human circles around trees so as to prevent loggers from cutting them down. Those women were the first tree-huggers, according to Guha, environmentalists who pitted the 'moral economy of provision in opposition to the political economy of profit' (Guha, 1993, p. 100).

Figure 10.3 Watchtower in Jim Corbett National Park, Uttarakhand, India, on the former site of a village that was displaced when the park was established.

Writing from the perspective of ecological economics, Martinez-Alier (2002) documented other cases of local resistance against environmental degradation in a range of global South countries including Ecuador, Colombia, Honduras and India. He made visible the SCS values of the poor, and showed how the destruction of their livelihood base was driven by asymmetries of power and wealth. Martinez-Alier among others popularized the concept of unequal ecological exchange, in which the negative environmental 'externalities' are borne by local communities whose livelihoods are often dependent on intact ecosystems that need to be resilient enough to provide services and NTFPs.

An example of where a deliberate attempt was made to involve local people is the CAMPFIRE program (Communal Areas Management Programme for Indigenous Resources) in Zimbabwe. The first program was established on the margins of Gonarezhou National Park and had fairly unique origins. Gonarezhou is a remote area on the border of Mozambique. During the civil war, it was a major route for fighters entering from training camps in Mozambique. The local people had been forcibly removed from the park in 1968, and were naturally resentful of the park management. The park was seen very much as an instrument of the government, and poaching was rife. The park is particularly known for its elephants, but groups of elephants would regularly leave the park to raid crops in the surrounding villages. The villagers retaliated by killing elephants. The founder of the Chilo Gorge Lodge, Clive Stockil, worked with local Shangaan elders to ensure that financial benefits from the park started flowing to the local communities, with the first project being the establishment of a school in Mahenye. A clinic, police station and mill have since followed, all funded from income earned through tourism. There have been many attempts to establish the CAMPFIRE program elsewhere, but not all have met with the success of the Chilo Gorge project. This is primarily because Clive Stockil was in a unique position, having grown up among the Shangaan and speaking several local languages, while at the same time being sufficiently persuasive to deal with the government in distant Harare.

Figure 10.4 Chilo Gorge Safari Lodge is located on a bluff overlooking the Save River in Zimbabwe. It is the site of the first, and arguably the most successful, CAMPFIRE program, in which part of the benefits from tourism are transferred to the local community.

Contemporary social values of forests

Employment and hence income derived from work in and from forests may explain the existence of some communities in or near forests. In addition to historical and spontaneous settlements, governments may create new or renewed settlements in order to stabilize viable rural populations and discourage urban drift. Two examples are in the tribal areas in northeast Thailand in the 1970s (Kunstadter and Chapman, 2002) and plantation forests in the UK in the 1940s and 1950s (Inglis and Lussignea, 1995). Private enterprises may hold logging concessions which require the creation or expansion of what are in effect single-company towns (Jari Florestal in Brazil; the former Crown Zellerbach at Ocean Falls in British Columbia, Canada). While Jari's Beiradão riverside shantytown has developed into the self-sustaining town of Laranjal, some of the pulp/paper towns in Canada and in Maine, USA, have suffered repeated booms and busts linked to the global business cycles that are the norm in the pulp/paper sector. Fiscal incentives and rules set by federal and provincial governments in Canada do not appear to have been able to counter the power of corporations, which own the businesses and thus in effect the towns, to close factories overnight and shatter the communities.

Traditional settlements in Sarawak, Malaysia, which opted to convert communal lands to rubber plantations in the 1950s and 1960s, sometimes suffered similarly from booms and busts in the pricing of natural rubber. From the 1980s, cropping systems have diversified away from traditional hill rice and NTFP collection into a variety of fruit orchards and some urban employment, aided by better access to urban markets through expansion of the road network (Cramb, 2007). Still, NTFPs and especially mast-fruiting species provide useful income supplements that also help to sustain social cohesion in local communities through inter-family collecting teams and processing of fruit on the verandahs of communal houses.

In more developed countries, further along the Kuznets 'forest transition' curve, diversification of sources of village income have helped to maintain economically viable rural villages and their associated diversified mosaic landscapes. The mosaic of habitats is attractive for tourism and recreation; economic analysis of the effect of the massive epidemic of foot-and-mouth disease in ruminant farm livestock (pigs, sheep and cattle) in England in 2002 showed that police-enforced closure of access to tourist areas was financially far more severe on rural livelihoods than loss of income from the culled livestock.

Conversely, the diversified habitats and landscapes including managed forests may stimulate conversion of low-yielding forests to financially high-yielding recreational resorts and holiday homes for the wealthy. Low-income employees of forest- and farm-based enterprises may be priced out of the villages in which their families have lived for generations, the houses purchased by relatively rich townspeople. While local governments may appreciate the increased income from local taxes on higher-valued properties, and while local construction and restoration enterprises may thrive on updating rural properties, both forest and farm enterprises may be forced into low-employment mechanization and the social hollowing-out of traditional villages. In much of Western Europe, both farming and forestry work are now carried out by mechanized contractors based in urban areas, leaving the countryside effectively depopulated. The associated more intensive cropping systems and forest management practices switch to a narrow range of commercially profitable practices. The traditional and often manual operations which contributed to the diversity of the landscape (such as hedge-laying, ditching and coppicing, and wood-pasture management) are abandoned as being relatively too costly, being sustained mainly by conservation NGOs and by the great family estates for whom the traditional landscapes are socially important as well as attractions to their paying visitors. So in the evolution of landscapes, there are trends which maintain and enhance social services and other trends which work in the opposite direction.

Also in Europe, regional funds to maintain and sustain the social viability of remoter areas have been very important in providing support to start-up family-based businesses often derived from traditional family activities but needing injections of capital investment to modernize and survive – for example, replacing human and animal power by small-scale mechanization and electric motors and wind power. Craftwork using forest-derived materials may be undertaken in family groups or community associations, thus providing social support as well as family income.

The public health benefits of a visible and accessible forest landscape, encouraging physical activity to keep healthy, and hastening recovery from illness have been

Figure 10.5 Forest interpretation centre and starting point for forest walks, Bennachie, Scotland, UK. The provision of such centres and signposted walks increases the number of people taking part in the opportunity to appreciate the ecology and history of forests.

recorded in many studies. Large economic benefits can be claimed from better public health in a forest environment. Specific and now well-defined measures are needed to make these benefits real to urban populations – easy access by road, the feeling of security, actual safety measures, signposts and information kiosks, introductory tours and guided walks, seasonal events well publicized, visible stewarding and patrols involving young children and emergency backup.

Mechanized forest operations have enhanced the social acceptability of jobs in forestry, with much lower accident rates than semi-mechanized (chainsaw) and manual forest operations with sharp-edged tools. Training and certification may set eligibility bars for poorly educated rural communities, hence the importance of linking educational and vocational training and mentoring in long-term support. The current need in North America is to replace an aging manual workforce with mechanically adept younger people, but in some cases a new manual workforce is being recruited from among immigrant and migrant workers. In addition the forestry sector is now more open to women workers because physical strength is not at a premium.

Contemporary cultural values of forests

Examples of the preservation of archeological sites in, near and underneath forests can be found on all continents: Mayan relics, including whole plazas, houses, roads and bridges in Belize and Guatemala; medieval villages in Europe abandoned during times of plague and never re-occupied; forest cemeteries and single tombs and chapels in many countries; and sites of battle shunned as homes of ghosts.

Sites of historical forest management, especially of activities no longer practiced, have become sites of cultural value. Examples are iron-ore smelting with forest charcoal ('blooming hearths') in the Weald of Kent, UK; coppice coupe boundary banks and dikes; pits for manual sawing of logs; charcoal kiln sites (Forest of Dean, England); and water flumes and mills in many countries.

Figure 10.6 Site of a bathhouse used by Chinese workers in the Bendigo goldfields of Victoria, Australia, in the mid-19th century. The site is preserved and one of the baths cleaned out, but there is no formal protection, and the signage had decayed to a point where it was no longer legible when this photo was taken in 2008.

Figure 10.7 The Washington winch at Nugong, in Gippsland, Victoria, Australia, is a heritage-listed site. It is the only intact high lead/skyline logging system within Victoria. The site also has a Washington Iron Works Yarding Engine, which may be the only one left in Australia.

Veteran trees, sometimes not just huge and old but also with specific historical associations, are landmarks of cultural value. Some examples from England are Major Oak in Sherwood Forest – about 1,000 years old and the reputed hiding place of Robin Hood, voted as England's tree of the year in 2014; Boscombe Oak in Gloucestershire, hiding place of future King Charles II; Rufus Oak in New Forest, where Walter Tyrell shot William II; and Tyburn Tree in London, where traitors were hanged.

Memorial clumps – trees planted in groves and fenced initially against cattle and deer, usually on hilltops, as commemorations of military victories – have become sites of cultural value. Some UK examples are Blenheim Clumps, Wellington or Waterloo Groves. Single trees – also to commemorate specific events – can become cultural markers. Some UK examples are Trafalgar (1805) and the Coronation trees of 1953. Some trees mark specific meeting places – a sycamore (*Acer pseudoplatanus*) in the village centre where farm labourers of Tolpuddle, Dorset, met to protest against the Black Acts in 1834 and which is claimed as the start of the trades union movement.

The culturally modified trees of the Northwest Coast and elsewhere in British Columbia are emblematic of the First Nations' ancient presence in their territories. Culturally important trees include those tapped for gums, latex and resin; coppiced and pollarded; and split planks of western red cedar (*Thuja plicata*).

Groves that are held under community or family tenure for fruits or nuts are also culturally esteemed. Examples are Brazil nuts (*Bertholletia excelsa*) and piquia (*Caryocar villosum*) in Brazilian and Bolivian Amazonia and illipe nuts (from *Shorea* spp.) in Borneo. Globally, trees and clumps are planted specifically as family insurance for weddings, dowries and funerals.

Forests are also culturally important as continued testimony to historical rights. In some parts of England, property owners have the right to let pigs forage for acorns, beechmast, chestnuts and other nuts (the right of pannage) and the right to collect peat (the right of turbary), and fuelwood and other woody material (the right of estovers). Some of these rights can be traced back to the Charter of the Forest of 1217. Today, claimed rights to manage and harvest NTFPs in many countries, with or without payment of royalty, are often contested. Indigenous peoples globally have challenged State claims that a customary

Figure 10.8 Plaque marking 'the Queen's Oak' planted by Queen Elizabeth II in 1979 to commemorate the 900th anniversary of the founding of the New Forest in England.

Figure 10.9 Culturally modified tree in the Haida Gwaii, Canada. A strip of bark has been removed from the western red cedar (Thuja plicata). Cedar bark had (and still has) many different uses for the Haida First Nation.

family right of collection has transformed into a commercial operation and so is subject to other regulations.

Increasing recognition of the SCS values of Indigenous and forest-dependent peoples in forest protection

As societies have evolved and changed, so have their SCS needs and values. Worldwide, there is continuing loss of unwritten traditions as oral transmission has weakened and as traditional practices and Indigenous knowledge are not passing to younger generations. The SCS needs and values of the majority populations of most countries have tended to override those of Indigenous and forest-dependent peoples, especially as national radio and TV convey predominantly the language and culture of the majority population.

External recognition of the SCS of Indigenous and local communities is important on a number of fronts. First, it ends the invisibility of IPLCs to the majority populations. Second, it affirms that traditional ecological knowledge (TEK) has value alongside Western scientific knowledge. Third, IPLCs can deploy this evidence of the validity of their claims to customary territories before national or international courts or in other fora in their efforts to get recognition of their native (or Aboriginal) rights to customary lands. In the case of Canada, Rajala (2006) has provided an in-depth treatment of settler governments' displacement of First Nations from their traditional, ancestral and unceded territories. In an influential article on the British Columbia context, Willems-Braun (1997) maintained that discourses of scientific forestry deployed in British Columbia were a form of colonial and post-colonial violence directed against First Nations people. Willems-Braun argued that the privileging of expert scientific and technical forest management, and the corresponding absence of references to the TEK of First Nations, who had never ceded their customary lands to the incoming settler governments, had perpetuated the exclusion of First Nations people from forests and submerged ongoing claims of First Nations to those lands. Willems-Braun suggested that dominant representations were 'abstracted and displaced from existing local cultural and political contexts' and acted to sever apparent ties between the forested landscape and any particular person or group (1997, p. 10). The *Delgamuukw* (1997) Supreme Court decision recognized TEK and oral traditions as demonstration of the continued practice of Aboriginal rights.

Measuring cultural and spiritual values

There are no commonly agreed metrics used to measure cultural and spiritual values. These values are generally labelled as incommensurable, in the sense that a dollar value cannot be used as a proxy for assessing their worth. Ecological economists have underlined the limitations of using non-market valuation techniques to assess the value of a natural resource or environmental service to Indigenous or local communities. As Martinez-Alier suggested, 'the poor sell cheap, not out of choice but out of lack of power' (2002, p. 30), including a lack of information about markets. However, Martinez-Alier further argued that 'incommensurability means that there is no common unit of measurement, but it does *not* mean that we cannot compare alternative decisions on a rational basis, on *different* scales of value, as in multi-criteria evaluation' (2002, p. 399).

Social scientists largely agree with this line of reasoning, but suggest that 'findings from ethnographic and qualitative research also may be useful in constructing economic behavioural models that are relevant to indigenous cultures' (Adamowicz *et al.*, 1998, p. 62). In this vein, leading social scientists in this field:

> advocate a multi-method and especially multi-metric approach. Likely key to this will be ability to either infer weights or preferences through choice surveys based on paired comparisons . . . or the actual construction of metrics through the use of subjective scaling when necessary . . . Such scales enable the assigning of value, ordinal ranking, or numeric tag to what are in large part intangible properties (such as awe in reference to spiritual value) . . . In the case of creating a metric for less tangible values using a multi-metric 'constructed' approach, the goal is best served by flexibility in the scales used.
> (Chan *et al.*, 2012, p. 15)

Figure 10.10 Prayer flags adorning a tree near Nyingchi in Tibet, China. Sites such as this are clearly of great spiritual importance, and forest managers need to take them into account.

Foresters also have advanced related arguments in favour of recognizing and measuring SCS values. Both governmental reporting processes and independent voluntary third-party certification schemes require active engagement and safeguarding of SCS needs and values.

Concluding remarks

Over the past half a century, both procedural and substantive gains in the recognition of the SCS needs and values of majority and minority populations have been achieved at national and international levels. In aggregate, however, those advances have not been sufficient to slow or reverse the steady conversion of the remaining natural world to more cropland or real estate development, or more degradation through natural resources extraction (timber, minerals and/or wildlife, for example). It is easy for governments and other powerful actors, who have legal authority over the majority of forest lands, to pay lip service to SCS needs and values. However, many governments are only forced to take action to uphold SCS values of their citizens as a result of sustained citizen protests, as demonstrated in Clayoquot Sound (British Columbia, Canada) from 1980 to 1994 (Satterfield, 2002). Similarly, court decisions, such as the requirement issued by the Supreme Court of Canada (SCC) in *Sparrow* (1990), *Van der Peet* (1996) and *Gladstone* (1997) that the Crown's representatives consult with and accommodate First Nations, have opened up spaces leading to the acknowledgement and protection of the SCS of First Nations. That recognition was taken to another level on 26 June 2014 when the SCC issued its watershed *Tsilhqot'in* decision, recognizing Aboriginal title over 40% (approximately 170,000 ha) of the traditional territory of the Tsilhqot'in Nation of British Columbia. At its core, *Tsilhqot'in* reversed the power dynamics between the provincial government and the Tsilhqot'in Nation. The SCC ruled that land which from the 1860s had been declared to be Crown land had been proved to be continuously used and occupied by the Tsilhqot'in Nation and was therefore covered by Aboriginal title, which held SCS at its core.

For their part, the majority of citizens may declare preferences for SCS. At the same time our consumerism has shown no signs of slowing down, and its negative externalities on the natural world proceed unabated. The process of globalization has accelerated these trends, so that few corners of the planet remain untouched. While biocentric and/or kincentric SCS needs and values are more vital than ever, they have to be complemented by national and international rules that compel us all to translate our facile protestations of caring for resilient ecosystems, on which all life on Earth is dependent, into concrete actions.

Further reading

Barton, G. A. 2002. *Empire Forestry and the Origins of Environmentalism*. Cambridge: Cambridge University Press.

Garforth, M., Mayers, J. (eds.) 2005. *Plantations, Privatization, Poverty and Power: Changing Ownership and Management of State Forests*. London: Earthscan.

Hayman, R. 2003. *Trees, Woodlands and Western Civilization*. London: Hambledon and London.

Langton, M., Mazel, O., Palmer, L., Shain, K., Tehan, M. 2006. *Settling with Indigenous People: Modern Treaty and Agreement-Making*. Sydney: Federation Press.

Mann, A. T. 2012. *The Sacred Language of Trees*. New York: Sterling Ethos.

Rajan, S. R. 2006. *Modernizing Nature: Forestry and Imperial Eco-Development 1800–1950*. Oxford: Oxford University Press.

Tindall, D. B., Trosper, R. L., Perreault, P. (eds.) 2013. *Aboriginal Peoples and Forest Lands in Canada*. Vancouver: UBC Press.

Venne, S. H. 1998. *Our Elders Understand Our Rights: Evolving International Law Regarding Indigenous Peoples*. Penticton, BC: Theytus Books.

Watkins, C. 2014. *Trees, Woods and Forests: A Social and Cultural History*. London: Reaktion Boos.

Chapter 11

Public participation in forest land-use decision-making

Howard Harshaw and Hosny El-Lakany

Introduction

Public participation is covered in the Montreal Process under Criterion 7, 'Legal, institutional and economic framework for forest conservation and sustainable management'. Indicators that include an element of consultation include:

7.1.a   Clarifies property rights, provides for appropriate land tenure arrangements, recognizes customary and traditional rights of indigenous people, and provides means of resolving property disputes by due process.

7.1.b   Provides for periodic forest-related planning, assessment, and policy review that recognizes the range of forest values, including coordination with relevant sectors.

7.1.c   Provides opportunities for public participation in public policy and decision making related to forests and public access to information.

7.2.a   Provide for public involvement activities and public education, awareness and extension programs, and make available forest related information.

7.2.b   Undertake and implement periodic forest-related planning, assessment, and policy review including cross-sectoral planning and coordination.

Forest land-use decisions involve the distribution of scarce resources; Smith and McDonough (2001) note that the issues that frame these decisions 'often involve limited resources, but multiple constituencies, creating a situation in which it is impossible for everyone to get what they desire' (p. 241). Land-use planning processes that incorporate, and are responsive to, the full range of social values are vital for the sustainable management of forested landscapes. This shift in the focus of forest management from timber to a broader range of values makes the representation of forest stakeholders' needs and desires an important consideration in forest land-use decision-making, and has paralleled the recognition of post-materialist values, an increased awareness of environmental values and issues, and a shift from forest management priorities being negotiated between governments and the forest industry to forest management practices and specific programs being challenged by environmentalists. This shift was motivated by an increase in public awareness of environmental and forestry issues. Carrow (1999) has characterized the evolution of public involvement in forestry issues as one that progressed from an atmosphere of hostility and antagonism in the 1960s to one

Figure 11.1 A practical consequence of adequate consultation is to reduce the number of protests, such as this one in British Columbia, Canada, by members of the St'at'imc First Nation objecting to a proposed skiing development.

that reflected greater degrees of local empowerment in the 1990s. This chapter reviews overall concepts and challenges of public participation to frame the representation of stakeholders in land-use planning.

Although there have been advances in the role that public participation has played in forest land-use decision-making, problems remain. Economic and market concerns have influenced the manner in which public participation has been incorporated into commercial forest management, as forestry companies have sought certification that their management and operations are ecologically, socially, and economically sustainable. In order to be seen as responsible corporate citizens and maintain market share, they have formalized the role of public participation primarily through the creation of public advisory committees. Although these committees make comments and recommendations about forest management plans and address some public concerns, they generally have no decision-making authority and tend to be heavily influenced by the company involved.

Public participation can also play a role in easing logistical tensions in land-use planning. The nature of choices available to decision-makers has changed as increased demands on a finite land-base, and decisions made under increased public scrutiny, have increased the likelihood of conflict and expectations of resolution. The success of land-use planning processes can be seen as a function of the intensity of mechanisms that are employed to involve the public, as more intensive mechanisms, such as collaborative processes, are more likely to succeed than less intensive mechanisms, such as open houses or advertisements in newspapers, in the current political and democratic climate (Figure 11.2).

As the intensity of the participation process increases, so too does stakeholder ownership of planning decisions; a consequence of this is that decisions that solicit meaningful public dialogue are more likely to have accounted for stakeholder interests, and people are more likely to understand the diversity of opinions surrounding a particular issue. Incorporating a broad range of publicly held values enhances the fairness, or equity, of the decision-making process. The results of that enhanced fairness are that

*Figure 11.2
Intensity of public
participation
mechanism and
likelihood of
success. (Adapted
from Beierle and
Cayford (2002).)*

the effectiveness and accountability of resultant decisions is greater, and that these decisions are more likely to reflect diverse public values and interests.

Additionally, interested and affected parties have knowledge and expertise – especially pertaining to local conditions – that can be lent to the planning process and help to improve the planning outcome. Forest management planning that is adequately informed by the views of affected people is more likely to be politically acceptable and socially beneficial to affected communities. In short, the effectiveness of decisions and the likelihood of successful planning outcomes increase in land-use planning processes that incorporate a high degree of public participation. This incorporation of representative stakeholders in land-use decision-making increases the likelihood of effective planning outcomes.

Socio-economic characteristics determining social values

To complement the detailed discussion of values in Chapter 10, this chapter addresses values as they pertain to public participation. Values can be considered to be 'cultural ideas about what are desirable goals and what are appropriate standards for judging actions . . . they are emotionally charged beliefs about what is desirable, right, and appropriate' (Tindall, 2003, p. 693). Values inform preferences; as values vary, so too will preferences. Inglehart (1977) proposed that Western public values have shifted from those that place importance on material well-being and physical security to values that emphasize quality of life and amenity values (termed post-materialist values), as economic and physical security have become more common in Western societies through structural change. With basic needs met, amenity values, such as outdoor recreation, may be pursued without compromising well-being. Inglehart (1977) suggested that a consequence of this is that a large proportion of the public has sufficient interest and understanding, as well as the necessary skills, to participate in decision-making at broader (national, international) levels, though not through traditional civic venues. There has been a shift from 'elite-directed' activities (i.e. traditional venues, such as political parties, labour unions, religious institutions) to 'elite-challenging' activities where the public are given a role in making specific decisions, such as formal intensive public participation processes. These two propositions reinforce one another. The change in values is a response to a decline in the legitimacy of traditional hierarchical institutions of authority. The shift in values has been influenced by a number of factors, including increased availability and use of higher education; shifts in occupational structure, such as the rise of the professional, knowledge, and service sectors; a broader and effective media; and general increased economic prosperity. People exhibiting or espousing post-materialist values are more likely to be younger, have achieved higher levels of education, have a sense of identity that transcends national boundaries (i.e. are more cosmopolitan), and are more comfortable with change and innovation. Women are less likely to hold post-materialist values than men (Inglehart, 1977). People holding post-materialist values have made increased demands for egalitarian-style decision-making.

The values that people hold about forested landscapes play a role in people's opinions about visual quality and aesthetic concerns in forestry. For example, people that highly value recreation experiences have different attitudes towards visual quality than do people that rate economic values highly; those that highly value recreation have similar attitudes to people that value ecology and aesthetics highly (Tindall, 2003). In a study of the role of social structure and identities in explaining the diversity of forest values, Harshaw and Tindall (2005) found that foresters, who are typically responsible for forest land-use decisions in British Columbia, Canada, had less diverse values than did non-foresters and concluded that knowing about the characteristics of social-psychological and social structural variables can be useful for identifying stakeholder representatives.

In their examination of the relationships between socio-economic characteristics, social influence, forest knowledge, forest values, and forest management preferences among motorized and non-motorized recreationists, McFarlane and Boxall (2000) found that the type of recreation that people participated in had an effect on their attitudes and values. Participants of appreciative, or non-consumptive, activities were more likely to have biocentric attitudes than were participants of mechanized or consumptive activities. While socio-economic characteristics like income, gender, and urban residence were not related to attitudes towards forest management, the authors found evidence to support the assumption that age, education, and place of residence are consistent predictors of environmental attitudes and beliefs. That is, women, people with higher levels of education, younger people, urban residents, people living in non-timber-dependent regions, and people with liberal political orientations were more likely to possess biocentric attitudes.

Perspectives on public participation

In North America, the need for the incorporation of public participation in land-use planning has increasingly been recognized since the 1970s. The sustainable forest management certification movement provides venues for discussions of the distribution and management of forest resources and amenities through the public advisory groups that represent public interests. Despite these opportunities for participation, the issues, concerns, and opinions of forest stakeholders are not necessarily being heard at forest land-use planning tables. This section reviews conceptions and challenges of public participation in an effort to identify those characteristics of planning processes that serve to benefit or constrain the interests and needs of outdoor stakeholders, to define what constitutes good participation and good representation, and to examine what may influence good participation and good representation.

Democratic traditions

Fundamental to democratic governance is the inclusion of citizens that will be affected by decisions in the decision-making process – it is a matter of fairness. In the context of decision-making, fairness can be conceived of as judgements about the legitimacy and relevance of a decision. People's perception of fairness influences how they evaluate the procedures that govern the decision-making process (i.e. procedural fairness), such that if the procedures are deemed to be fair, then it is more likely that resultant decisions (i.e. outcomes) will also be deemed to be fair (i.e. distributive fairness) (Lauber and Knuth, 1999). This conception of fairness requires that decision-making processes be open and transparent.

Procedural fairness
Procedural fairness has been defined as the 'adequacy of a procedure for producing an appropriate allocation of resources and obligations' (Lauber and Knuth, 1999, p. 20). Smith and McDonough (2001) summarize nine principles for assessing procedural

fairness: (1) consistency of process over time; (2) suppression of bias; (3) use of accurate information; (4) flexibility of decisions; (5) representativeness of the concerns of all recipients; (6) adherence to prevailing ethical and moral standards; (7) neutrality; (8) trust in the benevolent intentions of decision-makers; and (9) status recognition.

Lauber and Knuth (1999) identify a set of criteria that can be used to evaluate fairness. As these criteria are context dependent, a relevant subset is examined here: the influence that planning participants have over planning outcomes; opportunities for meaningful participation (i.e. process control); and a representation of all important viewpoints. Procedural fairness can be assessed in three ways. A review of relevant planning documents can provide some insight into the structure of the planning tables and intentions of the plan conveners. A survey of stakeholder representatives that participate in planning activities can record the opinions of those that played roles in the formulation of the plan. A further measure can help inform the examination of procedural fairness: the socio-demographic characteristics of the stakeholder representatives can be examined to gauge the degree to which the representatives are in fact representative of their constituencies.

Distributive fairness

Distributive fairness is concerned with the equity of planning decision outcomes. It can be assessed in two main ways: by reviewing actual planning documents and by surveying stakeholders for their opinions on outcomes. Three measures can be employed to examine distributive fairness: the accessibility of participation opportunities as a measure of the availability of participatory democracy; stakeholders' impressions of the abilities of proxies that are typically part of planning processes to represent their needs; and the feelings of representation among stakeholders – have they felt that their needs have been represented, for example in terms of outcomes?

Participatory democracy

Although there are many models of democracy, the democratic tradition can be conceived of as a continuum from participatory democracy to representative democracy wherein the degree of actual citizen participation and commitment varies. Participatory democracy requires the direct involvement of citizens in decision-making and assumes that citizens have both the capacity (i.e. knowledge, skill) and desire to engage in meaningful discussions about the issue at hand. The objectives of participatory democracy include involving a broad cross section of the public, building a sense of community, and developing a sense of self-reliance and self-worth among communities and their members. As Hemingway (1999) has noted, 'participatory democracy is in this sense a communicative process, and the citizen, rather than groups, is assumed to be the basic unit in it' (p. 153). The value of having these communicative processes has been examined by Delli Carpini *et al.* (2004), who argue that even the act of public deliberation, a formal (e.g. a town hall meeting) or informal (e.g. a discussion among friends) form of public participation that permits the development and expression of ideas, can lead to desirable outcomes such as increased social capital, since people who engage in public deliberation are more likely to become involved in other forms of civic engagement. In addition to being characterized as being communicative, participatory democracy is a cooperative process.

Representative democracy

Representative democracy, on the other hand, employs citizen proxies to represent and advocate for the interests of their constituencies, or stakeholders, with 'the basic acts entailed being the calculation of interest and the manipulative persuasion of others'

Table 11.1 Characteristics of participatory and representative conceptions of democracy.

Characteristics	Form of Democracy	
	Participatory	Representative
Citizen involvement	More	Less
Focus	Individual	Group
Style	Communicative	Adversarial
Orientation	Process	Outcome
Model	Developmental	Market
Demand on leisure time	Greater	Less

Adapted from Hemingway (1999, p. 152).

(Hemingway, 1999, p. 152). Representative democracy may have risen out of a need for proxy representatives, as citizens' time is typically constrained (e.g. biographical availability), and members of the public generally have limited capacities to understand the complexities of decisions. Other possible constraints on people's availability include their geographical distance from the locus of planning and the financial costs associated with participation. For example, in forest land-use planning, people may find it difficult to choose among alternative management scenarios and relate potential outcomes to personal preferences. As demands on citizenship are not great in representative democracies, it may be labelled 'weak citizenship', while participatory democracy can be labelled 'strong citizenship' (Hemingway, 1999). A comparison of typical attributes of representative and participatory democracies is presented in Table 11.1.

Hemingway (1999) has argued that 'democratic social capital grows out of leisure activity that fosters democratic norms like autonomy, trust, cooperation, and open communication', and that 'democratic social capital cannot emerge from activities administered in non-democratic fashions' (p. 162). Regardless of which democratic tradition frames a decision-making process, it is important that meaningful opportunities for public involvement be explicitly incorporated into that process. Hunt and Haider (2001) conclude that 'despite the different rationales for increased public involvement in decision making, the resonate echo is that people must be involved in these processes' (p. 874).

Approaches to public participation

Public participation in resource and land-use decision-making has been defined as 'any of several "mechanisms" intentionally instituted to involve the lay public or their representatives in administrative decision-making' (Beierle and Cayford, 2002, p. 6). There has been much discussion of the importance of incorporating opportunities for *meaningful* input into land-use planning processes, and the divestment of technical details from policy-making: 'rather than seeing policy decisions as fundamentally technical with some need for public input, we should see many more decisions as fundamentally public with the need for some technical input' (Beierle and Cayford, 2002, p. 75). Although it is now widely accepted that members of the public should be involved in environmental decision-making, there is less consensus about what that public involvement entails; for example, what degree of involvement ought to be accorded to individual members of the public?

Public participation constitutes a redistribution of power whereby the public are accorded opportunities to engage in decision-making through the sharing of information, knowledge, and ideas to effect change. For Arnstein (1969), public participation is an inclusive mechanism in decision-making that empowers citizens who typically do not have power: 'it is the means by which they can induce significant social reform which enables them to share in the benefits of the affluent society' (p. 216). An important distinction of public participation in forest land-use decision-making is that the decisions being made are about the allocation of scarce resources among a host of differing stakeholders, including commercial and public (e.g. outdoor recreation, aesthetics, access) interests. The key to effective public participation in decision-making is the degree of power (i.e. influence) that members of the public have; this has also been noted by Arnstein (1969), wherein 'participation without redistribution of power is an empty and frustrating process for the powerless' (p. 216). One typology of the degree of influence associated with public participation is presented in Figure 11.3.

	Citizen control
Degrees of citizen power	Delegated power
	Partnership
	Placation
Degrees of tokenism	Consultation
	Informing
	Therapy
Non-participation	Manipulation

(Y-axis label: Meaningful opportunities for public participation increase)

Figure 11.3 A ladder of participation. (Adapted from Arnstein (1969).)

In this typology, Arnstein (1969) distinguishes between eight different degrees of participation and influence and uses the analogy of a ladder of public participation. The bottom five rungs of this ladder of participation represent mechanisms that do not reflect meaningful opportunities for public participation; these mechanisms allow citizens to offer advice and be heard (at best), or provide opportunities for decisions to be explained and advocated for. These bottom-rung mechanisms that lack decision-making authority are generally the most frequently used forms of participation opportunities. The top three rungs of the ladder represent opportunities for the public to influence decisions in which they have a stake: these are meaningful opportunities for public participation. The benefits of meaningful public participation do come with responsibilities for participants, such as knowledge acquisition about the issue(s) at hand, the ability to communicate one's ideas and opinions, and effective negotiation for one's interests. These responsibilities borne by public participants contribute to the legitimacy of a planning process.

In an examination of resource planning in the USA, Daniels and Walker (2001) conclude that public policy is threatened by poorly defined direction and a lack of effectiveness, and may be characterized as lacking social legitimacy. These authors argue that effective policy 'is an adaptive process; uses the most appropriate science and technology; is implementable; [and] has low transition costs' (p. 2). To achieve social legitimacy, effective resource planning policy should be rational and solutions technically sound, and include affected publics through collaborative mechanisms.

Opportunities for public participation in decision-making have had an increasing role in land-use policy-making over the past 30 years and may be seen as climbing the rungs of Arnstein's (1969) ladder. In their review of 239 case studies of public participation in American environmental decision-making, Beierle and Cayford (2002) describe the transition of environmental decision-making from that of the managerial model, to decision-making processes that recognized pluralism, to an atmosphere that recognized popular democratic theory. The managerial model is a largely utilitarian approach to resource management that relied upon experts to identify planning goals and objectives and make decisions that would provide social welfare maximization; this approach to planning has also been termed 'rational planning' (Payne and Graham, 1993).

Managerialism gave way to the concept of pluralism, wherein it was acknowledged that the public good was relative to who the affected public was. Under this approach to resource planning, government administrators became the arbiters of the interests of the different constituencies that had stakes in the decisions. In the final stage of this transition, resource decision-making has adopted popular democratic theory, which 'stresses the importance of the act of participation, not only in influencing decisions but also in strengthening civic capacity and social capital' (Beierle and Cayford, 2002, p. 4). This inclusion and involvement of members of the public with an interest in planning outcomes throughout the planning process, including the identification and framing of planning goals and objectives, has been termed transactive planning (Payne and Graham, 1993). Although the transition to decision-making processes described by Beierle and Cayford (2002) is drawn from the American experience with environmental decision-making, it is strikingly similar to Canada's experience with the rise of public participation in land-use planning and social forestry.

Beierle and Cayford (2002) argue that planning success is a function of intensity of the mechanisms employed for involving the public in decision-making, and not the context of the planning. In their examination of public participation in environmental decision-making, the authors conclude that more intensive participatory mechanisms are more likely to succeed (see Table 11.2).

However, as the intensity of participation increases, the process moves away from the tradition of participatory democracy towards that of representative democracy.

Mechanisms such as open houses often involve large numbers of people, but may be perceived as hostile settings by some members of the public. Open houses can devolve into opportunities for announcing and defending decisions that have already been made, and may not provide meaningful opportunities for participation. On the other hand, more intensive processes, such as negotiations, engage fewer people and are less likely to be reflective of socio-economic characteristics, and are more limited in outreach to constituencies and communities. Beierle and Cayford (2002) note that more intensive processes 'demonstrate a strong tendency to reach consensus by leaving out participants or ignoring issues' (p. 48), and that 'as processes intensify, the range and

Table 11.2 Participatory mechanisms.

Mechanism	Feature			
	Selection of Participants	Type of Participant	Type of Output	Seek Consensus?
Public meetings and hearings	Usually open access; group size ranges widely	Average citizens	Information sharing	No
Advisory committees not seeking consensus	Small group of participants selected based on characteristics	Average citizens, interest group representatives	Recommendations to agency	No
Advisory committees seeking consensus	Small group of participants selected based on characteristics	Average citizens, interest group representatives	Recommendations to agency	Yes
Negotiations and mediations	Small group of participants selected based on characteristics or specific interests	Interest group representatives	Agreements to parties	Yes

(left margin, vertical:) Intensity of participation increases

Source: Beierle and Cayford (2002, p. 45).

representativeness of voices heard – as well as the social benefits of education, conflict resolution, and trust formation – tend to narrow down to the relatively small group of active participants' (p. 48). These authors also contend that more intensive processes can be more successful at overcoming pre-existing conflict among constituencies.

The degree of public participation that a decision-making process will incorporate should be clearly defined at the beginning of the process so as to avoid confusion and elevated expectations, and to provide a criterion for measuring the success of the process. Further, the participation of public constituencies should not be limited to the decision-making process, but should extend to implementing the plan and evaluating planning outcomes.

The situations and issues being confronted by decision-making processes that incorporate public participation, as well as the communities that are affected by the decisions, are unique and require specific approaches to the problem at hand. No single set of procedures can be applied universally to public participation processes, and local adaptations will be necessary. There are three common approaches to meaningful public participation: shared decision-making, consensus building, and interest-based negotiation. Shared decision-making is a type of decision-making that is based on consensus building, interest-based negotiation, and collaboration. Within this context, consensus building aims for win-win solutions and seeks to reduce power imbalances; consensus building has been characterized as being 'free from the tyranny of the majority' (Williams *et al.*, 1998, p. 864). Consensus building is dependent upon the willingness of all parties to pursue consensus and agree to outcomes, agreement that consensus is the most appropriate means of reaching agreement in a given situation, as well as the political will of government to embrace the consensus process and outcomes.

Interest-based negotiation focuses on stakeholder interests rather than positions. In this sense, interests are fundamental goals that are based on hopes, desires, concerns, and fears, whereas positions are desired end-states. Interest-based negotiation is premised on principled negotiation, and attempts to 'separate the people from the problem; focus on interests, not positions; invent options for mutual gain; and insist on using explicit objective criteria' (Williams *et al.*, 1998, p. 865).

Finally, collaboration 'occurs when stakeholders embroiled in conflict agree to explore their differences and seek solutions that go beyond their own limited vision of what is possible' (Williams *et al.*, 1998, p. 865). By encouraging planning participants to accommodate other stakeholder interests without necessarily compromising their own, shared decision-making outcomes can result in greater public ownership of decisions. However, there are limitations to collaborative planning methods including ideological and value differences among stakeholders, institutional resistance to change, a lack of trust among participants and between participants and government, and imbalances in power among stakeholders. Time is also an important factor when considering collaborative decision-making processes – due to the limitations already mentioned, the decision-making process can take longer than traditional processes. These three approaches to meaningful public participation need not be independent and can overlap; thus, for example, it is possible for collaboration to occur with different levels of consensus.

The inclusion of a broad array of stakeholders and ideas in planning processes that have high degrees of public participation permits the incorporation of experience and knowledge that can serve to produce innovative planning outcomes that are more reflective of community interests. Further, the sharing of knowledge through these collaborative processes can help to increase the knowledge and social capital of participants.

Among the benefits advocated for the incorporation of public participation into sustainable forest management decision-making are: (1) decisions that have been reached through public participation are more publicly acceptable and more likely to be implemented; (2) public participation improves the relationship between management agencies and the public; and (3) public participation in decision-making can reduce resource management conflicts (Lauber and Knuth, 1999). Although there are pragmatic reasons for adopting public participation mechanisms into land-use decision-making, such as increased ownership of outcomes by participants, Beierle and Cayford (2002) identify five goals of public participation that should also serve to produce more effective planning solutions: incorporating public values into decisions, improving the substantive quality of decisions, resolving conflict among competing interests, building trust in institutions, and educating and informing the public. A broader representation of interests at planning tables can also lead to a broader set of potential solutions being identified.

Challenges of public participation

Land-use planning is a complex undertaking. This complexity stems from bringing together multiple constituencies, each with their own deeply held values and worldviews, to discuss and resolve multiple issues, which are governed by multiple policies, that require a certain degree of scientific and technical knowledge in settings that are often plagued by scientific and technical uncertainty. Wondolleck and Yaffee (2000) characterized resource management as being predominantly utilitarian, brokered by technocrats who typically view under-use as wasteful. However, current land-use planning processes are an improvement over past processes, especially in terms of the degree to which information is made available, and the degree of opportunity that the public has to become involved. Despite these improvements, challenges remain.

One challenge to public participation is that participatory approaches are often less effective than representative approaches. That is, effective participation of the public requires knowledge of issues, skills (e.g. negotiation), and resources (e.g. time) that may be limiting factors in the participation of those citizens that would otherwise be inclined to participate. Thus meaningful public participation becomes the participation of public representatives – which in itself may not be a bad thing, provided that these representatives are advocating for the needs and interests of their constituencies and are in fact representative of their constituencies. Daniels and Walker (2001) introduce what they term the *Fundamental Paradox* of welfare maximization and interest group mediation: the juxtaposition 'between technical competence and open process . . . achieving one value may compromise our ability to achieve the other' (p. 4).

Other challenges to the effectiveness of public participation include imbalances in power and poorly defined accountability mechanisms of the process, which can exclude the public interest. Adding to these difficulties are uncertainties among participants about the process (both in terms of what the planning goals and objectives are as well as their role as participants) and the intentions of other stakeholders, and different assessments of acceptable outcomes or actions based on constituency needs and values.

Public representatives in planning processes have been referred to as members of civil society, or those people representing constituencies other than those of government or industry: 'These stakeholders are not professional natural resource managers, nor do they have any direct pecuniary interest in resource management decisions' (Finnigan *et al.*, 2003, p. 15). Civil society stakeholders typically have fewer resources than government and industry, including skills, funding, time, and experience. Such stakeholder groups also tend to have weaker organizational structures. These deficiencies can lead to what has been termed the 'two table' problem, wherein civil society stakeholders do not have the time or resources to participate at both the official planning table and their own constituency table, which can result in a compromise of stakeholder accountability to their constituencies. Consequently, two barriers for civil society stakeholders are commitments of time and financial resources that are required for effective participation and the power imbalances that exist among stakeholders.

It is important to foster effective two-way communication between stakeholder representatives and their constituents in order to maintain focus on a group's goals and objectives and to formalize the representative's accountability to constituency. However, such accountability, though necessary, increases the time demands of stakeholder representatives and may require additional skills to be developed. If forest land-use planning processes are to be representative of the range of stakeholders, then the processes ought to be accessible and be able to address socio-economic barriers to participation, both for stakeholders and their representatives.

The demands placed on discretionary time are typically higher in participatory forms of democratic tradition than in representative forms, as higher degrees of effort and commitment, such as the development and attainment of skills and knowledge, are required for the functioning of participatory democracies. In an examination of whether decision-making processes are representative of public values, Overdevest (2000) concluded: 'To the extent that participation in decision making is conditioned by unbalanced incentives, neither the participatory nor representative democracy models can succeed without overcoming structural barriers to participation [e.g. income, time, knowledge]' (p. 694).

In conclusion, characteristics of meaningful public participation (including the strength of representation) can be summarized as illustrated in Table 11.3.

As socio-economic representativeness is important in land-use planning, one criterion for evaluating the success of a planning process is the degree of representation present in the process. Thus, meaningful public participation is a function of the characteristics identified in Table 11.3. However, it would be unreasonable to expect all of these characteristics to be met in a single process. Meaningful public participation can occur when a majority of the characteristics are met. There is also a link between meaningful public participation and the strength of representation, as meaningful public participation in representative democracies is closely related to the effectiveness of the planning participants. Good representation is characterized as having the right people (i.e. representatives of all of the affected stakeholders) at the planning table who are competent and effective in their roles and in touch with their constituencies (Figure 11.4); they should be able to influence the planning process and outcomes through meaningful participation (as in Table 11.3).

Table 11.3 Characteristics of meaningful public participation.

Achievement of procedural fairness.	Structural barriers are recognized and addressed.
Achievement of distributive fairness.	
Open and transparent process.	People at planning table are representative of their constituency.
Opportunities to influence process.	
Opportunities to influence outcomes.	People at planning table have technical competence.
Opportunities for knowledge acquisition.	
Ability of participants to communicate ideas.	Accountability is clearly defined.
Opportunities to negotiate constituency's interests.	Imbalances in power among planning table participants are addressed.
Intensity of process sufficient to address issue at hand.	Communication between planning participants and their constituencies is strong.

Figure 11.4 Characteristics of weak and strong representation.

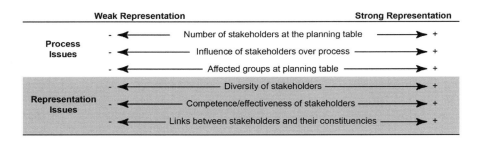

Although weak representation would typically be associated with poorly perceived representation, and strong representation with well-perceived representation, it is possible that disconnects may occur: strong representation may still lead to poorly perceived representation if communication between the representatives at the table and their constituencies is poor – for example, what would constituents base their perceptions of representation on if they were not aware of what occurred at the planning table? The effectiveness of planning participants is a key component to achieving strong representation. The importance of clear and open communication between the planning participant and their constituency cannot be overstated; this type of communication permits relevant issues to be identified and possible solutions to be discussed among affected recreation stakeholder groups. In the absence of this type of communication (and it is likely that, despite the best efforts of the planning participants, many affected stakeholders will not have an opportunity to engage in dialogue with their representatives; indeed, they may not even know that they are affected), it is expected that stakeholders may assess their perceived representation in part based on planning outcomes. In other words, if the needs of stakeholders are met, then they are likely to report high levels of perceived representation, even if they were not directly consulted by their representative(s).

The changing roles of forest stakeholders

Usually, the laws of countries mandate the degree of involvement accorded to stakeholders to participate in discussions about the governance and the management of natural resources, particularly forests. The participation of stakeholders in the consultative processes related to sustainable forest management has dominated the global forest policy discussion in recent years.

Stakeholders generally include individuals or groups that have an interest in forests, agriculture, and rural development. A stakeholder is any individual, social, or economic group or institution that is affected by and/or can influence decisions. Stakeholders are not necessarily formally organized groups, and may encompass many different interests (Davies *et al.*, 2011):

- Government or public sector (state, provincial, and municipal);
- Domestic civil society (NGOs, universities, research institutes, farmer organizations, Indigenous peoples' organizations, worker/trade unions, community organizations, and women's and youth organizations);
- Private sector (firms, cooperatives, and individual proprietors);
- Rights-holders (property owners, Indigenous peoples and tribal groups, communities or individuals that hold traditional or formally recognized usufruct rights to land or resources);

- The external community (some UN organizations, international financial institutions, bilateral donors, and INGO's, volunteer organizations and other parties recognized by inter-governmental conventions);
- Parties further invited to support capacity building and monitoring and assessment of sustainable forest management (SFM).

There can be no sustainable development in general, and SFM specifically, without the active participation of members of civil society. It is the role of these stakeholders to hold governments and businesses to account regarding their performance and honesty, to organize and mobilize communities and deliver services. Civil society organizations with interest in SFM include philanthropies that support science, research, and education and help poor communities. Other civil society organizations that defend the environment against degradation and other externalities include social enterprises, often with distinct legal status, that work on a business model yet do not pursue profit as their sole or main motive. Collaborating civil society organizations demand elaborate social and environmental safeguards before starting joint implementation of SFM projects.

Many countries are beginning to engage stakeholders in domestic SFM decision-making processes, often with support from, or sometimes at the request of, bilateral, multilateral, and non-governmental initiatives. There is general agreement that stakeholder participation is vital for helping decision-makers gather information needed to identify more effective solutions, mitigate risks with regard to potential conflicts, and ensure that the rights of affected groups are upheld. For example, the United Nations General Assembly has adopted the United Nations Declaration on the Rights of Indigenous Peoples, which emphasizes the full and effective participation of relevant stakeholders, in particular, Indigenous peoples and local communities, in actions related

Figure 11.5 Discussions underway among civil society delegates at the 2015 Conference of the Parties (COP21) to the United Nations Framework Convention on Climate Change in December 2015 in Paris, France. Approximately 40,000 are believed to have attended the conference, indicating the importance of civil society participation in such negotiations.

to natural resources management (United Nations, 2000a). As discussed in Chapter 10, Indigenous peoples are significant stakeholders who have their characteristics, ambitions, grievances, and approaches to plan and implement SFM policies and other forest-related initiatives. Some Indigenous peoples' organizations are well organized, powerful, and technically competent, while others are weak in some or all of these areas. Many Indigenous peoples' organizations that manage their own forests lack formal structure (such as umbrella associations), are overwhelmed with externally funded projects and initiatives, cannot agree on what representation entails, have different perspectives of SFM, and lack expertise to address specific technical SFM issues.

In addition to processes that engage stakeholders around specific forest policies, programs, or plans, there are many ongoing efforts to build the capacity of stakeholders to participate in existing or future SFM activities. The terms participation and consultation (the terms most commonly used to describe stakeholder engagement in SFM and related processes) are often used in different ways, which may obscure the extent to which stakeholders are involved in decision-making and plan implementation. Participation refers broadly to the involvement of stakeholders in decision-making or actions.

As shown throughout this book, SFM addresses a uniquely complex set of issues. It is often implemented in challenging environments that are largely outside the control of both managers and beneficiaries. Forest degradation, extreme poverty, insufficient infrastructure, weak governance, and social conflicts are prevalent in many situations where SFM is to be implemented. Furthermore, some implementation projects may face other challenges such as changes in government policies and bureaucratic delays in the transfer of funds. It is generally recognized that some of the challenges faced by stakeholders to perform a meaningful role in planning and executing SFM strategies stem from the ambiguity often associated with the terms used.

Partnership denotes a voluntary collective initiative to address common interests and goals wherein responsibilities for actions and outcomes are shared. In drawing national and local SFM strategies, these 'voluntary' arrangements have become almost obligatory. Usually, partners (stakeholders) agree to respect and understand each other's perspectives (Biermann *et al.*, 2007). Cross-sector partnerships have been proposed and practiced as an effective approach for implementing multilevel, multi-actor governance. Cross-sector partnerships involve different actors, with different levels of influence and power, who work together to implement policy.

Multi-stakeholder partnerships assemble a diverse assortment of societal interests, experiences, and perspectives. They provide meaningful opportunities for governments to practice their social responsibilities and support collaborative decision-making in the environmental domain. Collaborators are considered stakeholders when they are part of attempts to resolve certain problems and/or when they contribute certain experiences, information, knowledge, and views to the decision-making processes and perhaps later on during implementation (Poncelet, 2001).

National and international NGOs are generally active in the countries where SFM (including those initiatives that address climate change and social activities) is practiced. Environmental NGOs and various civil society actors have been instrumental in providing evidence of unsustainable land use and forest management practices and in helping develop new SFM standards. Their actions have frequently assisted both the demand and the supply sides of sustainable commodity production. However, for SFM to be effective on the ground it is necessary to influence and shift commodity investments that lead to deforestation, and to encourage new investments in sustainable practices that enhance value created in the forest sector (Hudson *et al.*, 2013). The overwhelming number of initiatives and projects related to SFM and climate change (especially REDD+) exerts additional stresses on many NGOs in several countries.

Although it is not uncommon that stakeholders have diverse perspectives of good forest management and governance and how they should be put into practice, it is generally accepted by all stakeholders that improving forest governance is vital in moving towards SFM that benefits people and nature (Broekhoven *et al.*, 2012).

Conclusion

A common perception of land-use planning is that it provides certainty. A consequence of this has been that land-use policy has centred on providing a stable environment for resource extraction activities that support the export of raw materials, such as timber, and that any use that is contrary to, or competes with, resource extraction is resisted, as economic certainty may be threatened.

An important principle for land-use planning participants is the legitimacy of the process (i.e. it should be open, democratic, and transparent), independent of the planning outcome achieved. An important component of legitimacy is the concept of representativeness: the representation of values, interests, and concerns. When this legitimacy comes into question, land-use planning may be seen as 'a cynical and undemocratic process in the guise of openness and participatory government' (Mascarenhas and Scarce, 2004, p. 31).

More intensive processes, such as negotiations, engage fewer people, are less likely to be reflective of socio-economic characteristics, and are more limited in outreach to constituencies and communities. This is consistent with Overdevest's (2000) observation that structural barriers to public participation do limit participation in land-use planning for the entire range of the public, as people with low incomes and little discretionary time are less able and less likely to become involved in land-use planning exercises, and participation in planning competes with participation for discretionary time (e.g. biographical availability). This lends support to the 'two table' problem, wherein all civil society stakeholders do not have the time or resources to participate at both the official planning table and their own constituency table, which can result in a compromise of stakeholder accountability to their constituencies.

Further reading

Beckley, T.M., Parkins, J.R., Sheppard, S.R.J. 2005. *Public Participation in Sustainable Forest Management: A Reference Guide*. Edmonton, Alberta: Sustainable Forest Management Network. Available at: http://cfs.nrcan.gc.ca/publications?id=26206.

Creighton, J.L. 2005. *The Public Participation Handbook: Making Better Decisions through Citizen Involvement*. London: John Wiley and Sons.

Jeanrenaud, S. 2001. *Communities and Forest Management in Western Europe: A Regional Profile of WG-CIFM the Working Group on Community Involvement in Forest Management*. Gland, Switzerland: IUCN.

Kusel, J. (ed.) 2003. *Forest Communities, Community Forests*. Lanham, MA: Rowman & Littlefield.

Larson, A.M., Barry, D., Dahal, G.R., Colfer, C.J.P. (eds.) 2010. *Forests for People: Community Rights and Forest Tenure Reform*. London: Earthscan.

O'Hara, P. 2010. *Enhancing Stakeholder Participation in National Forest Programmes: A Training Manual*. Rome: National Forest Programme Facility, Food and Agriculture Organization of the United Nations.

Scott, C. 2011. *Aboriginal Autonomy and Development in Northern Quebec and Labrador*. Vancouver: UBC Press.

Chapter 12

Sustainable forest management (SFM) from an international perspective

Hosny El-Lakany

Criterion 7 of the Montreal Process covers the legal, institutional and economic framework for forest conservation and sustainable management. Within this, there are a number of sub-criteria and indicators defined in the 2015 version of the criteria and indicators:

7.1.a   Legislation and policies supporting the sustainable management of forests
7.1.b   Cross sectoral policy and programme coordination
7.2.a   Taxation and other economic strategies that affect sustainable management of forests
7.3.a   Clarity and security of land and resource tenure and property rights
7.3.b   Enforcement of laws related to forests
7.4.a   Programmes, services and other resources supporting the sustainable management of forests
7.4.b   Development and application of research and technologies for the sustainable management of forests
7.5.a   Partnerships to promote the sustainable management of forests
7.5.b   Public participation and conflict resolution in forest-related decision making
7.5.c   Monitoring, assessment, and reporting on progress towards sustainable management of forests

Some of these areas have been addressed in Chapters 10 and 11 and will not be repeated here. As every country has its own unique set of laws and policies determined by its history and particular circumstances, it would be impossible to examine the forest laws of every country. Instead, this chapter focuses on the development of forest policy at an international level, since this generally reflects consensus views of the most important aspects of national forest policies.

Global policies to promote sustainable management of forests and woodlands

Brief historical background

Forests, covering nearly one-third of the Earth's terrestrial surface, have always been closely associated with human well-being worldwide. As an essential natural heritage and an important economic asset they are vital for sustaining life on Earth and stabilizing the global environment. Yet over hundreds of years, but especially in the past 50 years, humans have changed forest ecosystems rapidly, intensively and extensively. In recent history, there has been a net loss of forest cover of 129 million ha between 1990 and 2015 according to FAO's Global Forest Resources Assessment 2015 (FAO, 2015). Although there are indications that the rate of net forest loss (difference between gain in forest areas and loss to deforestation and conversion), is slightly declining, the rate of deforestation is still alarmingly high, particularly among natural forests.

From an international and national management and policy perspective, forests are complex ecosystems, politically sensitive and of cross-sectoral concern. Early approaches to forest management assumed that forests were isolated from broader social forces and had a singular, unambiguous purpose: timber production. Those approaches became insufficient in an era when forests are recognized as forceful features of the national and international social, economic and environmental fabric. Traditional approaches to forestry cannot cope with the modern intensity, variation and complexity of human activities and expectations. A new sense of scarcity and increased understanding of global functions are shifting attitudes of governments, civil society and the capital markets to the forest systems as valuable, diverse and vulnerable assets (El-Lakany *et al.*, 2007).

Forests emerged on the international policy and political agendas in the late 1960s and early 1970s following alarm raised by the environmental community and some governments about the unprecedented rates of deforestation and forest degradation, environmentally unsustainable forestry practices in many parts of the world, and the consequent loss of their multiple values and benefits. The 1972 UN Conference on the Human Environment held in Stockholm, Sweden, produced three major sets of decisions that included the Stockholm Action Plan, containing 109 recommendations for governments and inter-governmental organizations on international actions against environmental degradation. Although very few of the recommendations addressed forestry specifically, many governments accorded reducing deforestation high priority in their deliberations, and government regulations were seen as the key to offset these losses.

The UN Conference on Environment and Development (UNCED), also known as the Earth Summit, was held in 1992 in Rio de Janeiro, Brazil, with representatives from

Figure 12.1 Loss of tropical rainforest in peninsular Malaysia. Population pressure is resulting in rainforest being cleared for food production.

178 countries including over 100 heads of state and government and some 17,000 stakeholder participants. The continued deforestation and forest degradation at very high rates, among other unsustainable practices, were highlighted.

The principal output of UNCED was the Rio Declaration on Environment and Development, and a 40-chapter program of action known as Agenda 21. Despite wide differences of opinion on how to address pressing problems in the forest sector, the world forestry community gathered in Rio reached an agreement on a set of non-legally binding principles for the management, conservation and sustainable development of all types of forests, also known as the Forest Principles (United Nations, 1992a). This agreement represented a historical breakthrough in the global forestry debate by providing, for the first time, a common basis for commitment and action at international, regional and national levels to manage forests sustainably. Similarly, Chapter 11 of Agenda 21 was negotiated to specifically support efforts to combat deforestation and land degradation.

In the post-UNCED era, countries, intergovernmental organizations and several other stakeholders realized that the absence of a global forum dedicated to discuss forest policy issues seriously impeded progress in achieving sustainable forest management worldwide. The Intergovernmental Panel on Forests (IPF), followed by the Intergovernmental Forum on Forests (IFF), were created to facilitate intergovernmental forest policy deliberations between 1995 and 2000. That period witnessed confidence building, international dialogue and the forging of partnerships among stakeholders. Throughout those processes, the international community identified several pertinent issues and reached consensus on some of them.

The momentum created in Rio and the participation of additional stakeholders came to bear, and forestry was raised to new heights during the negotiation of the United Nations Framework Convention on Climate Change (UNFCCC) and the Convention on Biological Diversity (CBD), among other intergovernmental forums. The concept of forest management shifted from sustained yield forestry, aimed at producing a specific product(s), to managing the forests sustainably, a conceptualization that requires viewing forests as ecosystems that simultaneously provide multiple values and benefits to the environment and society. Civil society exerted powerful influence on forest policy and played important roles in mobilizing and pioneering innovative action on the ground.

Although the international community has made significant progress in the development and coordination of international forest policy since the 1992 UNCED, existing forest management plans in some countries remained limited to ensuring sustained production of wood, without due concern for other forest goods and services or their social and environmental values. In addition, many countries lacked appropriate forest legislation, regulation and incentives to promote sustainable forest management practices (Collaborative Partnership on Forests, 2012a), while in many others illegal forest activities, especially illegal logging, continued unabated. Forest governance and law enforcement remained weak in many developing countries, especially those who have inadequate financial and human resources for the preparation, implementation and monitoring of sustainable forest management plans and lack mechanisms to ensure the active participation and involvement of all stakeholders in forest policy and management planning and development. Many of the world's forests and woodlands, especially in the tropics and subtropics, are still not managed in accordance with the Forest Principles.

The United Nations Forum on Forests (UNFF) was established in 2000 by the Economic and Social Council of the United Nations to promote the management, conservation and sustainable development of all types of forests. The UN resolution (Resolution 2000/35) establishing the Forum further stipulated that UNFF

> should address forest issues and emerging areas of concern in a holistic, comprehensive and integrated manner; enhance cooperation and policy and programme

coordination on forest-related issues; foster international cooperation and monitor, assess and report on progress; and strengthen political commitment.

UNFF has also recognized the following seven thematic elements of sustainable forest management (SFM):

1   Extent of forest resources;
2   Biological diversity;
3   Forest health and vitality;
4   Productive functions of forest resources;
5   Protective functions of forest resources;
6   Socio-economic functions;
7   Legal, policy and institutional framework.

These thematic elements are based on the criteria of the nine ongoing regional/international processes on criteria and indicators for SFM, and are very similar to the first chapters of this book.

During the sixth session of the UNFF in 2006, member states endorsed four Global Objectives on Forests in support of SFM:

> (1) Reverse the loss of forest cover worldwide through SFM, including protection, restoration, afforestation and reforestation, and increase efforts to prevent forest degradation, (2) Enhance forest-based economic, social and environmental benefits, including by improving the livelihoods of forest-dependent people, (3) Increase the area of protected forests worldwide and other areas of sustainably managed forests, as well as the proportion of forest products from sustainably managed forests, and (4) Reverse the decline in official development assistance for SFM and mobilize increased new and additional financial resources from all sources for the implementation of SFM.
>
> (UN Forum on Forests 2006, pp. 3–4)

Figure 12.2 The Food and Agricultural Organization of the United Nations Committee on Forestry (COFO) meets every two years in Rome, Italy.

Based on the UNFF recommendation of 2007, the UN General Assembly (UNGA) adopted the Non-Legally Binding Agreement on All Types of Forests (NLBA) (United Nations, 2008, pp. 3–4), with the name being changed in 2011 to the UN Forest Instrument. The main objectives of that Instrument were to

> (a) strengthen political commitment and action at all levels to implement effectively sustainable management of all types of forests and to achieve the shared global objectives on forests, (b) enhance the contribution of forests to the achievement of the internationally agreed development goals, including the Millennium Development Goals, in particular with respect to poverty eradication and environmental sustainability; and (c) provide a framework for national action and international cooperation.

Meanwhile, at a special session of the UNGA in 2000, countries agreed on the UN Millennium Development Goals (MDGs) (United Nations General Assembly, 2000, p. 4), committing themselves to 'making the right to development a reality for everyone and to freeing the entire human race from want'. Although sustainable forest management was relevant to several of the MDGs, forestry did not feature prominently in the MDGs and remained splintered over many conventions such as the UNFCCC and the CBD. The only reference to forestry in the MDGs is forest area as one indicator under Goal 7: 'Ensure environmental sustainability'.

After the Rio+20 meeting in 2002, the United Nations set up several processes to develop Sustainable Development Goals (SDGs) to replace the 2000–2015 MDGs. The SDGs were agreed by the UNGA in September 2015. The SDGs not only frame global and national aspirations but also guide much of the investments from overseas development assistance (ODA), foreign direct investment (FDI) from the private sector, and the budgets of developing country national governments.

After almost 30 years of discussions and debates, sustainable development (SD) has taken centre stage in the new UN 2030 Agenda for Sustainable Development, which is focused on achieving a state of global sustainable development through a set of 17 broad global goals. The SDGs are accompanied by 169 targets and are being further elaborated through indicators focused on measurable outcomes. They are action oriented, global in nature and universally applicable.

Among the SDGs endorsed by the UNGA in September, 2015, Goal 15 is the main one that explicitly mentions forests, even though forests have a role to play in moving towards many of the other goals. Goal 15 reads: 'Protect, restore and promote sustainable use of terrestrial ecosystems, sustainably manage forests, combat desertification, and halt and reverse land degradation and halt biodiversity loss' (UNGA, 2015). Its targets are:

15.3 By 2020, combat desertification, and restore degraded land and soil, including land affected by desertification, drought and floods, and strive to achieve a land-degradation neutral world;
15.4 By 2030 ensure the conservation of mountain ecosystems, including their biodiversity, to enhance their capacity to provide benefits which are essential for sustainable development.

Forests and forestry are mentioned very briefly as well in Target 6.6: 'by 2020 protect and restore water-related ecosystems, including mountains, forests, wetlands, rivers, aquifers and lakes'. SDG 7 and others also link closely to forests and trees.

More recently, forests featured in the Paris Agreement adopted at COP21 of the UNFCCC in 2015 as a key climate change mitigation tool, while previous COP decisions for a framework to reduce emissions from deforestation and forest degradation (REDD+) were reaffirmed.

The Paris Agreement states in Article 5 (Forests): 'Parties should take action to conserve and enhance, as appropriate, sinks and reservoirs of GHGs as referred to in Convention Article 4.1(d) including forests'; and

> Parties are encouraged to take action to implement and support, including through results-based payments, the existing framework as set out in related guidance and decisions already agreed under the Convention for policy approaches and positive incentives for activities relating to REDD+, and alternative policy approaches, such as joint mitigation and adaptation approaches for the integral and sustainable management of forests, while reaffirming the importance of incentivizing, as appropriate, non-carbon benefits associated with such approaches.

In conclusion, nearly all the international political deliberations and decisions related to forestry have, so far, focused on reducing deforestation and forest degradation as the main objective of SFM. Often the debate addressed 'what' should be done, but little on 'by whom', 'who' would cover the expenses and 'how' to administer such global funds (if/when they materialized). A summary of some of the intergovernmental multilateral agreements is given in Box 12.1.

Box 12.1 Intergovernmental multi-lateral agreements pertaining to forest management.

Legally Binding Conventions

UN Convention on Biological Diversity (CBD) (UN, 1992a)	The United Nations Convention on Biological Diversity was one of the landmark agreements reached at the UN Conference on Environment and Development. The Convention produced two protocols dealing the biosafety (Cartagena Protocol) and access to genetic resources and benefit-sharing (Nagoya Protocol). By 2013 the number of the CBD parties had reached 193.
UN Convention to Combat Desertification (CCD) (UN, 1996)	This Convention built on long-term concerns: the United Nations Conference on Desertification (UNCOD), held in 1977, had adopted a Plan of Action to Combat Desertification (PACD), but this was largely ineffective. As a result, the United Nations Environment Programme (UNEP) reached the conclusion in 1991 that land degradation in arid, semi-arid and dry sub-humid areas was actually intensifying. The UN-CCD was adopted in 1994 and entered into force in 1996. It was intended to focus primarily on Africa, and the political will to push forward with it at UNCED may have been partly because African nations were looking for their own convention, as the UN-CBD and UNFCCC were primarily focused on other areas. As of 2013, 195 parties had signed the Convention.

UN Framework Convention on Climate Change (UNFCCC) (UN, 1992b)	The UNFCCC aims at limiting and reducing the emissions that are responsible for climate change. Although legally binding, the Convention itself drove little change in the behaviours of countries in limiting the emissions. Signing the Kyoto Protocol, which is a part of the process and sets specific goals for emission reduction, was a step towards strengthening the Convention. By 2013, 195 parties had signed the Convention; however, shifting commitment of some signatories questions the effectiveness of this mechanism.
Convention on Wetlands (Ramsar Convention, 1971)	An intergovernmental convention aimed at protecting the ecological character of Wetlands of International Importance, as well as sustainable use of all wetlands in signatory countries, the Ramsar Convention was developed outside the UN umbrella. At present, 164 parties have joined the Ramsar Convention.
Convention Concerning the Protection of the World Cultural and Natural Heritage (World Heritage Convention, WHC) (UNESCO, 1972)	The World Heritage Convention is aimed at preserving natural and cultural World Heritage List sites and finding the balance between humans and nature. Discussed at the Stockholm United Nations Conference on the Human Environment in 1972, the Convention combined documents developed separately for cultural and natural heritage. By 2013, 190 parties had joined the Convention.
Convention on International Trade in Endangered Species of Wild Fauna and Flora (CITES) (CITES, 1973)	Originated from the work of IUCN (the World Conservation Union), the Convention was ratified in 1973. The goal of the Convention is to regulate the trade in endangered species so that the survival of the species is not compromised. At present, the Convention unites 177 parties.

Non-binding UNCED Documents

Agenda 21 (UNSD, 1992)	Agenda 21 sets broad goals for eradicating poverty through understanding of the driving forces for environmental change, such as population, technology and consumption. This document did not result in developing a binding agreement; however, it raised issues that were later expanded in the Millennium Development Goals (UN, 2000b).
	The United Nations Commission on Sustainable Development (CSD) was established in 1992 as a functional commission of the UN Economic and Social Council. It arose from one of the recommendations made in Agenda 21, one of the agreements reached at the UNCED. It is the international body with the responsibility for reviewing progress made in the implementation of Agenda 21 and the Rio Declaration on Environment and Development, and also provides policy guidance on the follow-up to the Johannesburg Plan of Implementation (JPOI), an international agreement reached at a meeting held in Johannesburg, South Africa, in 2002 as a follow-up to the 1992 UNCED. It is the world's high-level forum for sustainable development within the United Nations system.

Forest Principles	The Principles appeared as Appendix 1 of the Rio Declaration. One of the most important principles is 2(b), which states 'Forest resources and forest lands should be sustainably managed to meet the social, economic, ecological, cultural and spiritual needs of present and future generations. These needs are for forest products and services, such as wood and wood products, water, food, fodder, medicine, fuel, shelter, employment, recreation, habitats for wildlife, landscape diversity, carbon sinks and reservoirs, and for other forest products'.
	The Forest Principles agreed in 1992, and progress made since then towards a non-legally binding Forests Convention, including the agreement reached in 2007 (UN, 2007).
Rio Declaration on Environment and Development (UN, 1992c)	The Declaration built on the ideas of the Stockholm 1972 Conference and described goals and responsibilities of nations for future development and care for the environment. It placed environmental protection as a starting point for economic prosperity. The Declaration was one of the provisions for the establishment of the CSD.
Non-legally Binding Instrument on All Types of Forests (UN, 2007)	The instrument builds on the global forest objectives agreed the previous year within the framework of the United Nations Forum on Forests. It places sustainable forest management centrally in the international discussion about forests, and has a set of elements closely related to the criteria set out in many documents in the 1990s and 2000s. The instrument was by the UNGA in 2007.

Internationally recognized interpretations of SFM

In its broadest sense, forest management encompasses the administrative, legal, technical, economic, social and environmental aspects of the conservation and use of forests. It implies various degrees of deliberate human interventions, ranging from actions aimed at safeguarding and maintaining the forest ecosystems and their multiple functions to favouring specific socially and economically valuable species or groups of species for the improved production of goods and services and maintaining the forest ecosystems.

'Sustainable forest management', 'ecologically sustainable forest management', 'forest ecosystem management', the 'ecosystem approach' to forest management and 'systemic forest management' are among the many terms used to describe concepts and practices that incorporate the three pillars of SFM – economic, environmental and socio-cultural aspects – to varying degrees. Recent discussions at international forest fora have focused on the similarities between SFM and the ecosystem approach as applied to forests, where they differ and how they could be integrated (Wilkie, 2003). Currently, SFM is regarded as inclusive of all other terms.

Sustainable forest management (SFM) and sustainable management of forests (SMF) are often used synonymously. For example, in 2007 the UNFCCC's Bali Action Plan

Figure 12.3 Selective harvesting in native forest in the Central Highlands of Tasmania, Australia. Despite definitions of what constitutes SFM, there is still controversy, with the most extreme position being that any form of intervention in natural forests should be prevented.

used SFM in a narrower context, relating to the productive functions of forests, but not including conservation functions and the enhancement of carbon stocks through afforestation, reforestation and forest restoration. However, it is generally agreed that the SFM concept encompasses both natural and planted forests in all geographic regions and climatic zones. Both terms provide for nearly all forest functions; management for conservation, production or multiple purposes, and the range of forest ecosystem goods and services at the local, national, regional and global levels (Collaborative Partnership on Forests, 2012b).

The need for global effective policies to promote SFM

While governance, planning and management reforms are gradually modernizing approaches to forest management towards sustainability (i.e. SFM), in many cases forests continue to be managed through conventional means with single or few objectives. These often fail to manage forests for their multiple functions and are therefore unable to adapt to, integrate or address the challenges facing forests today. The implementation of SFM, however, requires an effective policy and regulatory framework across sectors and institutions. SFM is often hampered by market distortions, a lack of ownership and secure forest land tenure, and governance failures (World Commission on Forests and Sustainable Development, 1999). Moreover, SFM can only work where there is strong societal recognition of, and demand for, the multiple functions of forests and a willingness among policy-makers to prioritize the long-term benefits of forests and SFM over short-term economic gains (Collaborative Partnership on Forests, 2012a).

Another obstacle to the implementation of SFM is the lack of valuation of many of the goods and services provided by forests (see Chapter 9 for further discussion). Forests benefit societies in many ways, but generally only a few such benefits, especially the provision of wood, are paid for. The failure to internalize the full range of benefits provided by forests – such as carbon sequestration and the protection of water catchments and soils – reduces the financial competitiveness of SFM versus forest conversion or unsustainable forms of forest management.

Although biodiversity and other key environmental services have traditionally been thought to be sustained through the establishment of protected areas, the wide range of competing uses for forests by diverse groups imposes constraints on how much can be achieved by protection alone. Improving forest management practices in production forests (forests where productive use is permitted and practiced) is an essential component of any strategy to protect vital local environmental services, in addition to efforts aimed at bolstering the effectiveness of management within protected areas.

Pressures on forests from poorly aligned strategies of other sectors such as agriculture, transportation, energy and industry, as well as from unsound macroeconomic policies, are major causes of forest loss and degradation. Cross-sectoral cooperation to coordinate policies is essential to avoid forest degradation and to ensure that forests are managed in a sustainable manner (World Bank, 2008). Conservation and production must coexist for the full potential of forests for poverty reduction to be realized. Although large areas of the world's forests must be preserved intact for their ecological and cultural values, much of what remains will inevitably be used for productive purposes. Consequently, a dual approach covering both protection and productive use is needed (Pokorny, 2013).

Criteria and indicators (C&I) of SFM (see Chapter 2) provide a framework to conceptualize, evaluate and implement sustainable forest management from local to global levels. Criteria define and characterize the essential seven thematic elements of internationally agreed SFM (see the previous section), as well as a set of conditions or processes by which SFM may be assessed. These include extent of forest resources; biological diversity; forest health and vitality; productive functions of forest resources; protective functions of forest resources; socio-economic functions; and the legal, policy and institutional framework. Indicators, periodically measured, reveal the direction and scale of change with respect to each criterion. Criteria and indicators of SFM can be used at the national or management unit level to report and assess progress towards achieving SFM. Nine international and regional C&I processes are operational across

Figure 12.4 Oil palms established by smallholders on recently cleared forest in Sarawak, Malaysia. Policies aimed at curbing deforestation need to be integrated with policies aimed at sustainable development and poverty alleviation.

various forest zones (boreal, temperate and tropical), including more than 150 countries (Hudson *et al.*, 2013).

Another global measure to support SFM is certification (see Chapter 16). Forest management certification is granted when independent inspection certifies that forest management meets internationally agreed-upon principles, criteria and standards of SFM or responsible forest management. Chain of custody (CoC) certification is granted when independent inspection tracks certified wood and paper products through the production process from the forest to the final product and to the consumer, including all successive stages of processing, transformation, manufacturing and distribution. The certified label ensures that the forest products used are from responsibly harvested and verified sources or forests under SFM.

Links between international, national and local forest management policies

SFM enables the creation of synergies, at the local and national levels, among policies for all forest goods and services including conservation and enhancement of biodiversity, carbon storage, soil and water productivity, livelihoods and other benefits provided by forests. SFM can also capture multiple benefits in a multipurpose approach spanning different sectors of national and local economies and achieving results that are greater than the sum of its parts. Under SFM, economic, ecological and social functions of forests can be realized and simultaneously pursued by setting a hierarchy of objectives at different spatial scales – from the landscape to the forest stand to single ecosystem components. The forest management unit (FMU) is usually the basic site for implementing national forest policies and SFM. However, there are organic links among local, national and international forest concerns. The interdependencies of policies governing all three levels of management have been demonstrated over many years.

Notwithstanding the sovereign rights of every country to control and manage its national natural resources including forests and woodlands, forests are of global concern. Forest issues are trans-boundary: forest dwellers, especially Indigenous people, usually move more or less freely across national borders; emission of GHGs, though initiated and exerting its effects locally and nationally, is a global issue, especially concerning adaptation to and mitigation of climate change; some countries share the same river catchment, and sometimes the upstream is located in one country while the downstream is in others; ecosystems and their biodiversity do not follow administrative boundaries; and forest fires, pests and diseases frequently cross political boundaries unabated. Moreover, there is a range of processing of and trade in timber and non-timber forest products from local to international levels.

Although the intergovernmental agreements, treaties and conventions such as UNFCCC, CBD, CITES and the NLBA on Forests are signed by national governments and ratified by national legislative bodies, they are implemented mainly at local levels, mostly through national forest programs (NFPs). In fact, the international policy dialogue on forests concluded that NFPs provide a sound framework for countries to lead forest policy development and implementation. At the fourth session of the Intergovernmental Panel on Forests in 1997, countries agreed that NFPs or similar national forest policy approaches to achieve sustainable forest management were long-term iterative processes, built on the principles of country leadership, broad participation, integration with national development strategies, and collaboration across sectors to address cross-cutting issues, including the need to reduce poverty and improve the livelihoods of people who live in and around forests.

*Figure 12.5
Fireweed
(Epilobium
angustifolium)
on the site of a
recent fire near
Tok, Alaska,
USA. Policies
allowing fires
in remote areas
to burn need
to be balanced
against the
carbon emissions
from such fires.
On some days
in 2015, forest
fires in Indonesia
emitted the
same amount of
carbon dioxide
as the whole of
the USA.*

When the UNFF reached an agreement on a non-legally binding instrument on all types of forests (NLBI) in 2007, NFPs were considered an important tool to make it operational (National Forest Programme Facility, 2012). There is an international consensus among stakeholders on many aspects of SFM including, for example: approaches to formulate national forest programs (NFPs) through a transparent, inclusive and participatory process; criteria and indicators (C&I) that characterize SFM in the context of diverse regional economic, environmental, social and political regimes; understanding the underlying causes of deforestation; the use of traditional forest related knowledge; conservation and rehabilitation of forests in countries with low forest cover; and fostering international cooperation (Maini, 2004). Nevertheless, some developing countries face major institutional, financial and political constraints in the development of their NFPs and the initiatives to mainstream NFPs into national development strategies and inter-sectoral coordination mechanisms as well as attempts to integrate international commitments into national forest policy formulation have been weak.

The national forest programs to implement SFM have, however, been, and to some extent still are, a challenge in terms of being fully understood by the state forest authorities. For example, 'national' is often understood in the sense that NFPs are government led and owned, but that they are also supposed to be developed, implemented, monitored and assessed at sub-national and local levels is still not clear. In addition, while 'forest' is understood as the framework for national plans for SFM, it is not fully understood that collaboration across sectors on governance, land management, environment, tenure, energy and poverty is fundamental to achieving SFM. That NFPs are not just about forestry is still a major challenge in several countries. Meanwhile, 'program' is often understood in the strict sense of the word, but that it is supposed to be a continuous process took time to be accepted. As of 2010, some 130 countries had reported that they have NFPs (FAO, 2010).

In order to further enhance the contributions of SFM to sustainable development, a number of coordinated actions and policies are still needed at the local, national and international levels. The range of policy actions include the promotion of cross-sectoral and cross-institutional collaboration, the further integration of sustainable forest management into national economic development strategies, and capacity building for systematic forest data and information, particularly on the non-cash and informal benefits that forests contribute to economic development (United Nations Forum on Forests Tenth Session, 2013). The policies of international development organizations are also increasingly significant sources of influence and inspiration nationally and sometimes locally. They give prominence to SFM and can often stimulate new thinking about the role and scope of forest laws and policies (Christy *et al.*, 2007).

Forest governance as an essential component of SFM

The concept and application of forest governance

The term 'governance' itself has no universally accepted definition. Sometimes it is used to mean simply the things that government does. Recently, though, there has been a tendency to use the word more broadly, covering informal as well as formal mechanisms and social and economic influences as well as official state actions (World Bank, 2009a). It is generally accepted that forest governance refers to the policy, legal, regulatory and institutional framework dealing with forests, and to the processes that shape decisions about forests and the way these are implemented. The practice of governance is based on fundamental principles of democracy, such as participation, fairness, accountability, legitimacy, transparency, efficiency, equity and sustainability (Broekhoven *et al.*, 2012).

The components of forest governance proposed by Kishor and Rosenbaum (2012) include the norms, processes, instruments, people and organizations that control how people interact with forests. They cover many areas such as traditional culture, current bureaucracy and private markets as well as public laws. It may be impractical to 'engineer' changes in some aspects of governance, like traditional culture, but understanding governance well requires detailed consideration of all its aspects.

Figure 12.6 Off-loading tropical rainforest logs at a mill in Bintulu, Sarawak, Malaysia. Identifying and tracking illegal logs in such situations is exceptionally difficult.

Forest governance has become more complex over the last few decades. The increase in the amount of goods and services that society expects forests to deliver has led to and is a result of an increase in the number of national and international actors and institutions involved in forest governance. Governments have had limited success in governing forests according to internationally agreed goals of sustainable forest management. And it is clear that without the involvement of non-government stakeholders, forest governance will not lead to achieving these goals.

The most obvious concern of forest governance is illegal logging. While legal exploitation (primarily timber and non-timber forest products and services) is a legitimate use of public forests, the findings from a number of recent studies on illegal logging and corruption are staggering. Illegal logging on public forest lands is estimated to cost governments of forest countries at least US\$ 10 billion to US\$ 15 billion a year – an amount greater than the total annual development assistance in public education and health (Collaborative Partnership on Forests, 2012c). Therefore, the importance of forest governance, from the international to the local level, cannot be overstated. And there are major challenges and opportunities associated with governance at each of these levels, and also with how different actors are involved in governance.

The corrosive effects of illegal logging, especially on governance, are not confined to forest operations. Illegal trade in timber and other forest products generates trade distortions and disrupts legitimate economic activities more generally. The illegal trade in wildlife from forests is massive and appears to be increasing, although fewer data are available than for illegal logging. To give an example, the number of rhinoceroses poached in South Africa increased from 13 in 2007 to 1,215 in 2014 (from a population of about 20,500). Approximately 30,000–40,000 elephants are poached annually in Africa, from a total population of about 470,000 (Wittemyer *et al.*, 2014).

Poor forest governance also 'empowers' criminals. Forest crimes such as illegal logging, illegal occupation and use of forest land, woodlands arson, wildlife poaching, encroachment on both public and private forests, and corruption thrive in an environment of poor governance (World Bank, 2009b). Forest loss and degradation resulting from weak forest law enforcement and poor governance have occurred at the expense

*Figure 12.7 White rhinos (*Ceratotherum simum*) in Mkuze Game Reserve, South Africa. Poaching of white rhinos in South Africa has reached epidemic levels, threatening the continued existence of the species. The problem can be directly related to poor governance.*

not only of national economies, but also of the rural people who depend on forest resources for their livelihoods. In short, in moving towards sustainability, the haemorrhaging caused by poor governance must be stopped. Understanding the strengths and weaknesses of forest governance can be useful to discover problems or to track the progress of forest policy reform. A broad diagnostic such as the one offered by Kishor and Rosenbaum (2012) serves to help discover problems and point towards options for addressing them. Subsequent targeted monitoring and assessment is useful to track reforms and improve implementation.

Improving forest governance is an essential prerequisite for promoting sustainable forest management and reducing deforestation and forest degradation. In fact, good governance is required of all sectors of society: governments, businesses and civil-society organizations. National and local governments need to build effective institutions and pursue sustainable development with transparency, accountability, clear metrics and openness to the participation of all key stakeholders. They should uphold and promote the rule of law as well as basic economic and social rights (Sustainable Development Solutions Network, 2012). Governments must also design financing strategies, help mobilize the necessary resources and provide the public goods needed for sustainable development. Such public policy decisions must be made on the basis of scientific evidence.

'Good' forest governance and law enforcement create the capacity for continuous learning and the ability to adapt to lessons learned among those engaged in the participatory processes of governance. This kind of social learning provides the dynamic and adaptive capacity of good governance. It also creates the stability and predictability necessary for all actors to make the long-term commitments necessary to achieve sustainable forest management that benefits people and nature. Transparency, communication and access to information, and multi-stakeholder engagement in deliberative processes (particularly the meaningful participation of disadvantaged groups) are essential ingredients in moving forward with forest governance (Broekhoven *et al.*, 2012).

Key features of good forest stewardship and governance include adherence to the rule of law, transparency and low levels of corruption, empowering stakeholders and seeking their inputs in decision-making, clear tenure rights, accountability of all officials, low regulatory burden and political stability (Kishor and Rosenbaum, 2012). Conversely, poor governance erodes institutions and spreads corruption across the economy through a corruption contagion effect.

The essence of the concept of governance is the many ways in which public and private actors (i.e. the state, private sector and civil society) work together in order to create capacity to make and implement decisions about forest management on multiple spatial, temporal and administrative scales. It is this mutual interaction that is the defining feature of governance institutions and arrangements. Forest governance institutions focus on five primary areas (Broekhoven *et al.*, 2012):

1   Creating coherence between various policies, laws and regulations, customs and practices, both in the forest sector and in other sectors that define ownership and use rights and responsibilities over forests;
2   Increasing the degree to which people respect and abide by these laws, regulations, customs and practices;
3   Enhancing the motivation of private actors to behave in a responsible manner that goes beyond regulatory requirements;
4   Equalizing the relative power and clarifying the mandates of stakeholder groups, as well as stabilizing the institutional arrangements that join them;
5   Enhancing the incentives, enabling conditions and capacity of organizations and individuals to engage in forest governance practices.

Forest governance is a complex endeavour that involves the active participation of a range of participants in civil society, not just forestry administrations. Voluntary market approaches (e.g. investment standards and forest and product certification) complement and implement regulatory goals by focusing on the behaviour necessary to achieve the goals. Thus, they often create incentives for corporate responsibility and opportunities for profit and interest-seeking behaviour to achieve desired public goals. However, voluntary and regulatory approaches depend on one another; neither alone is sufficient.

Trade-based regulatory approaches to forest governance reform include Forest Law Enforcement, Governance and Trade (FLEGT), voluntary partnership agreements, the EU Timber Regulation and the US Lacey Act, which all aim to decrease trade in illegal timber. They have the potential to be effective in that they tie forest governance reform to trade, thus uniting the private sector, governments and civil society around a common interest. In addition, these programs have a tangible benefit, such as continued market access, and contain measures that affect both producer and importer/consumer countries. Although it is generally understood that forest governance reform initiatives need to complement and reinforce each other – if only to avoid wasting limited resources and preventing unnecessary strains on limited capacities – in practice this is sometimes difficult to achieve. Avoiding overlapping or competing initiatives requires two measures, namely proactive strategies by the forest sector to ensure that the interests of the sector are adequately represented in cross-sectoral processes, and effective institutional settings for the forest governance reform process.

The capacity for multilevel and cross-sectoral learning is a distinctive feature of good forest governance. Much greater political commitment, institutional capacity building and strengthened law enforcement are needed to improve forest governance and make a significant and irreversible impact in reducing illegal logging, corruption, encroachment and violations of tenure and ownership rights. Reforms of forest governance will need to move away from the piecemeal approach towards more integrated approaches. Without good forest governance and promotion of legality and sustainability in the wider forest sector, achievement of the objectives of SFM, the European Union (EU)

Figure 12.8 Having an effective police force that is willing to go into forest areas to enforce forest legislation is a critical aspect of good governance. These guards from the Kenya Forest Service are responsible for enforcing local forest legislation, but they are limited in number and have large forest areas to cover.

Forest Law Enforcement, Governance and Trade (FLEGT) Action Plan and related programs will be jeopardized (Program on Forests [PROFOR], 2012).

Governance options towards greater stakeholder consultation and involvement, broader incorporation of diverse interests in decision-making, choices that include civil society, market-based, and hybrid instruments rather than regulations alone, and attention to attitudinal, behavioural, and institutional interventions also present substantial promise for enhancing the efficiency and sustainability of forests and forest management. Such governance options must also cover trees outside forests, and patch to forest areas at landscape levels. But research in the domain of forest governance is scanty, and for the most part is in its early phases. Far more work is necessary to identify the right mix of governance arrangements to address the incredible diversity of social, political, ecosystem and biophysical configurations in which forests are embedded.

The power of the market and consumer preferences has been used in the forest sector for the last 20 years through non-state regulation in the form of certification – and more recently in the form of legality verification through FLEGT, the Lacey Act and other measures – to promote better environmental and social outcomes. Comparable market power is now being applied to the production and trade of some agricultural commodities associated with deforestation including palm oil, soy, and to a lesser extent meat and leather.

Forest laws in support of SFM

In recent years forest laws have been significantly revised around the world in response to the need for implementing SFM and other international commitments. The visions of forests in, among other documents, the Forest Principles, Agenda 21, CBD and the NLBI on Forests have led many countries to move away from regulatory approaches focused primarily on government management and policing of forests as mainly economic resources. The policy changes have required a major reorientation of the law in such areas as local and private forest management, the environmental functions of forests, forest management planning, forest utilization contracts and trade (Christy *et al.*, 2007). Recent years have also seen particular attention paid to modifying heavily state-centric forest laws to accommodate, to the degree possible, customary tenure regimes or to enhance the rights of access, control and management of local communities.

While the regulation of forest management is a traditional core subject of forest law, new trends have emerged. For example, forestry laws now pay more detailed attention to planning and to a broader array of objectives than they did in the past. Increasingly, forest laws consider a wide range of ecological and social issues that might previously have been outside their purview. Planning procedures also provide an important point for the public to intervene in the design of SFM procedures. Many recent laws require public notices at various stages, opportunities to comment, public meetings and access to preliminary plans. New legal techniques have also begun to emerge for the more transparent and responsible allocation, pricing and monitoring of forest concessions and licenses.

Finally, forest laws have sought ways to use incentives to promote private forestry and to roll back the over-regulation of private forests that characterized many older forestry laws. The complementarities among law enforcement, certification and legality verification schemes can be realized only if promoted aggressively and attempts are made to better define and systematically harmonize the legality and sustainability standards, followed by better enforcement and monitoring systems.

Conclusions

Over the past 50 years, the global community has gradually realized the importance of foresters. While foresters themselves have long been stating the importance of forests, as witnessed in the declarations made at successive World Forestry Congresses, it has been and continues to be surprisingly difficult to raise political interest in forests. At the time when several intergovernmental conventions were drawn up on environmental issues (biodiversity, desertification and climate change), at the United Nations Conference on Environment and Development in 1992, forests were only accorded some non-legally binding guidelines for their sustainable management. Assessments and reports point mainly to a decline in the rate at which natural forests are being destroyed, suggesting that this represents a major success.

Developing universally accepted and sustainable international policies related to forests is extremely difficult. For many countries, forests represent a strategic resource, and they are also intimately tied to national development policies. Their functions vary greatly from region to region, and it is only with the realization of the importance of forests to climate change mitigation that there has been international agreement on the need to reduce the rate of deforestation and forest degradation.

A number of recent shifts are evident in the forest policy community. First, there is increasing interest in forest restoration. Large areas of degraded forest and deforested land are available for restoration activities, and this will have the added benefit of sequestering atmospheric carbon (Lamb, 2014). Second, there is recognition that to solve issues such as deforestation, a more holistic approach is required that views forests as a component within landscapes – the focus should be on the landscape itself rather than the individual components that make it up. Third, there is recognition that forests can play a significant role in sustainable development, and one of the Sustainable Development Goals agreed in 2015, as detailed earlier, makes explicit mention of forests (Goal No. 15: 'Protect, restore and promote sustainable use of terrestrial ecosystems, sustainably manage forests, combat desertification, and halt and reverse land degradation and halt biodiversity loss'). Fourth, a number of new actors have emerged. For example, the Asia-Pacific Network for Sustainable Forest Management and Rehabilitation, a major force in the Asia-Pacific region, is encouraging the sustainable management of forests and has strong political support. These all point to increasing attention being given to forests and forestry in the coming years.

Further reading

Arnold, F. E., van der Werf, N., Rametsteiner, E. 2014. *Strengthening Evidence-Based Forest Policy-Making: Linking Forest Monitoring with National Forest Programmes.* Forest Policy and Institutions Working Paper No. 33. Rome: Food and Agriculture Organization of the United Nations.

Christy, L. C., Di Leva, C. E., Lindsay, J. M., Talla Takoum, P. 2007. *Forest Law and Sustainable Development: Addressing Contemporary Challenges Through Legal Reform.* Washington, DC: The World Bank.

Dubé, Y., Schmithüsen, F. (eds.) 2014. *Cross-Sectoral Policy Developments in Forestry.* Wallingford: CABI Publishing.

Humphreys, D. 2006. *Logjam: Deforestation and the Crisis of Global Governance.* London: Earthscan.

Mayers, J., Morrison, E., Rolington, L., Studd, K., Turrall, S. 2013. *Improving Governance of Forest Tenure: A Practical Guide.* London and Rome: International Institute for Environment and Development and Food and Agriculture Organization of the United Nations.

McDermott, C., Cashore, B., Kanowski, P. 2010. *Global Environmental Forest Policies: An International Comparison*. London: Earthscan.

Pierce Colfer, C. J., Pfund, J.-L. 2011. *Collaborative Governance of Tropical Landscapes*. London: Earthscan.

Pierce Colfer, C. J., Ram Dahal, G., Capistrano, D. (eds.) 2008. *Lessons from Forest Decentralization: Money, Justice and the Quest for Good Governance in Asia Pacific*. London: Earthscan.

Tacconi, L. (ed.) 2008. *Illegal Logging: Law Enforcement, Livelihoods and the Timber Trade*. London: Earthscan.

Chapter 13

Addressing social, economic, and environmental objectives and values through decision support systems

Craig R. Nitschke, Anne-Hélène Mathey, and Patrick O. Waeber

Decisions about forests

Forest management in North America and some other parts of the world has been rapidly evolving since the 1990s as concerns for the environment, biodiversity, and the provision of ecosystem services have become increasingly considered along with the traditional utilization of forests for timber production (see Chapter 12). The evolution of forest management can be viewed as a paradigm shift from purely utilitarian management towards multipurpose management (Bengston, 1994). With this shift, ecosystem management (or integrated natural resource management) has developed, which promotes the application of management strategies that can achieve some future desired state over strategies that simply produce some desired mix of resource outputs over time. The development of sustainable forest management (SFM), which involves the integrated management of forests for social, economic, and environmental goals in a manner consistent with the Statement of Principles for the Sustainable Management of Forests and Rio Declaration on Environment and Development, has explicitly created a paradigm where potentially conflicting objectives must be managed, such as conservation of biodiversity versus maintenance of economic benefits.

SFM has also become increasingly complex due to the need for societal participation. The explicit inclusion and recognition of social and cultural values in SFM is increasingly being advocated (Gough *et al.*, 2008). As a result, an increasing number of natural resource management approaches have attempted to span various spatial and temporal scales while incorporating and addressing elements and principles that consider multiple disciplines and involve a multitude of stakeholders that represent a growing array of values and interests within the planning and implementation process. Today, forest managers must deal with these growing complexities that arise from increasing expectations of the forest expressed by a growing number of stakeholders and their values while additionally facing increasing risk and uncertainty due to changing frequencies or magnitudes of external forces, such as biotic and abiotic disturbance and economic or technological changes. Forest managers have to deal with the challenge of resolving these complex issues spanning spatial scales that range from stands of a few hectares in size to landscapes that are tens of thousands of hectares, and temporal scales that can range from hours to decades to centuries. Further complicating the decision making process are the decision makers themselves, since there are clear tendencies to tolerate unsolved problems rather than to accept solutions that are

not fully understood (Seidl and Lexer, 2013). This requires increasingly sophisticated methods for supporting decision making and developing and evaluating alternative forest management strategies that articulate outcomes in a clear and simplistic manner.

Decision making is central to forest management planning (Kangas *et al.*, 2008) and involves choices between distinct alternatives based on three elements: information, alternatives, and preferences. According to Belton and Stewart (2002), decision-aid or support processes consist of three phases: problem structuring, model development, and using the model to inform decision makers and challenge alternative thinking.

Due to the increasing number of values, forest uses, and changes in forest stressors, decision making is becoming increasingly multi-dimensional and characterized by an increased number of unknown elements. Unknown elements are divided into two specific groups that we need to distinguish between: risk and uncertainty. Under risk we have no knowledge about what is going to happen next, but we do know what the distribution of the risk looks like. Under uncertainty we have no knowledge about what is going to happen next, and no knowledge about the possible distribution of this uncertainty.

Uncertainty can be found in all parameters of a decision problem, and the future outcomes and consequences of some or all of the parameters may as a consequence also be uncertain. The consequences of actions are determined by the type of action, which is pre-determined by the decision alternatives and by the external forces or factors that are beyond the decision makers' control. Decision problems are therefore related to both risk and uncertainty.

Achieving truly sustainable SFM requires the incorporation of both risk and uncertainty into long-term planning (Nitschke and Innes, 2008). The complexity of social and ecological systems, however, presents a challenging mix of uncertainty and risk for forest managers. Uncertainty does exist in any decision making process related to forest management, which means that we have to accept a certain level of risk regarding the outcome of our actions. Again, this is caused by the nature of the complex system we are dealing with as forest managers, in which we have yet to find ways to estimate the

Figure 13.1 Tree shelters used to aid the development of a new stand in the Haida Gwaii, British Columbia, Canada. Shelters such as these involve a significant financial outlay, and the costs of this must be assessed against the risk of browsing damage.

probability of natural events – such as disturbances – or natural states that influence the forest planning process. The assessment or predictions of risk for various events are therefore an important element for successful forest management.

In response to the diversity of forest management actions and the ensuing complexity and uncertainty that they represent, an array of tools for decision support, the so-called decision support systems (DSS), have evolved. DSS are developed and used to improve the robustness and transparency of the decision making process. They are computer-based systems that aim to provide support to solve complex decision problems and help evaluate the various courses of action in terms of the desirability of their outcomes. As such, DSS must be able to accomplish at least three user-defined tasks: (1) estimate the consequences resulting from each decision alternative; (2) enable the correct definition and incorporation of what constitutes a 'desirable outcome'; and (3) offer a means of identifying some alternatives as more desirable than others.

In the context of forest management, decisions can range from harvest scheduling to land conservation, the sustainable yield of timber to water and to a factorial of land use and management actions. The consequences and desirability may be of interest to government, industrial operators, non-governmental organizations (NGOs), Indigenous groups, and community forest managers. In the latter case, stakeholders could have a larger and more diversified set of values which would call for a different decision support process than the industrial operator which may have a relatively simpler set of values. At broader landscape scales, the consequences and desirability of management outcomes can incorporate the interests and values of many of the aforementioned groups. The number of possible combinations of decisions, consequences of interest, and sets of values that forest management encompasses can therefore range from a few to boundless. A DSS may not be appropriate for every decision making context, as decision support tools have different properties which affect their aptitude for addressing any given problem.

The next section identifies the elements of decision making problems and decision-support methods that are used in decision support.

Figure 13.2 Decision-support tools must be capable of helping a land manager determine the management strategies for an entire landscape, not just the forests, Central Highlands, Costa Rica.

293

Problem structuring

Decision alternatives: defining the decision problem

The purpose of a DSS is to guide an individual or group through a series of tasks related to an identified problem, including problem analysis, the design of alternative management solutions, and ultimately the selection of a preferred solution that best meets the goals and restrictions of a particular situation. Articulating the stakeholder's goals, preferences, and management paradigm lies at the core of decision support. Based on these attributes, management alternatives and the type and scope of consequences that need to be investigated can be identified. These attributes also form the basis for a differentiation between management alternatives and their desirability. A DSS therefore attempts to express some belief about the 'world', including its complexities and inherent uncertainties.

One common method is to use scenario-based approaches. Scenarios are simply possible futures under certain circumstances. For example, in climate modelling, different scenarios are developed for what might happen to global emissions, and the future climate is modelled based on the expected emissions expected under a particular scenario of world development. Scenario-based approaches have been used to address and deal with complexity and uncertainty and are particularly useful when used in the context of non-linear and paradigmatic changes. Scenario planning is a technique that has been used for many purposes, mostly aiming to develop results of practical applicability, as a guide for the development and filtering of ideas for projects or management, and for the evaluation of existing or new concepts and strategies.

The nature of the decision problem in question also defines the time frame to be covered by all planning activities that become part of our decision making process. The greater the planning time frame (i.e. the further into the future we look), the greater the number of possibilities or possible futures, with some of the alternatives being more probable than others. The most desirable future may not even overlap with possible futures. As robust as scenario-based approaches are as DSS, as spatial and temporal complexity increases, their successful application becomes extremely difficult. To address these challenges, collaborative or participatory decision making can be incorporated in DSS.

The temporal and spatial contexts are critical elements in the initial definition of a problem. The near future bears more certainty; as such, outputs from a DSS focusing on the near future are more reliable than those focusing on more strategic time frames, as they are based on existing relationships rather than probable ones. This means that for shorter time frames decisions can be made under greater certainty. The paradigmatic shift towards SFM, however, has led to an expansion of planning time frames. With this evolution the existing structural stability is replaced by complexity and certainty by uncertainty. One approach used to address these issues is to structure a complex problem into different hierarchies.

Typically, hierarchical structures have been organised into three levels: the strategic level, the tactical level, and an operational level (Bettinger *et al.*, 2009). At the strategic level, the basic goals for achieving what is wanted or desired are identified and formulated over longer time frames (e.g. usually more than 20 years). The tactical level is used to define and describe the proximate mechanisms on how these goals are achieved (usually 5–10 years), and the operational level details the local actions on the ground (one to five years). For example, with the goal to 'maintain a sustainable timber yield', decisions could be made that range from determining a long-term sustained yield in volume per year (strategic) to where in a landscape harvest activities will take place in a given period (tactical), to how harvesting will be done at the

*Figure 13.3 A strategic decision might be to improve forest resilience. Tactically, it might be decided to do this through various mechanisms, including selective thinning and surface fuel reduction. The operational decisions in this scene (ponderosa pine [*Pinus ponderosa*] forest near Cranbrook in British Columbia, Canada) would cover when the operations were performed and how much would be removed from the site.*

stand-level (i.e. silvicultural system; operational). Which choices are more desirable depends on rules and regulations in the jurisdiction, the preference of stakeholders regarding sustainability, impacts on other forest values, and increasingly public acceptance of management actions.

Whether the forest is perceived as a set of distinct elements in juxtaposition or as an integrated system has further ramifications on framing the forest management problem and on which tools need to be drawn upon to link decisions to outcomes. The former relates to the use of simple decision support tools that predict single values and assume that the states of nature (soil, climate, biotic circumstances) are expected to remain unchanged. The focus on a single value, typically growth and yield of timber, dates to the beginnings of forest management in the 1700s (see Chapter 1); however, the reality is that current issues in forest management are complex because they are related to ecosystems and not merely to its sub-components (Kimmins *et al.*, 2008). The acceptance that natural resources are derived from complex systems with overlapping components has led to three main implications. First, all components of a system are interconnected and necessary. Second, these inferences have led to the expansion of the forest management focus from an anthropocentric view (e.g. sustained yields) to emphasizing the maintenance of ecological integrity of the forest ecosystem along with the other, more human-centred values (e.g. ecosystem-based management). The third is that most ecosystems have such a complex web of interacting environmental drivers and ecological processes that they exceed the capacity of mental models and contemporary forest decision support tools.

Consequences of actions: computing and evaluating decision outcomes

Once the decision problem is framed, a DSS can then be used to assess whether possible actions will lead to the desired outcome. A DSS must be able to represent the causal and interdependent relationships between decisions and outcomes. The ability

*Figure 13.4
Fuel reduction
underway in
Cache National
Forest, Idaho,
USA. Much of
the forest area of
the western USA
is considered
unhealthy
because of the
fuel that has
accumulated
(as shown by
the trees in the
background).
Improving the
health of these
forests through
fuel reduction
is considered a
priority, but is a
huge task.*

of the DSS to provide a strong versus weak test of SFM must also be considered: a DSS that can provide an explicit evaluation of a wide range of indicators that represent multiple values (i.e. timber, water, biodiversity, social, cultural, and economic benefits) with minimal assumptions about forest condition and values will provide a strong test. A DSS that provides a weak test of SFM is one that focuses on one or few values and assumes that the maintenance of forest structure over time and space will maintain other forest values. Addressing multiple criteria and multi-stakeholders, often with conflicting interests, requires a more flexible and versatile DSS than traditional simulation and optimization tools. A weak DSS will likely be inexpensive and easy to apply while a strong one is likely to be expensive and time-consuming (Reynolds, 2005). Nonetheless, in the era of ecosystem management and SFM, DSS that provide a strong test of SFM are essential for assisting forest managers in balancing an increasing diversity of resource objectives.

DSS that are capable of providing a strong test of SFM usually include methods such as multi-criteria decision analysis (MCDA) and/or optimization and/or simulation modelling tools that incorporate multi-criteria analysis (MCA), such as the analytic hierarchy process (AHP), and knowledge-based systems that provide a framework for collaborative and/or participatory engagement. To describe variation between alternatives, information from additional models (e.g. growth simulators), information from geographic information systems (e.g. LiDAR data, or Light Detection and Ranging), simple spreadsheets, or databases are used within a DSS framework. For a DSS to deliver robust and useful results relevant to the decision making process a reliable scientific assessment of alternative scenarios, ideally developed through participatory approaches and applied in a collaborative modelling process needs to be the basis. The selected DSS must be able to accommodate the complexity of coupled human-natural systems, while at the same time being able to communicate outcomes with clarity and simplicity. Bohnet *et al.* (2011) classified the use of modelling in decision making as projective or predictive approaches and participatory approaches as an explorative approach to decision making. Individually, both approaches cannot provide enough information for a holistic decision making process. The combination of predictive and explorative approaches is therefore advocated.

Predictive, projective, and explorative approaches to decision support

Predictive and projective approaches

Traditionally, forest planning has been conducted with the aid of decision making tools that focused on optimizing economic (e.g. net present value) or wood production objectives subject to constraints. The focus of research into decision making was concentrated on developing linear programming, integer programming, or heuristic models that attempted to maximize timber flows while satisfying non-timber objectives. The initial models were timber supply models that utilized simple habitat constraints. Many forest planning models now incorporate wildlife, habitat, biodiversity, carbon, water, and timber values, although one value is typically set in the objective function with other values represented as targets, goals, or constraints. The best or optimal value is usually defined in terms of highest volume harvested or most profit obtained (Boyland *et al.*, 2006). Many of the timber supply/harvest scheduling models have evolved into large, multiple-objective forest-level models that simulate landscape patterns and structure. The use of heuristics has facilitated the solving of computationally difficult planning problems that linear and integer programming approaches are unable to solve; however, heuristic approaches still have difficulty in formalizing spatial objectives (i.e. connectivity, reserve formation) during the planning horizon (Mathey *et al.*, 2007). Decentralised and bottom-up modelling approaches, such as cellular automata, which represent discrete dynamical systems whose behaviours are influenced by local relationships, have been advocated for improving the ability of forest planning models to capture the spatial and temporal interconnectedness of forest management parameters (Mathey *et al.*, 2008). The more recent development of agent-based planning models using co-evolutionary algorithms has overcome these spatial limitations and allows for spatial objectives and global objectives to be robustly modelled, although

*Figure 13.5 Harvesting hoop pine (*Araucaria cunninghamii*) in Allan State Forest, Queensland, Australia. Most decision support tools focus on timber supply, and often issues such as biodiversity and riparian zone protection are treated as constraints. New generation models are taking a more balanced approach.*

the approach requires further work before it can be applied to real-world situations. Despite the evolution from single to multi-objective and aspatial to spatial contexts, the decision process within these models is still focused on the timing/location of management activities (i.e. harvest scheduling) by algorithms designed to seek the best or optimal solution to a perceived problem. However, in reality, an optimal or best solution is unlikely to exist (Mendoza and Prabhu, 2006).

Today, there are two forms of decision making models: normative (optimization by exact or heuristic methods) and descriptive (simulation). Within the forest planning model paradigm, simulation models predict the consequences of harvest policy on sustainable timber volume as well as landscape attributes and structure given a set of constraints and simple harvest rules while optimization models find the combination of management actions that produce the highest level of a given objective. The pros of optimization approaches are that they directly address the objectives and provide precise solutions; the cons are that they are constrained by the large size of planning problems and the non-linearity of spatial relationships. For optimization techniques the complexities defining forest management problems suggest that optimal solutions are unlikely to exist. The pros of simulation modelling are that they can accommodate the non-linear patterns that emerge as the consequence of multiple distinct events; they are more holistic and flexible in modelling forest dynamics and management and can efficiently process large planning problems. The con is that this approach does not address management objectives directly but rather describes the effect of decision alternatives on the modelled forest.

To address issues of dynamism in the environment, process-based simulation models have been increasingly advocated as a method for addressing multiple environmental factors. Simulation models have been developed to assess the cumulative impact of stand-level forest management activities on economics, nutrient balances, and biodiversity indicators through the simulation of forest operations, tree growth, regeneration, mortality, and decay of dead wood at the landscape scale. Landscape-level disturbance and succession models (LDSM) have also been developed that simulate the impacts of disturbance (such as fire, insects, drought, and wind), harvesting, and climate on forest composition and spatial pattern. Although these models can address issues of environmental uncertainty, they only provide metrics to assess environmental outcomes of management; they do not address social or economic values. Despite these issues, in the debate between optimization and simulation, landscape simulation models are often regarded as the best tools for predicting future forest conditions and the long-term, large-scale outcomes of management actions. Regardless of model choice, the singular application of these models in a decision making context is predictive and will likely provide only a weak test of SFM. The complexities facing forest managers highlight that forest-level planning models are just one component that is needed to address the shift to multiple-objective forest management planning and provide a strong test of SFM.

Explorative approaches

Forest managers can better align their policy and practices with public perceptions if values are incorporated in the planning process, as values and beliefs are important underlying factors in social acceptability judgements. The development of management scenarios is challenged, however, by the social context and ecological and economic dynamics of the system. For example, when information is inconsistent with one's personal beliefs about the consequences of a particular management action, it may not be used to judge if a management action is socially acceptable (Ford *et al.*, 2009). Despite these issues, Bengston (1994) describes three benefits to managers for understanding social values: (1) aid in establishing forest management objectives by increasing the understanding of the relative importance of different outcomes of management to the public; (2) help managers predict how people will react to particular forest practices; and (3) analyze values to help in understanding and managing value conflicts. To aid in the development of management objectives and scenarios, a range of approaches have been applied to recognize and incorporate social objectives and values. For a comprehensive review of scenario development techniques see Bishop *et al.* (2007). The application of participatory models is increasingly occurring in SFM as it provides a method for incorporating local communities and stakeholders into the planning process. Participatory models are typically broad and strategic in nature rather than operational or tactical, and focus on understanding the problem rather than solving it. Participatory-based approaches can assist local communities with diverse stakeholder interests to develop sustainable strategies for their regions by facilitating the incorporation of local knowledge, perspectives, and priorities which in turn can lead to the development of more effective, locally based, and long-term strategies for addressing complex problems. As such, participatory methods have become an important tool for incorporating stakeholder involvement and for defining and balancing different objectives (Mäkelä *et al.*, 2012).

The use of participatory approaches is critical for SFM, as determining the relative importance of objectives is a significant challenge but critical to scenario development. While participatory methodologies are strong in terms of inviting participation, they typically lack in their ability to provide a structured framework from which management alternatives and strategies can be analysed and evaluated in a robust and quantifiable manner. The quantification of alternative management scenarios and strategies is an important step because it allows for the desired future conditions to be defined, which in turn form the basis of the decision analysis. Multi-criteria analysis (MCA) techniques have been widely adapted to structure and implement criteria and indicators (C&I) for the assessment of SFM.

MCAs provide a structured process for developing scenarios based on a collaborative process that combines expert evaluations and stakeholder input to weight

Figure 13.6 Capitol State Forest, Olympia, Washington, USA. This forest has been used extensively to determine public reactions to particular forms of silviculture such as clearcuts and shelterwood systems.

criteria and/or management objectives. There are numerous approaches for eliciting judgements and weights for objectives which are described in detail by Mendoza and Martins (2006). Common approaches are voting models, aggregation of partial utility/value functions, Paretian analysis, game theory, vote-trading models, interactive approaches, public value forums, and analytic hierarchy processes. Role playing has also been found to be effective for defining scenario objective weights, as has public multiple criteria analysis. One of the most frequently applied decision support techniques in natural resources management is the AHP (Saaty, 1977). An AHP consist of three components: (1) problem structuring, where goals and criteria are identified and defined in a hierarchical structure, (2) evaluation of the problem by stakeholders through the judging and comparisons of elements at each hierarchical level (each element of the AHP structure is rated by identifying the relative importance of the elements in the criteria and sub-criteria); and (3) synthesis (calculation of relative weights and local priorities to global and overall priorities). Elements within an AHP can include objectives, scenarios, events, actions, or outcomes that best describe the criteria listed within the hierarchy. The final result of the AHP is a numeric value that indicates the relative priority of each management alternative in achieving the management goal.

Waeber *et al.* (2013) used a participatory-based approach in combination with an AHP analysis to develop alternative management scenarios for forest managers of the southwest Yukon, Canada. The participatory approach involved a local reference group to define desirable future conditions (management themes), tactics, and local forest values and an expert group to characterize the importance of each forest value given the management theme and tactics available. Hajehforooshnia *et al.* (2011) used an AHP to assign weights to criteria for land-use zoning and linked these to a GIS-based MCDA approach (weighted linear combination approach) to create spatial maps of criteria and constraints which were then used in to delineate management zones in a wildlife sanctuary in Iran.

The approach used by Waeber *et al.* (2013) enabled the exploration (participatory) and quantitative comparison (AHP) of alternative strategies. However the approach, as with most explorative approaches, did not explicitly test outcomes in time and space. The approach used by Hajehforooshnia *et al.* (2011) combined the weighting of values by stakeholders but did not incorporate management alternatives, nor did it explore the predictive outcomes of zonation on land-use values. These approaches, while inherently important to the decision making process, do not provide for a strong test of SFM, as they do not allow the consequences of the alternative actions on environmental and social values to be explicitly tested in time and space.

Combining explorative and projective approaches

To address the limitations of explorative approaches, the outcomes from participatory studies are being combined with decision making tools that enable a rigorous selection of the most preferred alternative when several criteria are considered simultaneously. Many of today's complex forest management planning challenges are addressed using MCDA methods, each with its own set of characteristics and techniques for a specific decision situation and planning context. Mendoza and Martins (2006) provide a comprehensive review of MCDA approaches and frameworks. MCDAs are frameworks for analyzing complex multi-objective decision problems and as such are widely used in forest management planning. MCDAs are highly complex mathematical models that permit the evaluation of multiple C&I and the exploration of trade-offs between competing objectives, and can quantify uncertainty to a degree (Mendoza and Prabhu, 2000). MCDAs are projective in nature as they forecast likely land use and management scenarios based on past trends and actions.

The application of an MCDA typically involves the evaluation of alternative management scenarios across a range of different C&I, creating a matrix within which the performance of each scenario is assessed. Linear optimization, multi-objective optimization, and goal programming are common methods used in MCDA to evaluate alternative management scenarios across a range of different objectives and indicators. When the C&I are based on participatory approaches, an MCDA becomes a decision-support system that has clear applications for SFM planning. Goal-driven MCDA systems can support strong tests of SFM as they can allow for the effects of change on multiple states and processes to be assessed. In combination with AHP techniques, MCDA frameworks can used to define objectives, identify criteria to measure and assess the stated objectives, and thus provide support for a strong test of SFM, which in turn will assist in the development and selection of decision alternatives.

A good example of a MCDA model linked with AHP is provided by Stirn (2006), who developed a DSS based on fuzzy, dynamic, and multi-objective models which can select the sequence of decisions that jointly maximizes economic, ecological, and social objectives while maintaining constraints. Another example is provided by Vacik and Lexer (2001), who developed a spatially explicit DSS to aid in silvicultural planning of forests managed for sustained yield of water resources based on a combined AHP-MCDA approach.

Given that optimization and goal programming approaches are commonly used in MCDA models, the same pros and cons that apply to normative models that have been developed for forest planning problems pertain. While MCDAs can directly address the objectives and provide precise solutions, they are constrained by the non-linearity of spatial relationships (Mathey *et al.*, 2008). They also fail to incorporate sufficient representation of key ecological processes and conditions to make plausible forecasts of changes in ecological and social values at relevant spatial and temporal scales. In addition, the complexities of current forest management problems suggest that 'optimal' or 'best' solutions are unlikely to exist. They are also limited by their inability to reconcile both aspatial goals (e.g. harvest or cash flows) and spatial goals

Figure 13.7 Styx Valley in Tasmania, Australia. Decision tools have been developed that help determine the likely impacts of forestry activities on water supply from forested catchments.

such as dynamic changes in the resources on the land base. The normative structure of MCDAs is also problematic when dealing with climate change, because they are limited in their ability to address the novel conditions of the future as they are based on relationships from the past. The ability for MCDAs therefore to provide a strong test of SFM under dynamic and uncertain futures is weakened by their projective/ normative nature.

Combining explorative and predictive approaches

To achieve a strong test of SFM, ecologically based forecasting tools (i.e. ecosystem management models) are needed that incorporate sufficient representation of key ecological processes and conditions that can make plausible forecasts based on both ecological and social values at relevant spatial and temporal scales (Kimmins, 2002). Models can no longer focus strictly on the management of individual trees or stands, rather they must be able to consider stands in the context of the landscapes in which they exist. The need to consider climate change has also highlighted a further challenge to DSS as the environment that characterizes model inputs and behaviours is dynamic, not static as traditionally assumed. Models therefore need to move from empirical approaches (normative) to mechanistic approaches (simulation) to address the novel conditions of the future (Gustafson, 2013).

Process-based simulation models provide a method for addressing multiple environmental factors and issues of dynamism and uncertainty in the environment. Landscape-level disturbance and succession models such as LANDIS-II that can simulate the impacts of disturbance (e.g. fire, insects, drought, and wind), harvesting, and climate on forest composition and spatial pattern are important tools for supporting strong tests of SFM. Another approach being increasingly used is to develop toolkits that contain different models that operate on different scales, spatially and temporally as well as structurally. The combinations of models into toolkits are commonly referred to as meta-models, and when used in the context of DSS they are referred to as hierarchical decision support systems that can enable the modelling of stand dynamics, stand productivity, habitat, biodiversity, hydrology, and the visualization of management outcomes, among others.

Hierarchical meta-models such as UBC-FM developed by Seely *et al.* (2004) are critical components of a DSS that can support a robust test of SFM. Combining MCDA approaches with simulation approaches is another meta-modelling approach which can allow for the interactions between ecosystem components to be addressed within a normative modelling framework. Hjortso and Straede (2001) combined a simulation model with a multiple criteria decision making (MCDM) model to explore alternative management scenarios on forest harvesting and the provision of non-timber forest values. The simulation model was used to model relationships between forest structure and non-timber forest value functions while the MCDM addressed the multiple management criteria as objective functions and constraints. Although these MCDA meta-models can address issues of environmental uncertainty and provide metrics to assess environmental outcomes of management, they do not directly address social values and have not explicitly addressed climate change.

For a meta-model to be applicable to SFM it must be able to represent the different aspects of ecosystem functioning and the societal values that are represented in the criteria and indicators of SFM (Mäkelä *et al.*, 2012). While meta-models provide outputs that aid in the biophysical and economic implications of forest management aimed at addressing multiple objectives, social values are usually addressed in terms of a scenario-based approach with alternative management scenarios applied within these decision support models. The development of alternative scenarios that address/

Figure 13.8 Beekeeping operation in a forest developed primarily to combat desertification in Chifeng, Inner Mongolia, China. The bees require the shrubs associated with the plantation. Some decision support tools can factor in the trade-offs between the production of non-timber forest products such as honey and the more traditional measures such as canopy closure and optimization of timber production.

incorporate social and economic dimensions is therefore critical for achieving outcomes that are relevant to SFM. The DSS developed by Bohnet *et al.* (2011) is a good example of a meta-model that combines predictive approaches with explorative approaches to aid in decision making; likewise, the collaborative modelling framework present by Sturtevant *et al.* (2007) provides an approach that links simulation metamodelling with scenario development and evaluation, in turn fostering local participation. The UBC-FM described earlier links model outcomes to visualizations to engage stakeholders in scenario evaluation, which links predictive outputs with explorative outcomes to facilitate decision making. The key to using these predictive meta-models is transparency and active participation in scenario development and model analysis.

Judging social acceptability: the communication of model outcomes

Irrespective of the evaluation approach used to assess alternative management actions/strategies, it is important to continue the participatory approach through the analysis stage to ensure a strong test of SFM is delivered. Meta-modelling approaches could integrate ecological and forest planning models that derive the biophysical attributes with visualization systems that can effectively communicate the outcomes of alternative approaches to stakeholders and the public. Communication of knowledge from forest managers to the public is an important step for increasing public understanding and gaining support for management actions. Information on indicators of consequences and animated sequences showing changes in forest structure through time are important attributes required for judging social acceptability. Communicating outcomes in this manner is important, as the process of addressing visual management objectives or scenic beauty constraints in a DSS will not insure public acceptance as there are deeper social reasons that underlie public perceptions of sustainability. For these reasons the visualization of alternative management strategies within a participatory framework is important to the DSS process and has been found to convey complex information quickly and relatively simply, which in turn can provide a comprehensive, engaging, open, and accountable decision making process.

Fitting it all together

Current knowledge in landscape ecology, biology, economics, and management sciences demonstrates that forests are complex co-evolutionary systems with changing functional controls in the ecosystem, in the economy, and in the society (Holling *et al.*, 1998). Decision makers should aim at developing or utilizing participatory decision support systems that are underpinned by spatially explicit simulation models. Some key characteristics of a DSS required to provide a robust and strong test of SFM are:

1 A clearly structured decision making process;
2 Broad representation of stakeholders;
3 A collaborative planning and evaluation process;
4 Scenario development to reflect alternative management objectives and/or actions;
5 Multi-attribute analysis methods such as AHP, MCA, and MCDM structured around systems of criteria and indicators;
6 Spatially explicit and temporal forecasting of ecological, social, and economic values over tactical and strategic time periods, which can incorporate environmental change and uncertainty;
7 Clear and simple communication of information on indicators of consequences and visualizations showing changes in forest values and structure to aid judgements of social acceptability of management alternatives;
8 Provision for adaptive management within the decision making process.

Step 8 is important, as it allows for the selected alternative to be implemented or for the decision making process to cycle back to the problem identification, scenario development, or alternative selection steps. As new decision support tools become available, existing tools are improved, or environmental, economic, or social context changes, having a DSS that is flexible to allow for the re-evaluation of management objectives as well as criteria and indicators is critical for ensuring selected management actions and policies will continue to deliver SFM.

Incorporating DSS into the development of a sustainable forest management plan

The decision making process in SFM comes down to choosing and applying a course of action(s) from myriad alternatives to a landbase that incorporates the values, desires, and potential states of nature that exist within a system. The final result of the decision making process is a management plan which recommends a course of action, including predictions of the consequences of implementing the plan (Kangas and Kangas, 2005). In the context of SFM planning, there often needs to be a participatory decision support system that is underpinned by a spatially explicit simulation model (especially on public land). For management problems that are tactical in context and have singular or limited objectives that do not require spatially explicit problem solving at large scales, normative MCDA approaches are still applicable if linked to explorative approaches. The steps that can be followed to incorporate participatory DSS into the development of SFM plans are illustrated in Figure 13.9. The framework incorporates pathways for utilizing classic MCDA and spatially explicit simulation models/meta-models. MCDA and other normative modelling approaches are best suited for simple problems that are aspatial or have limited spatial constraints and are tactical in nature, while spatially explicit simulation modelling/meta-modelling is better suited for spatially complex problems that are strategic in nature, as they will provide a stronger test

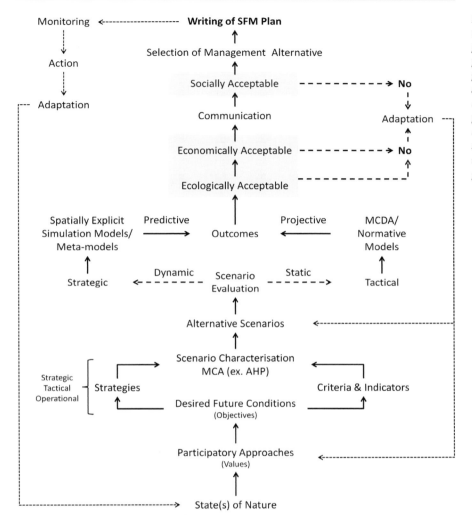

*Figure 13.9
Conceptual
framework
for combining
participatory
and quantitative
decision support
processes for the
development of
sustainable forest
management
plans.*

of SFM under environmental change and uncertainty. Adaptive management loops are provided (small dashed lines) to facilitate changes to management scenarios and values within the scenario evaluation process and also to inform management following the implementation of a SFM plan. Participatory approaches and multiple criteria analysis are critical for scenario development, while collaborative analysis and communication are key elements for scenario selection. The steps outlined in Figure 13.9 should provide for a strong test of SFM that is adaptive and accommodating of environmental, social, and economic objectives in a dynamic world.

Conclusions

In an age where large amounts of information are being generated, more and more stakeholders are getting involved in planning and decision making, and multiple forest values need to be maintained, DSS have become an indispensable part of the planning process. A range of different techniques exist, and a forest manager using such tools must decide on the best combination of methods. While most of the early DSS were focused on optimizing timber harvesting, more recent tools have been able to incorporate a range of forest values. However, no system will give perfect answers, and it is still up to individual managers to decide which actions will best achieve the desired outcomes.

The use of LiDAR and other remote sensing techniques, together with developments in field technology, mean that data are more available than ever before. The challenge for the decision maker is to make best use of these data. This also includes making the data available to all those involved in the decision making process in a form that they can readily understand. As co-management becomes increasingly common, it also means ensuring that all relevant information is shared. In some cases, this may be difficult, such as with culturally sensitive, place-based information. In such situations, decision makers will need to work together to ensure that any decisions are based not only on knowledge, but also on trust and respect.

The DSS that have been developed are becoming increasingly sophisticated and, as a result, difficult to use. There is a danger that they become 'black box' systems, producing outputs that may not be readily understood. To avoid this, managers need to use such outputs with care, and use their knowledge of the forest to interpret the outputs in a meaningful way.

Further reading

Bettinger, P., Boston, K., Siry, J. P., Grebner, D. L. 2009. *Forest Management and Planning*. Burlington, MA: Academic Press.

Buongiorno, J., Gilless, J. K. 2003. *Decision Methods for Forest Resource Management*. San Diego: Academic Press.

Harding, R., Hendriks, C. M., Faruqi, M. 2009. *Environmental Decision-Making*. Sydney: Federation Press.

Kangas, A., Kangas, J., Kurttila, M. 2008. *Decision Support for Forest Management*. Berlin: Springer.

Schmoldt, D. L., Kangas, J., Mendoza, G. A., Pesonen, M. 2001. *The Analytic Hierarchy Process in Natural Resource and Environmental Decision Making*. Dordrecht: Kluwer Academic.

von Gadow, K. (ed.) 2001. *Risk Analysis in Forest Management*. Dordrecht: Springer.

Chapter 14

Assessing the quality of forest management

Anna V. Tikina

The quality of forest management is often measured by the attainment of the outcomes desired by the society. While in the 1940s the desired outcome was timber production in industrialized areas around the globe, societal interest in forest values beyond timber was reignited in the 1970s and never ceased in others, such as with subsistence food or fuelwood collection in South America. The quality of forest management oriented solely on timber was questioned, and governmental reporting on the outcomes of forest management became insufficient (Upton and Bass, 1996). As discussed in Chapter 12, the international concern over forest management did not produce a legally binding mechanism and instead led to the non-legally binding Forest Principles. In the absence of a strong binding international agreement on forest management, non-governmental approaches to the assessment of the quality of forest management gained dominance in the 1990s (Rayner *et al.*, 2010).

Without international legal mechanisms to improve forest management, regionally or nationally governments assure the quality of forest management through legal and regulatory requirements. Beyond legally binding requirements, the best management practices approach links mandatory and voluntary efforts to improve the quality of forest management. This chapter examines the different approaches to assuring the quality of forest management and focuses on voluntary forest certification. Since the 1990s, forest certification, one of the dominant non-governmental approaches based on a set of comprehensive criteria and indicators, has become a tool for the assessment of forest management and for communicating its quality.

Approaches to effectiveness evaluation

Effectiveness evaluation is basically concerned with how well something is achieving its objectives. In this sense, the effectiveness of forest management is directly linked to the objectives that a manager has defined for the forest, and how these are reflected in the outcomes. The theory of effectiveness evaluation in environmental governance embraces multiple approaches. To assess the effects of a program/management, there is a need to:

- Differentiate outputs, outcomes (results) and impacts (effects of implementation and maintenance) of a mechanism;
- Identify modifications of behaviours that lead to elimination or mitigation of the problem;

- Argue the necessity of comparisons with 'control' situations when the mechanism is not applied;
- Describe separate criteria or aspects of effectiveness;
- Assess the 'beyond compliance' performance outcomes of the program.

Mandatory versus voluntary assessment

Given the importance of forests and forest management to the economies of many countries, governments should be a major contributor to the assessment of the quality of forest management. The governmental efforts often include not only rule-making but also audits of compliance with and enforcement of legal and regulatory requirements related to forest management. The role of governmental compliance and enforcement (C&E) oversight is especially important where forest management is carried out by private organizations (e.g. industrial corporations). On the other hand, where the government itself is involved in forest operations, the self-policing of forest management activities by governmental audits have raised doubts about its effectiveness. IFMAT (2013) has emphasized the need for external bodies to assess the quality of forest management, proposing an independent national-level commission of experts to periodically review government-managed forest operations.

Besides the policing role of C&E audits, audits inform governmental decision-makers on how well their regulations and rules are meeting the intent they were developed for. For example, the British Columbia Forest and Range Evaluation Program (FREP) delineates four types of evaluations, based on the types of monitoring developed by Noss and Cooperrider (1994):

1   Compliance evaluation (analyzing the degree to which the requirements are met);
2   Implementation evaluation (recording progress to a specific goal);
3   Effectiveness evaluation (analyzing whether the program is meeting its goals);
4   Improvement/validation evaluation (establishing causal linkages between forestry practice and outcome).

Some governments have experimented with combining their own legally required (hard law) C&E audits with third-party verifications for voluntary mechanisms (soft law) based on considerations of efficiency and cost-effectiveness, but this practice is not widespread.

Soft law (i.e. rules with no legally binding force but with possible practical effects) may fill the absence of a legally binding rule. In forestry, these mechanisms are generally developed to define and assess the quality of forest management. The proponents of a 'beyond compliance' approach (i.e. following a voluntary mechanism to improve environmental performance beyond what is required by law) advocate enhanced flexibility of the soft law mechanisms; they tend to focus on impacts and cost-efficiency. Voluntary mechanisms have shown some positive results, but their overall effectiveness to ensure the quality of environmental performance is still being debated.

The influence of soft law has been greatly affected by pre-existing governmental regulations. A mechanism can be successful in certain regions or a certain part can be more effective than the rest of a mechanism. Regions with less stringent environmental governance may experience greater influence of an introduced soft law mechanism (Tikina and Innes, 2008). This, however, depends on the stringency (which can be considered as the ability of rules to regulate behaviours) and the 'beyond compliance' potential of the soft law mechanism. The two-dimensional grid of the effects of a soft law mechanism (Figure 14.1) can be augmented by a third dimension pertaining to company philosophy and efforts on leadership, often expressed by corporate social responsibility (CSR) activities. Multiple sources discuss CSR in forestry (e.g. Vidal *et al.*, 2010), and the topic is only briefly mentioned in this chapter.

Independence of the assessor

The assessments of quality of forest management originate from different sources (Table 14.1). In some assess-

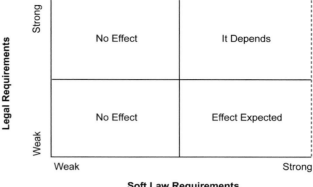

Figure 14.1 Expectation of change related to the stringency of soft law and governmental regulation.

ments, agents of the quality assessment, the auditors, come from the same organization or company. This is the case when the government checks forest operations performed by a governmental corporation or company (as occurred with the former Lesprom-khozes in the USSR). In the private sector, this pertains to internal audits the results of which are rarely publicized, or to company statements about achieving publicly announced CSR commitments. This is usually called first-party certification.

Second-party certification is performed by a customer or an association that a company belongs to. Certification may be a requirement of membership in such an association and carried out through the association. An independent assessment (third-party certification) is carried out by an independent organization that has documented procedures that are accepted by an accrediting body, and is accredited to carry out assessments. In forestry, this category of forest management certification is now the most widely used, as first- and second-party certifications have little credibility.

The difficulty of quality assessment in forest management lies in the long-term delay in the effects of forestry activities. Positive or negative effects of forest management may not be apparent for decades, and in some cases, centuries. Therefore, any type of assessment only takes a snapshot of the current state of management, and long-term continuous monitoring of the effects is required to ensure the appropriate quality of forest management.

Table 14.1 Categories of quality assessment by auditor-auditee relationship.

Categories	Examples
First-party (self-certification) – an internal assessment, claims about own performance	Government certifying governmental operations. Forest procurement policies that have not been certified otherwise.
Second-party certification – an assessment by an arm's-length organization	Before the development of a third-party process, Sustainable Forestry Initiative (SFI) certification originated as a second-party certification: American Forest & Paper Association (AF&PA) used to certify its members.
Third-party certification independent assessment	Standards by Forest Stewardship Council and standards endorsed by the Programme for Endorsement of Forest Certification.

Difference between process- and performance-based approaches

International environmental regulation includes many soft law instruments. The standards developed by the International Organization for Standardization (ISO) improve industrial environmental performance in countries with weak governmental regulatory control. However, the impact of ISO 14001 in minimizing the negative environmental impacts of forest management has been debated, because the focus of the standard lies on management systems rather than on management (Bass *et al.*, 2001). The standard is considered process based, and lacks performance measures directly linked to forestry. The absence of required field visits during ISO 14001 certification audits and the absence of ecological measures are other criticisms of this process-based approach.

Performance-based standards (i.e. standards containing criteria and indicators of forest management) are a much stronger method to assure the quality of forest management. Examples of such standards have been developed by the Forest Stewardship Council (FSC) and its regional standards, and the Programme for Endorsement of Forest Certification (PEFC) and the national standards that it has endorsed. Although early versions of some current performance-based standards (e.g. Canadian Standards Association CSA Z809 of 1999) had much greater reliance on ISO 14001, later revisions of the CSA standard has led to the inclusion of ecological and forest-related measures.

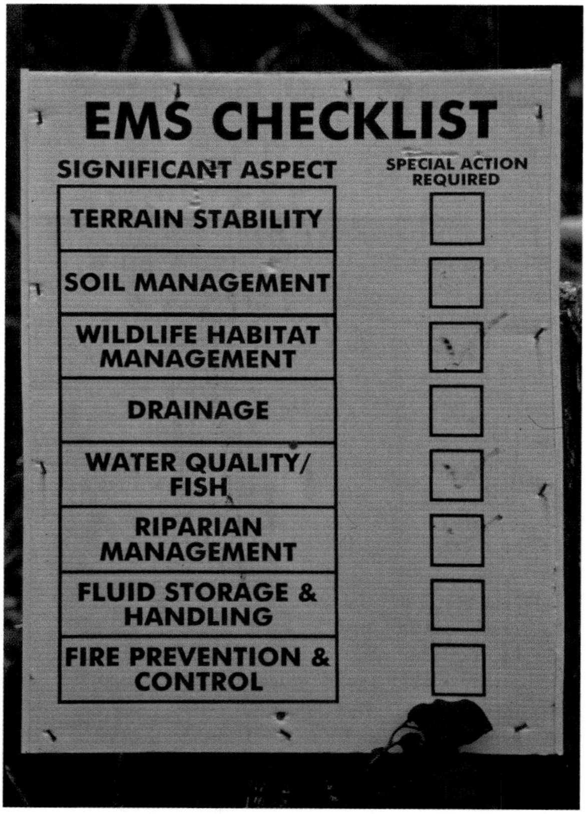

In forest management, ISO 14001 is described as a viable tool to update forest operations, and was used, for example, by the Canadian Certification Coalition to report on ISO 14001 as one type of forest certification in Canada. While the ISO 14001 standard does not contain any forestry requirements, the Canadian forest industry sees it as streamlining forest operations and thus making it easier to obtain certification by a performance-based forest management standard (Tikina *et al.*, 2010).

ISO 14001 certification has also been adopted as an additional guarantee of quality forest management by provincial governments in Canada: the government forestry program in Saskatchewan has been certified to the standard, as have been many regional branches of the provincial British Columbia Timber Sales (BCTS) program. The positive impact of ISO 14001 has been observed in other parts of the world, such as in China (Zhu *et al.*, 2012) and Mexico (Blackman, 2012).

Figure 14.2
Checklist used onsite to ensure compliance with an Environmental Management System, an important part of ISO 14001 certification.

Best management practices (BMPs)

Best management practices (BMPs) are defined as guidelines developed by a government or a designated agency on technological, economic, and institutional means to

Figure 14.3
Best management
practices are often
concerned with
water quality and
the maintenance
of riparian
vegetation. They
are particularly
used in the USA
where many
practices remain
voluntary,
whereas in other
countries, such
as shown here
in Victoria,
Australia, codes
of practice are
legislated.

control pollution 'at levels compatible with environmental quality goals' (Ice *et al.*, 2010). Some BMPs in the USA have legally binding force (regulatory BMPs in 16 states), some have a blended approach (voluntary and mandatory), and 22 states have chosen developing non-legally binding BMPs.

BMPs vary in their level of detail and also in their rigour. Some represent a basic minimum standard that would be expected, whereas others, such as those published by the Minnesota Forest Resources Council (2005), are much more detailed and provide a comprehensive guide to good forest management. Similar guidelines have been published under other names: FAO (2006) for example provides voluntary guidelines for the management of planted forests in what amounts to a best practices guide.

Voluntary BMPs appear to be as successful as regulatory programs (Sugden *et al.*, 2012). Usually, voluntary best management practices guide industrial performance where the governmental governance alone fails to mitigate or eliminate an environmental problem. Under such conditions, voluntary BMPs may bring about positive change in forestry (Ice *et al.*, 2004), although care needs to be taken over the choice of best practices. The difficulty with voluntary BMPs is that they cannot be enforced, and so they are often accompanied by comprehensive education programs aimed at persuading stakeholders to adopt the BMPs in the interests of leaving the forests in a better condition than they found them. This is a basic ethic of forestry, but not all adhere to it.

Evolution of forest certification

Originating from concerns about biodiversity and the livelihoods of people relying on forests (Upton and Bass, 1996), forest certification is aimed at verifying that forest management complies with a set of principles, criteria and indicators of sustainable forest management. The concept of forest certification appeared in the mid-1990s. Its purposes have been to identify responsible forest companies and to assure the consumers of forest products about their quality of forest management in areas where the products are procured (Vogt *et al.*, 2000).

Forest certification is intended to drive irresponsible forest companies off the market through the consumer choice of 'greener' products, and to help responsible companies to emerge or stay in the market. The Forest Stewardship Council (FSC) certification, which emerged in 1994, was particularly promoted by environmental non-governmental organizations (ENGOs), whereas other certification organizations have evolved from a range of different stakeholders, including industry, private landowners and governments.

Since the mid-1990s, forest certification has become a major component of sustainability efforts in forestry and an additional non-governmental verification of the quality of forest management. Since the adoption of the FSC standard, the development of alternative standards has resulted in competition between standards, which at times has been highly acrimonious. Examples of alternative standards include the Sustainable Forestry Initiative (SFI) developed in the USA and widely used in the USA and Canada, the Canadian Standards Association forest management standard (CSA Z809 for large companies and CSA Z804 for small non-industrial forestry), the China Forest Certification Council standard, the Indonesian Forestry Certification Cooperation standard, the Pan African Forest Certification System Gabon and the Australian Forestry Standard.

Besides forest management, chain of custody (COC) also can be an object of certification assessment. Although COC certification is not directly linked to the evaluation of the quality of forest management, only this type of certification can lead to labelling certified products.

The label can be used to inform the consumer about the quality of forest management in a certified forest. This could be helpful to the end users of forest products in identifying responsible forest management, if it can overcome issues such as consumer confusion over labels, fraudulent labels and misused labels. Such problems have rendered the use of the forest certification labels problematic for assurance of the high quality of forest management. The number of COC certificates issued in a country is useful for understanding its legitimacy among the stakeholders involved into the supply chain of forest products: the greater numbers refer to greater adoption and greater involvement in trade of forest products, especially internationally (Table 14.2).

Table 14.2 Adoption of forest certification globally and in selected countries with high forest area in January 2014.

	World	USA	Canada	Russia	Brazil	Indonesia	Malaysia
Number of national/regional FM standards							
FSC	40 in 30 countries	1	3	1	2	1	0
PEFC	34 in 32 countries	2	2	1	1	0	1
Number of forest management certificates							
FSC*	1,260 in 81 countries	124	76	109	100	27	6
PEFC	511 in 28 countries	99	74	3	16	0	10
Number of COC certificates							
FSC	27,246 in 114 countries	3,263	964	248	1,030	191	164
PEFC	10,078 in 64 countries	302	184	13	59	17	273
Area (million ha)							
FSC*	190.8 in 81 countries	14.6	62.2	39.2	7.0	1.7	0.5
PEFC	258.1 in 28 countries	34.3	119.9	0.6	1.6	0	4.6

Note: * FSC numbers are reported for all countries – with and without national standards.

Source: FSC (2014); PEFC (2014a, 2014b).

In 2015, the largest area of certified forest was under the various schemes endorsed by the PEFC, and Canada had the largest area of certified forest.

Many factors can shape the development of a national forest certification standard. These may include political as well as market influences. Indonesia and Malaysia tend to export to similar forest markets, but their past choice of certification system (FSC in Indonesia and PEFC in Malaysia) is explained by other drivers (noting that Indonesia now has a PEFC-endorsed scheme). In Russia, difficulties of establishing a national PEFC body were one of the reasons for the huge difference in the legitimacy of the systems and a discrepancy between the current FSC-certified versus PEFC-certified area.

When a forest certification is introduced to a country, the growth in the number of certificates and certified forest area generally follows an exponential pattern (Figure 14.4). The rapid growth is normally followed by a relative stability once all opportunities to expand forest certification have been taken up. At that point, when all larger forest companies are certified, the changes in the number of certificates and area of certified forest are caused by company mergers and acquisitions. This can be upset by circumstances. For example, in 2015, the PEFC withdrew their endorsement of the Russian forest management certification standard, meaning that no forest could claim to be producing PEFC-certified wood.

The generic FSC standard first developed in 1994 can be used for certification in countries where no national FSC standards exist. The standard is based on 10 FSC Principles, which pertain to a range of areas from legal compliance to the well-being of Indigenous people (as discussed in Chapter 10) to forest operations and monitoring. In 2014–2015, a new set of International Generic Indicators was under development: details on progress in the adoption of these are available from the Forest Stewardship

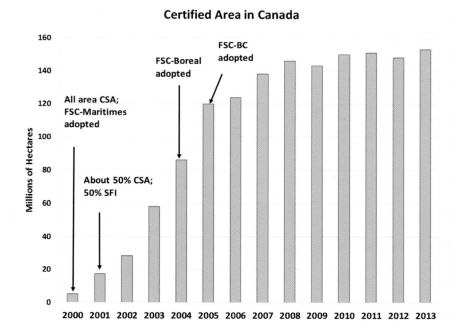

Figure 14.4 The trend in certified forest area in Canada 2000–2013, overlapping FSC and PEFC areas excluded. (Source: FPAC (2014).)

Council website (http://igi.fsc.org). Other certification systems may rely on their own criteria and indicators (C&I; e.g. SFI), or may be based on an internationally adopted set of C&I.

FSC certification is usually introduced into a country through the application of a generic FSC standard. In Russia almost half of the currently certified area was originally certified to a generic FSC standard before a national standard was adopted. In 2014, more than 30 million ha were certified to an FSC standard in countries without a national standard. The use of the generic standard helps in testing the applicability of the standard to forest management and gives a start to the national FSC body. However, it has also caused criticisms that the certification outcomes can be negotiated by local auditing companies or that the standard lacks in national or regional detail.

The generic FSC standard can only be loosely related to by local/regional conditions. To avoid this issue, the national or regional FSC body normally develops one or more national or regional standards. An example of the use of regional standards is provided by Canada, where there are three regional (and very different) FSC standards: FSC-Maritimes (2000), FSC-Boreal (2004) and FSC-British Columbia (2005). A fourth Canadian FSC standard, FSC-Great Lakes-St. Lawrence, covering the southern parts of provinces of Ontario and Quebec, has been in a draft form since 2006. The complicated regional peculiarities of relations with Aboriginal peoples in the Canadian boreal forest and the coastal region of British Columbia deferred FSC standard development in these regions. The slow development of regional FSC standards in Canada, particularly FSC-BC, precluded the fast growth of the FSC system in the early 2000s (Figure 14.5).

The regional specifics of a standard were aimed at capturing the peculiarities of forest management in a particular region. This good intention has led, however, to issues for companies operating in similar forest ecosystems on different sides of national borders. For example, the requirements of the FSC-Maritimes standard in Canada differed from the requirements of FSC-Eastern standard in the USA. This discrepancy

Figure 14.5 In 2003, 18,400 ha of the Pskov Model Forest in western Russia was certified to Forest Stewardship Council standards by the SmartWood Program of the Rainforest Alliance, using the Draft Smartwood Baltic Interim Standard. This was subsequently replaced by the Smartwood Interim Forest Management Standard for Pskov Region in Russia, which in turn was replaced by a national FSC standard for Russia. The Pskov Model Forest project concluded in 2009.

created problems for a cross-border company, J. D. Irving, and it ultimately decided to forgo FSC certification (Lawson and Cashore, 2003).

The development of FSC standards for certifying plantation forestry (e.g. in Brazil) and Principle 10 (Plantations) in the generic standard triggered a heated debate about the sustainability goals of the certification. In response, the current guidelines of the FSC International require integration of plantation considerations into all FSC Principles and Criteria. Since the global demand for forest products continues to increase, an increased reliance on plantations may be more sustainable than the continued extensive forest management of primary forests.

Another current direction of the FSC International is the development of national standards as opposed to regional ones (FSC, 2014). While several countries (e.g. Canada, Brazil, Bolivia) need to align their regional standards with this policy, the newly developed standards only focus on the national level.

The recognition between FSC and other national standards was widely discussed in the 1990s, but the mutual acceptance between FSC and non-FSC national standards has been elusive. Some non-FSC standards arrived at mutual recognition (e.g. SFI and CSA reached such agreement), but FSC has maintained its position of not entering into similar agreements. This has resulted in some companies seeking, and obtaining, certification under both the PEFC and FSC systems. However, certification by a different standard may help with the verification of legality of FSC-certified wood products (e.g. by labelling 'responsible sources'). Most alternative standards now seek endorsement from the Programme for Endorsement of Forest Certification (PEFC).

The PEFC abbreviation initially stood for Pan-European Forest Certification. Its origins lie in European certification based on the Helsinki C&I (see Chapter 2). PEFC subsequently became an international system for the endorsement of national certification standards in Europe. Further development of the system allowed for its extension beyond Europe as the trade in wood products became increasingly international. It is now an umbrella organization that includes a number of national forest certification initiatives, ensuring that they achieve an identical minimum of requirements. The standards endorsed by PEFC rely on national processes that aim at sustainable forest management, such as the national initiatives following the international C&I developed by the Ministerial Conference on the Protection of Forests in Europe (MCPFE), the Montreal Process, or the Lepaterique Process in Central America. Some FSC supporters claim that its international FSC office also acts as an umbrella organization for national and regional standards, but given the possibility of FSC certification to a generic standard in many countries that do not have their own national standards, this proposition is untenable.

Convergence of certification schemes and international recognition

The competition between FSC and PEFC – 'the war of the standards' – was in full swing in the 1990s. Supporters of both camps produced critical reports condemning the opponents of greenwashing, obscurity of certification requirements and low incentives for improvement in forest management (e.g. the 'Don't Buy SFI' campaign by Rainforest Alliance Network; 'FSC-Watch', http://www.fsc-watch.org/). The criticism, however, led to a considerable degree of convergence among the systems and standards, including the following:

- *Development of similar requirements.* The Australian Forestry Standard (2013) that is endorsed by PEFC refers to the FSC Principles and Criteria as one of the sources of its requirements.

- *Many requirements are based on ISO procedures*. Although this does not directly translate into the assessment of forest management, the reference to internationally adopted ISO requirements on auditing procedures and auditor qualifications aligned standards in both PEFC and FSC systems.
- *Auditing companies serving both FSC and PEFC*. Originally, auditors usually specialized in only one system. With a growing demand of companies wishing to certify, larger auditing companies obtained accreditation to certify to both systems. This further unified procedures used in auditing.
- *Comparable legitimacy*. Both systems gained support from governments, industry and industry associations, buyers' groups and to various extents from different non-governmental organizations (although a few environmental groups actively oppose certification). Acceptance of the systems is evident from the growth of the number of certificates, the growth of certified area, and governmental or association commitments.

The majority of large forest companies in North America are certified to one or more of the standards, initially in the hope of higher prices for certified wood products. Others were striving to be ahead of their competitors, and a few were concerned about demonstrating their quality of forest management (Vogt *et al.*, 2000). Forest certification is now widely viewed as a cost of doing business (Chen *et al.*, 2010), and certification systems compete for expansion to more companies and more countries.

Nonetheless, voluntary forest certification constitutes another means of assessing the quality of forest management in addition to governmental (mandatory) measures. In countries where governmental oversight and law enforcement are incapable of ensuring sustainable forest management, forest certification is an additional mechanism to react to stakeholders' concern about sustainability.

Conclusions

The assessment of management performance can be undertaken in a number of ways. The most obvious is direct compliance with the laws, policies and best management practices applicable to a forest area. This requires some form of authority that is in a position to assess the degree of compliance. Such compliance assessments can be undertaken before work is undertaken, as in the approval of forest management plans, or afterwards, when on-the-ground assessments are made of the extent to which the operations were in compliance legally and with the tactical plans.

Frequently, this is a government department with responsibility for land management. The capacity (and willingness) of government to do this is often limited, however. The process is also prone to corruption in many parts of the world. Further concern is that when a non-compliance is discovered, there must be appropriate mechanisms in place to deal with it. For example, the watchdog appointed by the government of British Columbia to oversee forest practices has recently reported that government officials have been able to do little to control forestry developments that have put local environmental and community values at risk (BC Forest Practices Board, 2015).

Certification grew out of concerns about poor forestry practices, particularly in tropical countries. The greatest adoption has been in temperate and boreal countries by companies seeking to demonstrate their environmental credentials. Two major schemes have emerged, one through the Forest Stewardship Council (FSC) and the other through the Programme for the Endorsement of Forest Certification (PEFC). Over time, the two schemes have shown increasing levels of convergence, and while there have at times been bitter arguments between the two, both are aimed at recognizing good forest stewardship. For many companies, obtaining certification is seen as a demonstration of their corporate social responsibility, and customers, while unwilling to pay a premium for certified forest commodity products, are increasingly expecting that the products that they buy come from well-managed sources. Consequently, it appears that the certification of forest management will not only continue, but will continue to evolve, covering in the future all the goods and services provided by forests.

Further reading

Nussbaum, R., Simula, M. 2004. *The Forest Certification Handbook*. 2nd edition. London: Earthscan.

Rockwell, W., Levesque, C. 2007. *Forest Certification Auditing: A Guide for Practitioners*. Washington, DC: Society of American Foresters.

Tollefson, C., Gale, F., and Haley, D. 2008. *Setting the Standard: Certification, Governance, and the Forest Stewardship Council*. Vancouver: UBC Press.

Chapter 15

Adaptive management

A case study of how some of the principles of SFM can be applied in practice

Fred Bunnell

Introduction

Knowing the types of things that a forester should be valuing is important, but at some stage this knowledge has to be put into practice. As explained in Chapter 3, there is a great deal of uncertainty surrounding many aspects of forestry, and this means that management needs to be flexible. Adaptive management presents an opportunity to formalize this flexibility. Adaptive management has been adopted in many different situations and with varying levels of rigour. It has been expanded to include adaptive co-management, which brings a stronger element of public participation, particularly from local people (Armitage *et al.*, 2007). The treatment of adaptive management in this chapter focuses on the challenges of sustaining forest-dwelling biodiversity. The reason for this focus is that, although many other aspects of forestry can and have profited from adaptive management approaches, sustaining biodiversity in managed forests is particularly challenging. The examples of lessons learned that are presented here are most often derived from efforts to sustain biodiversity but are generic and can be applied to many different contexts.

Forest biodiversity and wicked problems

Among terrestrial ecosystems, forests contain the largest and most long-lived components. The long lives of trees, the large spatial extent of forests and their planning, and the social perceptions of forests combine to create particular difficulties in managing forests to sustain biodiversity. Bunnell *et al.* (2009a) recognized five, as follows:

1   For historical and social reasons, forest practitioners bear an uncommon responsibility for sustaining biodiversity. In western North America, for example, both forestry and agriculture provide highly valued crops, but only forestry is consistently charged with sustaining biodiversity.
2   More species reside in forests than in any other plant community. That challenges monitoring and encourages a search for surrogates.
3   Forestry is usually planned over large areas that rarely are homogeneous. Any area large enough to permit a sustainable harvest contains both natural disturbances and forest practices that create stands of different structure and, often, composition. That variation encourages species richness, but complicates setting objectives and planning.

4   Forestry must be planned over long periods. Natural disturbances in forests are highly variable in extent, frequency, duration, location, and intensity. Some forest-dwelling species occur primarily in stands older than the typical rotation age.

5   Public goals and perceptions change quickly, but consequences of forest practices are evaluated slowly. Trees can grow large (changing the environment as they grow), but grow slowly.

These difficulties and the values provided by forests combine to make management of forests a 'wicked problem'. There is no moral sense to this use of 'wicked', but a recognition that some problems defy tidy solutions no matter the effort and good intention directed to them. Society seeks a variety of values from the forest, some of them competing. Management of forests is a challenge of dealing with organized complexity and directing clusters of interrelated parts towards specific ends. Ackoff (1974) termed problems that cannot be resolved by treating such clusters in relative isolation 'messes'. Wicked problems arise when the boundaries of messes expand to include sociopolitical and moral-spiritual issues (King, 1993). The most succinct definition of wicked problems is that presented by Churchman (1967) quoting a seminar by Rittel, who described them as a: 'class of social problems which are ill-formulated, where the information is confusing, where there are many clients and decision makers with conflicting values, and where the ramifications in the whole system are thoroughly confusing'.

*Figure 15.1
No moral sense
is implied in the
term 'wicked
problems'.
(Figure prepared
by I. Houde.)*

The original formulators of the term 'wicked problems' (Kunz and Rittel, 1970; Rittel, 1972) recognized several key attributes, among them:

*   There is no one way of formulating the problem – they can be described in different ways that have different solutions;
*   Each problem is unique;
*   There is always more than one plausible explanation for outcomes;
*   There is no single correct test for a solution (i.e. there is no 'stopping' rule);
*   Solutions cannot be true or false, but can be more or less effective.

The term wicked problem has been applied to many resource management issues, but Rittel's definition of a wicked problem is a perfect fit with the problem of managing forests to sustain biodiversity.

Adaptive management, wicked problems, and managers

Adaptive management is a formal process for continually improving management practices by learning from the outcomes of operational and experimental approaches (seminal works are Holling, 1978, and Walters, 1986; see also Walters, 1997). Four elements of this definition are key to its utility. First, it is *adaptive*, and intended to be self-improving. Second, it is a well-designed, *formal* approach that connects the power of science to the practicality of management. Third, it is an ongoing process for *continually improving management*, so the design must connect directly to the

Figure 15.2 Tropical rainforest near Manaus, Brazil, after selective harvesting. Given how little knowledge there is about the impacts of selective harvesting in a forest like this, a manager needs to monitor impacts carefully and adjust practices if unacceptable effects are recorded.

actions it is intended to improve. Fourth, although experimental approaches can be incorporated into adaptive management effectively, *operational* approaches and scales are emphasized to permit direct connection to the efforts of managers.

The relationship between adaptive management and wicked problems is threefold. It is a system for continually improving management and acknowledges that there is no stopping rule. It explicitly accepts that there is no single true or false solution, but seeks to make management more effective. And, it serves to connect ways of addressing wicked problems, such as managing for biodiversity, directly with the multi-faceted desire to change or manage in a better way.

There is nothing easy about implementing adaptive management well. The design itself is difficult to conceive and implement, particularly in forested systems (see Huggard *et al.*, 2009). Major barriers include effectively linking the different worlds of science and management, and developing concepts and approaches that permit feedback from findings to changes in action. At its best, adaptive management links science with management so that each responds to the needs and information of the other. Neither science nor management is consistently subordinate; rather the priorities of each are reconciled through the design of the process. In this way the tensions within wicked problems are reduced. Ideally, managers help establish the program's direction so that it connects to management issues and they commit to using the results of the monitoring. Ideally, researchers understand management issues well enough to design sampling and experiments that connect directly to management questions, and can help to interpret findings and their implications to managers. Even when this ideal is attained, the process may fail and not be adaptive because institutional structures or philosophies become barriers to implementing changes (Bunnell and Dunsworth, 2009).

There are four major components to an effective adaptive management program:

1   Clearly defined objectives;
2   Planning and practices to attain those objectives;
3   Ways to assess proximity to the objectives;

4   Ways to modify practices if objectives are not attained or are changed (as by links to management actions; Figure 15.3).

Bunnell *et al.* (2009a) noted key features of these components if they are to be successfully applied. Those features are summarized here:

Specify clear objectives: Objectives must be sufficiently clear that means of assessing proximity to objectives are likewise clear. Objectives cannot be rigid, but must respond to both changes in values and information. In sustainable forest management, major objectives are increasingly being described as criteria for success, while indicators are used to measure outcomes that should be evident if a criterion is successfully attained. Examples of criteria and indicators have been provided in earlier chapters.

Select planning and practices: Once objectives are clear, the planning and practices likely to attain those objectives can be chosen. In forests, effects of practices at the stand level both affect and are affected by forest planning over broader areas, and vice versa. That is particularly true for effects on forest-dwelling biodiversity. Planning and evaluation of success in sustaining biological diversity must occur at a variety of scales.

Assess proximity to objectives: Unlike objectives for visual quality, the objective of sustaining biological diversity cannot apply to specific cutblocks, but must be evaluated for broad areas. Assessments everywhere are too costly, so must be designed in a fashion that can be 'scaled up' (as by computer projection) to reflect larger areas. We learn fastest, or reduce uncertainty fastest, when we organize our current information in a way that can be directly challenged by new information – that is, by making predictions and evaluating their accuracy. Where prediction is difficult, comparisons remain useful. A key feature is to provide a structure for learning.

Ways to modify practices: New information is of little use to practitioners if it does not link to management practice. Linkage is provided by both the evaluation system developed for each indicator of success and descriptions of potential management actions that are expected to help correct any failures or strengthen areas of weakness. Ability to change is enhanced by a formal mechanism for accepting results and associated management or policy changes.

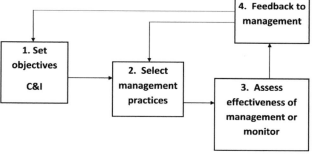

Figure 15.3 Major components of a successful adaptive management process. C&I represent criteria and indicators used in some forest certification systems.

For adaptive management to work, these four components must influence each other (Figure 15.3). Appropriate feedback to management is essential because it can affect all other elements of an adaptive management program. A major advantage that adaptive management brings to the challenge of managing forests is that the primary elements of a successful adaptive management process (Figure 15.3) correspond directly to the major questions managers ask when confronted with any challenging task. The four major management questions are:

1   What do we want?
2   How do we get that?
3   Is our approach correct?
4   What do we do if the approach is wrong?

Table 15.1 Why adaptive management can be successful.

Managers questions	Elements of adaptive management
What do we want?	Clearly defined objectives
How do we get that?	Planning and practices to attain objectives
Is our approach correct – right direction?	Ways to assess proximity to objectives
What do we do if the approach is wrong?	Ways to modify practices – links to management action

*Figure 15.4 Red and silver beech (*Fuscospora fusca, Lophozonia menziesii) *forest, Mount Ruapehu, New Zealand. Concerns about the impacts of logging of native forest in New Zealand, particularly on birds, resulted in it being effectively halted. Adaptive management provides a way to test and modify practices if they are having adverse effects.*

These questions are a near perfect match with the four elements or tasks of an adaptive management program (Table 15.1).

The types of questions we ask and why adaptive management fails

Adaptive management is about learning. We learn by questioning, but during management we ask very different kinds of questions. These questions form a typology of monitoring. They are ordered in the following list as we might ask them as we grow up and assume more responsibility.

1   Have we done what we were told to? *Compliance monitoring.*

When you do not have, or do not want, responsibility, the first question is sufficient. Compliance monitoring assesses actions relative to external rules, regulations, or targets and can be a useful regulatory device provided the rule will attain the desired outcome. Compliance monitoring assumes that the regulation or standard is appropriate, does not evaluate effectiveness, and contributes almost nothing to the growth of knowledge. Moreover, reliance solely on compliance implies rules for everything. Only the rudiments of responsibility are assumed.

2   Have we done what we said we would? *Implementation monitoring.*

Implementation monitoring accepts more responsibility because specific goals or targets are created and acknowledged. The monitoring may record rates of adoption of new practices and whether they were implemented as planned. Implementation monitoring is all about progress towards some internally defined goal. It also contributes nothing to evaluating whether the target was a good one, but it is good for measuring progress, especially when the goal truly is desirable.

3   Did our actions achieve our objectives? *Effectiveness monitoring.*

Acceptance of more responsibility is evident as recognition of the possibility of error. Effectiveness monitoring is used to determine whether the plans or practices implemented do actually meet the anticipated outcomes; that is, are they effective? To be effective and informative, effectiveness monitoring must address what the most common or extensive practices are achieving as outcomes; that is, assess outcomes over the largest area affected. Effectiveness monitoring is a necessary part of operational feedback in adaptive management, but because it assesses common practices, it provides limited opportunities to refine current standards or future operations. Typically, it selects the best practices among *existing* practices, unless novel practices have been introduced. It is relatively simple to select the best practice for attaining an outcome from among existing practices, when that is the chosen measure of 'most effective'. However, some standard, comparison, or target that defines effectiveness must be specified. Knowing one management approach provides two large snags per ha over 30% of the area and another provides four over 30% of the area indicates which is better (so is good), but still does not reveal whether either approach will be effective at sustaining large, primary cavity excavators.

   We can monitor natural benchmarks to create a comparison for effectiveness monitoring. Some term that baseline monitoring, but because the sampling regime for the baseline must be similar to that in the managed area, it is actually merely one part, albeit an important part, of effectiveness monitoring.

4   Can we achieve our objectives better, faster, or more cheaply? *Refinement monitoring.*

More responsibility is assumed when we believe we have enough experience with effectiveness monitoring that we can refine it. We already have figured out how to determine what is effective. Refinement monitoring samples beyond common practice, usually with very specific questions in mind. Creating learning opportunities through experimental treatments beyond the normal operational range is a part of refinement monitoring. In this case, the approach is synonymous with research.

   Both effectiveness monitoring and refinement monitoring are part of adaptive management. Whereas effectiveness monitoring evaluates current practice, refinement monitoring asks questions like, 'Do I have sufficient confidence in the underlying relations to apply them in novel ways?' or 'Can I attain the same ends more cheaply?' Designing an approach to compare practices satisfactorily is difficult enough; validating the underlying mechanism or generality of a relationship is still more complex and more costly.

   When adaptive management fails, it often fails because questions were not answered well. Three reasons are common: too many things were measured, it cost too much, and it was inadequately linked to management (often taking too long for results to

*Figure 15.5
A cutblock with
retained patches
of trees such as
this one in the
central coast
area of British
Columbia,
Canada, may
look appropriate
immediately after
harvesting, but
its effectiveness
is compromised
if the patches
are too small to
be of value to
wildlife or they
blow over.*

appear). I noted that the four big questions managers ask and the four elements or tasks of adaptive management correspond closely (Table 15.1). In sections following, I summarize major lessons learned while applying adaptive management to the challenge of sustaining biodiversity in forests, grouping them under the managers' largest questions. Most of these lessons apply equally to efforts to sustain other forest values.

What do we want?

Answering this question well requires clearly defined objectives. By specifying objectives we also specify outcomes for which we accept responsibility. Many forest tenure holders follow a criteria and indicators (C&I) approach. The Canadian Council of Forest Ministers (CCFM) recognized the importance of biodiversity by making it Criterion 1: the variability among living organisms and ecosystems of which they are part. The wording is consistent with the Convention on Biological Diversity. The expanded criterion includes three elements: ecosystem diversity, species diversity, and genetic diversity. Similar priority is given to biodiversity in the Montreal Process and in a number of other C&I schemes.

While working with individual forest companies, my co-workers and I developed a C&I approach for biodiversity that connects directly with forest planning and practice. The version that proved helpful for several companies follows. Much of the input to developing a criterion and associated indicators comes from practitioners, but scientists are helpful in exposing implications, particularly for monitoring.

Criterion: Biological diversity (native species richness and its associated values) is sustained within the tenure. The criterion has three indicators of success. Indicator 1: ecologically distinct ecosystem types are represented in the non-harvestable land base of the tenure to maintain lesser known species and ecological functions. Indicator 2: the amount, distribution, and heterogeneity of stand and forest structures important to sustain native species richness are maintained over time. Indicator 3: the abundance, distribution and reproductive success of native species are not substantially reduced by forest practices. In this example, Indicator 1 is the coarse filter, Indicator 2 is a medium

*Figure 15.6
Varied thrush
(Ixoreus
naevius), a
characteristic
species of the
coastal forests
of British
Columbia,
Canada, where
the work
described in
this section was
undertaken.*

filter of broad habitat types, and Indicator 3 addresses the fine filter of organisms (see also Chapter 3).

Lessons learned include the following.

Ensure that management objectives are clearly specified. Because C&I specify the obligations practitioners have agreed to assume, they need to be clearly specified and understood.

Identify major issues. The three indicators of the example specifically address three levels of the scale problem (ecosystems, habitat, and species) and two of the three elements of the CCFM criterion (ecosystem and species diversity). Typically, genetic diversity is evaluated primarily by such indices as seed sources. Species and subspecies are the only self-replicating sources of genetic variability. Their persistence and the exchange of genetic material are addressed by ecosystem representation in Indicator 1, habitat distribution in Indicator 2, and Indicator 3 as an ultimate outcome.

Forestry has increasingly become a social exercise (Chapters 9, 10, and 11). Omitting or ignoring social issues can haunt future effort. For example, questions at public meetings frequently address species on the forest tenure that do not inhabit forested land. It is helpful to anticipate such questions. It is for this reason that the Species Accounting System of Chapter 3 includes species of non-forested habitats. Such anticipated questions should not increase the number of indicators, but may have to be included within the monitoring (e.g. simple GIS exercise for species of non-forested habitats, more complicated for climate change).

Connect directly to forestry practices. As worded, the three indicators connect directly to forest planning and practice. It is important that they do, because it is the planning and practices for which foresters have responsibility.

Bound the indicators. It was always true that forest-dwelling organisms were responsive to phenomena other than forest practices. That is particularly true now, with climate change encouraging species to shift ranges and altering the relative competitiveness of tree species onsite. Climate change is likely to become a major challenge to

*Figure 15.7
A black-throated
grey warbler
(Setophaga
nigrescens).
Changes in
this species'
wintering habitat
in Mexico
and Central
America may be
influencing its
abundance as a
breeding bird in
western North
America.*

forestry planning and practice. Trends in climate change are still too poorly known to merit a separate indicator, but monitoring should include measures that help anticipate changes in tree species composition, disturbance regimes, and forest health. Similarly, forest practitioners responsible for habitat on migrating birds' breeding range should not be held responsible for events on winter range or along migration routes.

How do we get that?

This managerial question is simply a rephrasing of that step in adaptive management that identifies planning and practices to attain objectives (Table 15.1). The criterion specifies what success looks like broadly. The indicators identify measures that will be used to assess whether success has been attained. Planning and practices must thus connect directly to those indicators in a manner that works towards successful outcomes.

The initial step for Indicator 1 is to determine amounts and distribution of ecosystem types by ecosystem representation analyses (e.g. Wells *et al.*, 2003). The rest is planning. If ecologically distinct ecosystem types are to be represented in the nonharvestable land base, it is likely that some areas will have to be set aside. There are four questions: (1) What is 'distinct'? (2) If there are choices, where to set them aside? (3) How much should be set aside? (4) Should I be responsible?

A well-developed system of ecosystem classification is usually available for public lands and should be employed to assess ecosystems. Those systems will specify widely agreed-upon definitions of what is distinct. The finest divisions of ecosystem classification can often be aggregated for vertebrates, less often for plants. Answering the question 'Where?' can be guided by a GIS depiction of a given ecosystem type's location. General guidance is to seek areas least subject to disruptive disturbances and to acknowledge that in many instances exposure to chance events suggests that reserves often serve better when distributed rather than concentrated. Private holdings are often too small for ecosystem representation to apply for all but the smallest ecosystem types. How much to set aside usually is a corporate decision recognizing the tensions between shareholder concerns and the desire to attain social license. We found it useful to recognize 'responsibility' defined as the proportion of an ecosystem type within the forest tenure (Huggard and Kremsater, 2009). For example, if a particular ecosystem is well represented within a park, other rare and distinct ecosystems within the tenure have higher priority for inclusion in the non-harvestable land base.

Indicator 2 specifies that the amount, distribution, and heterogeneity of stand and forest structures are an important route to success. Both natural disturbances and the aging and growth of trees change amounts and character of stand and forest structures. Forest planning and selection of practices imposes additional future changes to amounts and kinds of structure as well as their distribution. The widespread increase in forest fire frequency is both good news and bad news. The good news is that the habitat of species seeking early seral stages will increase and will not be a constraint. There are two major types of bad news: (1) if harvest is to be sustainable, it will have to proceed at a lower rate than only a few decades ago, and (2) any targets derived from historical stand structures may no longer apply, and the challenge of a moving target is increased. In aggregate, even the widespread mortality of lodgepole pine is primarily good news for biodiversity, because relatively few species seek out lodgepole pine stands (Bunnell *et al.*, 2004). The loss does, however, reduce the sustainable allowable cut, so must be incorporated into planning.

Both planning and practices determine success in attaining Indicator 3. Practices determine the variety or heterogeneity of stand structures. Planning determines the distribution and amounts of different types of stand structure.

Effects of global warming extend to practices as well as planning. The consequences of increasing warming and drought suggest that the heterogeneity of stand structures normally created by natural succession and age will be reduced (more early seral stages, more older stands burned or killed by pathogens). Together the consequences imply

Figure 15.8 Part of the area burned by the 2003 Robert Fire in Glacier National Park, Montana, USA. Fires represent an important ecosystem process, and without them early seral species would not survive in a landscape.

that foresters will be less able to simply let natural processes change stand structure and will need to intervene creatively more often than they have in the past. Fortunately, foresters have proven creative in the past and have a good understanding of how different practices change the structure of stands. There is ample experience to build on. Nature will create more areas analogous to clearcuts. In the reduced areas of older stands, more practice will have to be directed to creating structures more often provided by older stands. Practitioners will also be developing approaches that increase climate resilience within stands. Specific suitable practices will have to be region specific, but there are example efforts to increase resilience to draw on (e.g. Wang *et al.*, 2010).

The goal of appropriate distribution of a variety of stand structure is more challenging than the degree of variety. Issues created by the 'scale problem' are noted in Chapter 3; challenges to monitoring are noted in the following section. A simple general guideline is to avoid concentration of any specific practice or age class over large areas.

There is sufficient experience in applying forest planning and practice that while new and helpful specifics were acquired during monitoring, we encountered few new lessons during this step of implementing adaptive management. We already had learned one of the biggest: don't do the same thing everywhere. There were three inadequately anticipated outcomes that provided lessons that appear general.

Novel practices make monitoring more expensive. That may seem obvious. We often underestimate the additional cost of going from effectiveness monitoring to refinement monitoring (see the section 'The types of questions we ask and why adaptive management fails'). Introduction of novel practices may be necessary, but frequently creates unanticipated costs.

Discuss the implications of informed choice versus future guidance early. Whether the adaptive management process is intended to select among existing practices or work towards continual improvement has large implications to the length and costs of the program (see the section 'Lessons learned from designing adaptive management'). The decision needs to be discussed early to avoid unrealistic expectations.

Monitoring will cost more than expected. Assessing Indicator 2 is more costly than assessing Indicator 1, but costs of assessing Indicator 3 can dwarf both, simply because

Figure 15.9 Monitoring is expensive and time-consuming. A temperate rainforest such as this one at Carmanah in British Columbia, Canada, presents many challenges. Monitoring the effects of management activities in a tropical rainforest is even more difficult.

there are so many different species requiring so many different kinds of monitoring. There is no completely satisfactory way around this condition, but the Species Accounting System of Chapter 3 and the cross-design described in the section 'Is our approach correct?' are significant aids. The major implications to the question 'How do we get that?' are that there inevitably is a choice between assessing outcomes of a few forms of planning and practice well or more forms less well. The primary challenge is that even an incomplete assessment of some practices can provide an originally unknown amount of general guidance. An additional challenge for both researchers and practitioners is expectation management and potentially promising too much. Combined, these challenges emphasize the importance of lessons learned while asking the next question.

Is our approach correct?

Having selected ways of meeting the indicators of success, the next step is to determine ways to assess how close you are to success or proximity to objectives (Table 15.1). This step is more simply termed 'monitoring'. Experience suggests several key lessons.

What would you do with the data if you had them? This is the overarching question. Monitoring is always the most expensive step in an adaptive management program. That results partly from the fact that many processes influence the sustainability of species and partly from the variety of species with somewhat different requirements in any forest. One result can be a modest tug of war between practitioners and scientists about the specific data to be collected. Answering this question effectively can save future grief. The goal is to keep monitoring as simple as is credible.

Do the data connect directly to planning and practice? This question is a corollary to the previous one, but merits emphasis because of the expense of monitoring. The process of adaptive management has four major components (Figure 15.3); one is feedback to management. Data are collected to guide practitioners' selection among approaches to planning and practice – they must be connected to management. In some cases, the data can lead to changes in objectives. That is rare when 'What do we want?' has been addressed well, but could become more common with global warming. Some objectives believed possible to attain may become impossible.

To connect to planning and practice while ensuring a practical monitoring design, species selected for direct monitoring should be forest dwelling; be sensitive to forest practices employed; be practical to monitor in terms of sampling, identification, and cost; and provide information useful in guiding forest practice.

We did not consistently follow these guidelines, and sometimes monitored species (e.g. slugs and snails) for which sensitivity to forest practices was poorly documented. Findings were informative. One slug new to science was discovered and named (*Staala gwaii*); clearcuts retained the fewest small snail species; larger retention blocks (0.8 to 1.2 ha) retained gastropods at similar levels to those found in control sites (Ovaska and Sopuck, 2008).

There will be more questions than can be asked effectively. There is always a cost constraint. Moreover, any combination of practitioners and scientists can create a lengthy list of questions they would like to have answered. Managing to sustain biodiversity will always be a wicked problem, with few tidy yes or no answers and no handy 'stopping rule'. The simplest approach is to identify major questions, rank questions, identify data needs to answer highly ranked questions, and develop the design for answering them.

A useful device for focusing the monitoring design is the 'what and how' matrix of Huggard *et al.* (2009). The 'what' aligns specific forest elements or organisms believed to reflect success in sustaining biodiversity with the various 'hows' of practice

Table 15.2 A 'what and how' matrix to help focus monitoring on the most desirable comparisons.

Stand level indicator priority	Retention type & amount	Older stands	Edge effects	Patch size	Silvicultural system	Harvest method	Non-harvestable scrub
	VH	VH	VH	H	H	M	M
Habitat elements							
Standard + heterogeneity	1	1	1		1	1	1
Rare (e.g. large snags)	3	2	3		3	3	2
Surface soil structure	4				4		
Habitat integrators							
Ecological classification unit	1	1			1		1
Microclimate	3	4	3		3		
Organisms							
Vascular plants	1	1	1		1	1	
Bryophytes on forest substrate	2	3	2		2		3
Lichen epiphytes	2	3	2		2		
Ectomycorrhizal fungi		2	2				4
Gastropods (+ amphibian by-catch)	2	2	2	3	2		3
Carabid beetles	2	2	2	3	2		3
Aquatic breeding amphibians	4			2	4		4
Common songbirds	2		4	1	2		3
Selected woodpeckers and owls	3				3		

VH: very high; H: high; M: medium. Modified from Huggard *et al.* (2009).

as reflected in stand level measures. Table 15.2 is an abbreviated example of such a matrix used to guide design for stand level comparisons. (Indicators in the table are those noted in the section, 'What do we want?') In this instance, the major issue was the introduction of large-scale variable retention to sustain biodiversity to coastal British Columbia and for the first time anywhere. A more complete table would reveal organisms that fell off the list: conk and bracket fungi (unclear what would be gained), terrestrial breeding amphibians (expensive for information gained), squirrels (pilot study revealed that caches were inadequate and expensive telemetry would have to be used), and others. The priorities assigned reflect priorities of the questions and the information acquired by measuring the combinations of items listed would provide to answering those questions.

There never is sufficient funding to address all interesting questions or organisms. The 'cross-design' of Table 15.2 helps focus monitoring by using a few critical variables to make all selected comparisons, while the highest priority comparisons are made with a more extensive array of variables. Benefits include reducing the complexity of biodiversity, basing the most revealing comparisons on a more complete representation of biodiversity, and increasing program cohesion and efficiency by focusing a number

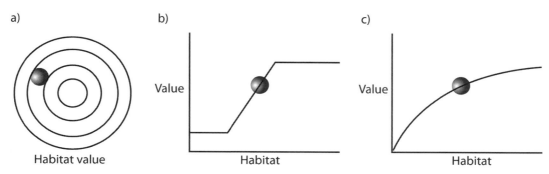

a) Habitat value

b) Value / Habitat

c) Value / Habitat

*Figure 15.10
Responses of
organisms to
habitat elements:
(a) there are no
unequivocal
'bullseyes' for
habitat values in
nature; (b) sharp
thresholds would
clearly guide
management;
(c) nature has
few sharp
thresholds.*

of projects on the same basic questions at the same sites (for a more complete list of benefits see Huggard *et al.*, 2009).

Design monitoring to search for thresholds, but don't expect tidy thresholds. There is a misconception, particularly among the public who influence social license, that somewhere in nature useful targets exist. It is true that nature provides the only unequivocal 'targets', but these rarely appear as a tidy 'bullseye' or even a sharp threshold (Figure 15.10).

Bunnell and Dunsworth (2004) noted that even for small groups of similar species, responses appear more curvilinear than sharply defined. The 'habitat alteration' threshold of Guénette and Villard (2005), for example, comprises at least four quite different variables. That is, responses most commonly follow the pattern of Figure 15.10c and rarely are guided by a single variable. Habitat has a marginal value – increasing the number of large snags from 5 to 6 per ha contributes much less than increasing them from 1 to 2 per ha. When axes of marginal-value curves, such as that illustrated in Figure 15.10, are reversed they represent a marginal-risk curve. In either form, they have some utility in guiding definition of a threshold for action, but are still burdened by the complexity of forest-dwelling biodiversity. Different curves exist for different species and species groups and assume different shapes depending on the kind and number of habitat attributes included.

An important consequence of this pattern in nature is that most monitoring will be directed to comparisons of different practices, rather than a search for a specific target. That is particularly true when novel or uncommon practices are being implemented. Comparisons yield no single correct answer (a major attribute of wicked problems), but allow distinction between better and worse.

Specify the management responses during design. A very helpful exercise is to specify at the outset: if the data show this we will do that. Such a process not only helps ensure that the data are practical and informative but reduces confusion when it comes to changing management action in response to feedback from monitoring. For example, several forms of feedback can be anticipated from monitoring ecosystem representation. The most general of these include guiding the intensity of harvest, focusing finer scale monitoring, and simplifying operational decisions. The intensity of harvest can be guided by the likelihood that relying on the non-harvestable land base will be sufficient or insufficient to maintain all organisms. Gains are made when stand or structural retention is allocated to poorly represented ecosystem types. Favouring forms of retention harvest within poorly represented ecosystem types also focuses habitat monitoring (medium filter) and specific species monitoring (fine filter) on those habitat types. Ecosystems represented relatively well in non-harvested types are of less concern, although they may provide 'benchmark' examples of natural conditions. The task of distributing monitoring throughout the management area can be better guided and focused. Operational decisions are simplified when a group of species appears well supported in non-harvestable areas, so need not enter operational decisions for the harvested land base.

Don't expect tidy answers about habitat distribution. The problems of scale and confusion in terms used to evaluate habitat distribution are discussed in Chapter 3. These will not go away. Nor are models that address the gradients of nature readily available. It is certainly simpler and often just as accurate to interpret GIS-based maps visually than use models with inaccurate assumptions.

What do we do if the approach is wrong?

This step involves the links to management action. Because managing to sustain forest-dwelling biodiversity is a wicked problem and full of a variety of species and interactions, there will be few clear answers. That is one reason why identifying major questions and then ranking them is so important.

Conflicting recommendations are inevitable. What form of retention during harvest is best, or even adequate, for large cavity nesters will not be the same as for mycorrhizal fungi or carabid beetles. These conflicts are not a failing of the adaptive management program, but a product of how nature is put together, so will not go away. There is no single best approach so choices have to be made.

Several values will determine success. The problem would be wicked enough if only nature were involved, but economic and social features are a significant part of it. As long as forestry is intended to make money and employ people, costs will remain a concern for managers. They may be as simple as various forms of trade-offs between contiguous reserves and scattered retention patches. Such choices also are not a failing of the program, but a product of us being part of the problem as well as the solution. It helps greatly to anticipate and specify management responses to particular findings early in the program, but that cannot anticipate all economic and social change.

*Figure 15.11 Cutblock in 40-year-old flooded gum (*Eucalyptus grandis*) forest at Nymboi Binderay, in New South Wales, Australia. Harvesting of native forest in Australia is under increasing pressure, with some arguing that it should be completely replaced by plantation forestry.*

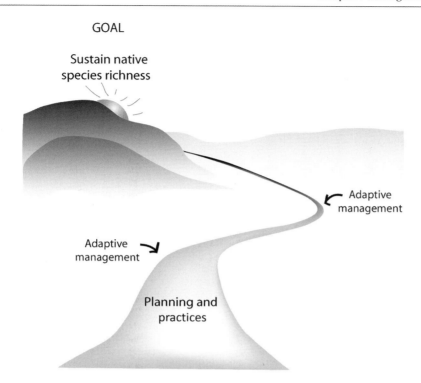

GOAL

Sustain native
species richness

Adaptive
management

Adaptive
management

Planning and
practices

*Figure 15.12
Adaptive
management
is a process
of repeatedly
making the
better choice.
(Figure prepared
by I. Houde.)*

Different measures will rank different practices differently for different values. The most obvious and pervasive example is that different species respond to the same stand treatment differently. The relative costs of treatments also differ. However well the program is designed, there is no way that the rankings of stand treatments or their relative amounts can be consistent across species and areas. The problem is increased because thresholds, when they exist, are seldom tidy. Choices have to be made.

There is no single correct path, but you know what is worse and what is better. This is the single redeeming feature of a wicked problem and the feature that adaptive management is intended to exploit. There is no single 'best' way of managing forests, but adaptive management can keep forest planning and practices from going off the road and into the ditch for any specified goal (Figure 15.12).

Examples of adaptive management in forestry

The roots of adaptive management are in the Pacific Northwest of North America, specifically at the University of British Columbia. They can be traced to 'the myth of the omniscient forester' (Bunnell, 1976), who noted 'acknowledging existing uncertainty at the initiation of a management decision helps eliminate the pathological avoidance of paying for mistakes, which often leads to foreclosure of options' (p. 150). The formal approach for avoiding paying for mistakes was first described by Holling (1978) and Walters (1986), also then at the University of British Columbia. It is likely for this reason that the most inclusive and comprehensive examples of adaptive management applied to forestry are from the Pacific Northwest.

The Survey and Manage Program of the Northwest Forest Plan (NWFP) employed adaptive management principles in an attempt to sustain little-known species

Figure 15.13 Coastal temperate rainforest near Quinault Lake, Washington, USA. The management of old growth forest such as this presented a wicked problem for managers that could not be resolved by reserving large areas as national parks.

associated with late-successional and old-growth forests on more than 7.7 million ha of federal lands in Washington, Oregon, and northern California (Molina *et al.*, 2006). Over 400 species of amphibians, bryophytes, fungi, lichens, mollusks, vascular plants, arthropod functional groups, and one mammal were listed under this program, because viability evaluations indicated the plan's network of reserve land allocations might not sustain the species over time. Among the important lessons learned was much about specific survey techniques, that it is unlikely that the reserves alone suffice to provide habitat and geographic protection for all rare and poorly known species associated with late-successional or old-growth forest in the Pacific Northwest, and that the problem remains 'wicked' – there are inherent difficulties in balancing broad, regional conservation goals with the other social and economic goals of the Northwest Forest Plan.

The adaptive management program described by Bunnell and Dunsworth (2009) covered 1.1 million ha on Vancouver Island, portions of the adjacent mainland, and Haida Gwaii. It was equally expansive in the species studied (songbirds, owls, carabid [ground] beetles, gastropods, bryophytes, vascular plants, epiphytic lichens, ectomycorrhizal fungi, and aquatic-breeding amphibians) but was not limited to late-successional and old-growth forests. It focused directly on potential advantages of introducing wide-scale variable retention. The utility of reserves was examined, but did not dominate program design.

Although adaptive management must be well formulated to attain its promise, it does not have to include all the questions that arise to be informative. Examples of findings developed from smaller adaptive management programs in northeastern British Columbia are presented in Chapter 3. Those programs were limited in scope and expense, but were able to guide significant improvements in practice (e.g. location of anchor points for variable retention, diameter limits of retained trees and down wood) and to enhance social license.

Many examples of adaptive management in forestry focus on biodiversity because sustaining biodiversity receives the most public attention and is usually the most challenging part of forest management (see the section 'Forest biodiversity and wicked problems'). Other features of forest management are at least as amenable to adaptive management. The program described by Bunnell and Dunsworth (2009), for example, included consequences to regeneration of variable retention and the potential to increase the rate of at which old-growth attributes might be attained by appropriate management of riparian areas. Bunnell and Kremsater (2012b) note the role of adaptive management in exposing and responding to the challenges of climate change.

Lessons learned from designing adaptive management

Adaptive management programs provide abundant opportunities to learn about individual species and practices. Some lessons learned while asking the major management questions are summarized earlier. Here I summarize more general lessons about the design and implementation of adaptive management. These lessons are summarized from experience of larger projects (e.g. Molina *et al.*, 2006; Stankey *et al.*, 2003) and smaller projects (Bunnell *et al.*, 2009e, 2010). They are collected under three broad groupings: organizational structure, design, and feedback to management. Lessons can be pertinent to more than one grouping; connections to those described earlier are noted.

Some lessons pertain primarily to organizational structure. Attempts at adaptive management fail less often in design of the work than in closing the loop so that management actions are taken in response to findings (Figure 15.3). The two large examples noted struggled more with this challenge than the smaller programs in northeastern British Columbia. There were two broad reasons: (1) naiveté with regard to the world of management decisions and (2) failure to implement a shared vision within the world of researchers. Lessons on organization structure include the following.

Commit time and resources to communication. Inadequate communication among scientists helping design the monitoring program, those overseeing the program, and researchers acquiring data is common. The need for a shared vision needs to be clear to all participants at the start; adoption of that vision should be assessed early and encouraged within contracts to consultant researchers.

Diverse funding is a challenge. Diverse funding competes with a shared vision. Often both corporate and public funding is involved. This can produce a program partially driven by interests of individual researchers and funding agencies rather than being designed and rationalized explicitly within the context of long-term monitoring. Expectations of annual 'stand-alone' products are encouraged by external funding sources, annual reviews, and traditional expectations of researchers themselves. Diverse funding sources also greatly increase issues of intellectual propriety that can thwart the analyses and syntheses necessary to design an effective monitoring program. Conflict between short-term expectations or demands and longer-term requirements will not go away. It should be acknowledged at the outset and the overall design carefully evaluated to determine where focus can produce significant short-term contributions within necessarily longer-term efforts. Funding sources and award systems can then be matched more appropriately with intended products. We did not anticipate this challenge at the outset of the program described by Bunnell and Dunsworth (2009). By the time we were working in northeastern British Columbia we had developed an approach characterized as the following.

A long-term approach AND short-term delivery. The costs of an adaptive management program can be large. If the program is to be sustained long enough to answer management questions, there has to be some short-term delivery of useful results. One approach we found useful is the Species Accounting System described in Chapter 3.

Because it helps to focus questions and is largely based on GIS, interim or even near final answers can be offered early, while data for more challenging questions are being accrued. Ideally, the program should be designed to deliver some answers early and often. Usually, it can be.

Recognize different reward systems. Many creative researchers are somewhat maverick in nature, and nearly always have a different sense of reward and of risk than do practitioners. Academic researchers often find that pilot study work will not produce publishable papers, while conservation-oriented researchers often do not feel that pilot studies directly support their interests. Pilot studies for a long-term monitoring program have a different purpose (thus a different design) than the shorter-term research projects with which biologists are familiar. A major difficulty in implementing long-term monitoring programs has been the mismatch between the need to repeat the same monitoring over several years to obtain useful results and demands for short-term output from the work. Part of this difficulty derives from funding sources with short-term objectives and part from the very diversity of questions that monitoring is intended to address. The latter diversity encourages the tendency of researchers to chase an interesting new question, rather than building up the results needed to answer an older question well. Communicating pilot study goals and finding researchers willing to address those goals is critical to an efficient long-term monitoring program.

Other lessons can be grouped under design issues. Designing and implementing an effectiveness monitoring program teaches a lot of lessons about appropriate sampling methodologies for different organisms, statistical blocking factors, questions more accessible to experimental approaches rather than operational comparisons, conditions that encourage pre-treatment measures, and more. It also provides generic lessons.

Identify major questions and stay with them. It is easier to get the first half of this correct by winnowing, then ranking questions and assessing which potential comparisons are most relevant to which questions and ranking those as well. You can then match indicator variables with comparisons through a cross-design, using a suite of indicators for the highest priority comparisons, while ensuring that each important comparison is assessed with at least one or two suitable indicator organisms (design in Table 15.2). That determines the basic monitoring framework. The challenge is staying with the framework. Large programs in particular depend on diverse, sometimes opportunistic, funding sources that distract from focus on the highest priority questions and comparisons. That challenge emphasizes the importance of clarity in and commitment to a shared vision, and of efforts to match reward systems to activities. We found it easier to stay focused in smaller programs where the vision of practitioners and scientists was sufficiently shared that they worked together towards funding and implementing the cross-design.

Begin modelling early; you know you'll have to stay late. Typically several types of models are required simply because the over-arching question of whether sustained habitat is provided for biodiversity must be addressed over large areas and long periods of time. These include: stand projection models to project habitat elements of stands and consequences on regeneration if novel practices are introduced; landscape projection models to evaluate timber supply, habitat distribution, and visual quality objectives; and habitat suitability models at both the stand and landscape level.

Major advantages are gained by question- or model-oriented sampling. Development and refinement of useful question-based models is a long-term exercise that must be initiated early. Testing modelling components has implications for design of a monitoring program, particularly the relative role of comparisons and mechanisms and the allocation of funding to them, so must begin early. That is particularly true of habitat models and models projecting habitat elements, as discussed next. Effort must be sustained if all advantages are to be gained.

One size does not fit all. The kinds of data required to assess the effectiveness of forest planning and practice at sustaining biodiversity are diverse, but most involve relating organisms to their habitat. Relating habitat elements to the needs of a range of organisms through empirical habitat models gathers many habitat elements into fairly simple, easily understood indices. Developing such habitat relationships often requires organism monitoring to use study designs and measurements that differ from those needed to make direct comparisons. Similarly, models required to project habitat through time require information on processes, such as growth and mortality of retained and regenerating live trees, decay and fall of snags, and decay of down wood. The interdependency of models and sampling emphasizes the need to develop the framework of explanatory models early in the program and to continue to evaluate progress on the modelling framework. It also means that if researchers are not closely involved in model formulation they must, at the least, understand why and how their data can inform modelling.

Monitoring and research are not discrete. We tend to view monitoring and research as discrete and different activities. Monitoring is often viewed as repeatedly (thus boringly) recording conditions, while research asks more immediate questions – typically about explanatory mechanisms. Such distinction ignores the important complementarity between the two activities, particularly within an adaptive management program (Noss and Cooperrider, 1994). Monitoring is necessarily a longer-term activity, while a well-designed research question may be answered relatively quickly. Usually, both ways of asking questions are required: making comparisons and evaluating mechanisms. The former is often emphasized because comparisons of operational sites are fruitful in addressing near-term management questions. Explanatory mechanisms, however, are necessary to project long-term consequences of present management decisions, particularly for habitat elements. Explanatory mechanisms are first to succumb to pressures of designing a cost-effective system of comparisons (e.g. Table 15.2). That impedes the ability to answer key questions about larger-scale, longer-term questions.

Where new practices are introduced, some questions are better asked in experimental than in operational sites and require pre-treatment measurements. There are difficulties inserting experimental sites or active adaptive management into operational forest activities, which can mean that anticipated contributions to operational comparisons are not made for some organism groups. Experimental sites have strong appeal to scientific reviewers, they help involve operational staff directly in adaptive management, and they may produce more definitive information when the scattered replicates are eventually completed. However, the design of the overall monitoring program must recognize that there is a substantial cost to monitoring experiments. With dispersed replicates and implementation over many years, there is a long delay before results from experimental sites become available, and considerable uncertainty about how meaningful they will be.

Scale or context is essential. Even precise measurements are unhelpful without context or scale. Regulatory targets are often in flux and inconsistently supported by data; natural benchmarks in older forest provide less equivocal context (Figure 15.14). When economic conditions were good, my co-workers and I found that companies were willing to accept comparison of practices with each other or to benchmarks and to change practice (e.g. diameter limits of leave trees). The implication to monitoring design is that resources must be allocated to such benchmarks – the 'baseline' monitoring of Noss and Cooperrider (1994).

The third group of lessons learned concerns feedback to management. The purpose of adaptive management is to improve management through information gained on actions taken. Feedback to management can take several forms. Experimental sites

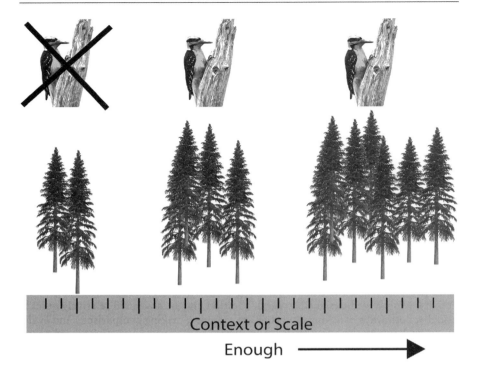

Figure 15.14 Context is necessary; natural is the context. (From Bunnell et al. (2009b) with permission of UBC Press.)

are necessary for some questions, but the kinds of planned comparisons summarized in Table 15.2 are always cheaper and still informative. They can be designed to reveal areas of relative weakness where improvements to management would be most effective. For example, direct comparisons of the amounts and type of variable retention can inform decisions about appropriate mixes of practices. Feedback also can be based on comparisons to thresholds or targets established externally by government, certification groups, or requirements of organisms. Other mechanisms linking monitoring to management are formal review of monitoring results and recommendations from scientific, operational, and public groups.

One lesson closely associated with feedback was noted earlier. The fact that there are few tidy thresholds in nature must be clearly communicated when feedback is offered, and leads directly to an additional lesson.

Simpler is not always better. In the program described by Bunnell and Dunsworth (2009), most recommendations from monitoring to management involved selecting better practices, and thus gradually increasing performance as assessed by the indicators. Specific targets based on ecological thresholds did not play a role in this feedback, although natural benchmarks provided scale or context. Comparing retention results to benchmark sites to identify and focus on the weakest points in retention was a useful approach to simplifying the message for practitioners. In the smaller projects of northeastern British Columbia (see Chapter 3) feedback also derived from comparisons of practice to natural benchmarks.

Such simple comparisons are most effective when the goal is to compare responses of one or a few elements among a few alternative treatments. It does not evade the complexity inherent in relations among forest-dwelling organisms and their environment. Focusing on individual elements can lose sight of the fact that we want to maintain a diversity of habitat, not just a few focal elements.

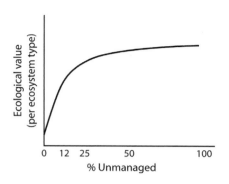
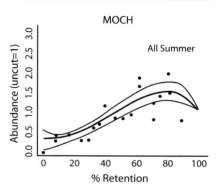

Figure 15.15

*Illustrative examples of marginal value curves. Left: Changes in ecological value with changes in the percent of area unmanaged. Right: Response of mountain chickadee (*Poecile gambeli*) to amount of forest cover retained; outside lines are 5% and 95% quantiles; 100% represents the natural benchmark. (Adapted from Bunnell et al., 2009c, with permission from UBC Press.)*

Likewise, instead of searching for discrete thresholds, describing response or marginal-value curves indicating the benefit of incremental gains is our best option for presenting information (Figures 15.10c and 15.15). The converse, showing the marginal cost or risk to various organisms of potential options, also is informative. In each case natural benchmarks provide the scale for the relationship. We have such curves implicitly in mind when we evaluate information. Making them explicit helps inform decision-makers.

For example, going from 40% unmanaged in an ecosystem type to 50% is good, but it is not nearly as good as going from 1% to 11% (e.g. Figure 15.15 left). The response of mountain chickadee (*Poecile gambeli*) to amount of forest cover retained is illustrated in Figure 15.15 right. It is one example from a likelihood-based meta-analysis of 69 bird species from over 50 North American studies. As expected given the variety of bird species, the curves assume most conceivable shapes, including negative responses to increased forest cover. They illustrate why simpler is not always better. During feedback to management, it is important that creation of such curves be transparent and that a number of curves for different elements of biodiversity be incorporated in any decision-making process.

Informed choice or future guidance. The concepts of informed choice or future guidance determine the way information is presented to decision-makers. The dichotomy between the two is always implicit in the design of the monitoring program. Informed choice presents decision-makers with predictions of the effects of different clearly specified management actions on a number of valued components based on the best available science. Selection of the Northwest Forest Plan ('President's Plan') for forest practices in the American Pacific Northwest is an example. Although informed choice is helpful when specific management alternatives can be defined, it does not provide general guidance to management questions, such as 'What should we be doing more or less of or doing differently?' or 'What issues will we be facing in 10 years?' The distinction can be characterized as a short-term length of view (choosing between currently available options) or a long-term length of view (providing guidance towards an ever-increasing ability to attain the goal of sustaining biodiversity). An alternative and equally relevant distinction is whether the program is to 'mitigate' (find the least bad

Figure 15.16 Harvesting in plantation forests near Braemar, Scotland, UK. The results obtained in the British Columbian studies may be directly applicable here, as many of the species are the same. However, the applicability to sub-tropical woodlands or tropical rainforest remains untested.

current option) or 'do good' (work toward a better condition). The latter incorporates information beyond common current practice.

To reduce disappointment, practitioners and researchers need to discuss this distinction and expectations of the program early, because it determines the design of the program and its length or cost. Given the wealth of species, making comparison among current practices is itself expensive; designing and monitoring novel practices can be significantly more expensive and lengthy. Any adaptive management program attempts to identify potentially deleterious practices, but also is designed to help improve existing practice. A major challenge to a program is to maintain a commitment to long-term guidance and improvement while addressing short-term choices and attempting to detect deleterious effects quickly enough to mitigate them.

Expectations for monitoring are often naïve. Monitoring or feedback to management is critical to the success of adaptive management, but expectations of monitoring are often naïve simply because tidy answers are necessarily elusive. Monitoring is a form of risk management that can expose unintended or unanticipated consequences early enough that costs of correction are not large. The findings or feedback to management do not eliminate risk, but can lessen consequences of the inevitable exposure to risk (e.g. keeping out of the ditch; Figure 15.10).

It is naïve to expect consistent and direct short-term responses to monitoring results in the face of strong economic pressures on forest managers. It is likewise naïve to think we can fully implement a long-term monitoring program as designed. Instead, programs will evolve at a far faster rate than they will provide answers. Even though approved by President Bill Clinton, the Northwest Forest Plan was not implemented as designed (Molina *et al.*, 2006). As with the program described by Bunnell and Dunsworth (2009), the primary difficulty was monitoring organisms. At best, people overseeing a monitoring program can provide selective forces that help keep the monitoring useful in the long term.

Experience suggests that the appealing idea of a simple adaptive feedback loop – modifying management based on results from comparing options – may be unrealistic

Figure 15.17 Harvested Acacia mangium *plantation in Sarawak, Malaysia. In many parts of the world, forestry is not accompanied by detailed research or monitoring, yet such work is essential if forestry is to provide any credence to its claims as a 'green' industry.*

with management and response variables as complex as forest practices, particularly when organisms other than trees are included. The issue is too thoroughly an example of a wicked problem to permit simple solutions. Such loops can work for simple operational questions, especially when single, small-scale factors affect individual species of direct management concern. Changes in forest management are more likely to be based on information received less directly, as recommendations from people who integrate many information sources, such as market groups or policy-makers. Because such recommendations are partly informed by science, a monitoring program can make an important contribution by generating sound scientific results. These may ultimately have more impact on improving management – by whatever indirect route – than monitoring by attempting direct, immediate feedback to management decisions. For a company, conducting sound scientific projects (even if not directly applied) can be seen as 'rent' for the use of the broader scientific knowledge that is the fundamental basis of informed management decisions. With an ever-changing organizational environment for monitoring, and the inherent interests of most researchers, good scientific contributions may be the main legacy of a long-term monitoring program.

Conclusions

This chapter has presented how a forest manager can go about managing a forest in the face of considerable scientific uncertainty. It has focused on the uncertainty surrounding the effects of management on forest biodiversity, but the methods can be applied to other values associated with forests. The work was undertaken on Vancouver Island in British Columbia, Canada, but is applicable (with appropriate modification) in other parts of the world.

The process of adaptive management remains our best and cheapest way of guiding improvements in management with data or knowledge, in part because we don't stop doing but learn by doing. The process, however, is not easy. There are three common reasons why adaptive management fails. Two of these (measuring too many things and high costs) can be corrected by limiting efforts to the most important questions the budget can address.

The third reason for failure, failure to link well with management, also can be addressed, but not as effectively. Specifying how data acquired are anticipated to change management at the outset helps significantly. Management, however, responds to changing economic and social environments as well as the natural environment. The problem remains wicked. The significant advantage of documenting what is better and what is worse also persists.

Further reading

Armitage, D., Berkes, F., Doubleday, N. 2007. *Adaptive Co-Management: Collaboration, Learning and Multi-Level Governance*. Vancouver: UBC Press.

Bunnell, F. L., Dunsworth, G. B. (eds.) 2009. *Forestry and Biodiversity: Learning How to Sustain Biodiversity in Managed Forests*. Vancouver: University of British Columbia Press.

Gauthier, S., Vaillancourt, M.-A., Leduc, A., de Grandpré, L., Kneeshaw, D., Morin, H., Drapeau, P., Bergeron, Y. 2009. *Ecosystem Management in the Boreal Forest*. Québec: Presses de l'Université du Québec.

Messier, C., Puettmann, K. J., Coates, K. D. 2013. *Managing Forests as Complex Adaptive Systems: Building Resilience to the Challenge of Global Change*. London: Routledge.

Walters, C. J. 1986. *Adaptive Management of Renewable Resources*. New York: Macmillan.

Chapter 16
Conclusions

John L. Innes

Humans have always interacted with forests, having evolved from forest-living animals. Early hominids in Africa likely lived in savanna ecosystems; as they spread out from this ancestral base, different species lived in forests and woodlands. Woodlands were particularly attractive as they were easier to pass through, and food was easier to locate and hunt. These early hominids probably used fire to manage the woodlands much as Aboriginals in Australia were doing, a practice that is now emulated by modern land managers (Russell-Smith *et al.*, 2009). Forests and woodlands were a major source of food, both animal and plant, and were a source of firewood used in cooking, to keep warm and to ward off predators. The very small populations of hominids are unlikely to have had a major influence on forests, except perhaps through deliberately set fires.

This changed with the evolution of modern humans. Early humans were hunter-gatherers and, as they developed tools, they started having an impact on forest ecosystems. The relative importance of climate change and humans in the extinction of the megafauna that existed in much of the world during the Pleistocene is strongly disputed, but the loss of Australia's megafauna about 45,000 years ago, shortly after the arrival of humans, and the loss of the North American megafauna about 13,000 years ago and the South American megafauna about 500 years later, again shortly after the arrival of humans, strongly suggests that hunting pressures were involved. One of the best documented cases is the loss of moas, giant flightless birds, in New Zealand following the arrival of the Maori people; this occurred through intensive hunting and through habitat loss caused by deliberate burning of the forest cover. The impacts were both direct and subtle: species were lost, but the loss of those species seems to have created changes in forest vegetation. For example, in Australia, the loss of a megafauna that consisted mainly of browsers led to an increase in vegetation density, which in turn resulted in an increase in fire frequency (as shown by layers of charcoal). The increased fires in turn led to the dominance of sclerophyllous vegetation over rainforest species. This last dynamic is still occurring – when fire is excluded from sclerophyll vegetation, rainforest species can successfully grow.

Through time, people living in or near forests have accumulated a wealth of information which today is recognized as 'traditional ecological knowledge'. This included all sorts of information about the ecology and use of forests and the species occurring in them and, in most cases, was closely tied to the cultural and spiritual use of forests. It is impossible to say how much of this knowledge has been lost over time: knowledge

*Figure 16.1
Rainforest
species growing
under a eucalypt
canopy in
the central
Highlands
of Tasmania,
Australia. The
rainforest species
can only survive
as a result of fire
exclusion.*

is continually gained and lost, but is recognized today as being vitally important to forest-dependent peoples.

The development of agriculture freed people from their dependence on forests and woodlands. It enabled towns and cities to develop, and these population centres also became centres of resource use. Resources around these population centres gradually became depleted, with wood being in particular demand because of its many uses. In some cases, resource depletion was so severe that it led to the collapse of the cities and, in some cases, entire civilizations. Under such circumstances, it became increasingly important to manage the resources, and there are records of such management dating back thousands of years.

The demands of towns and cities for resources, and the continuing existence of people directly dependent on forests, led to conflict between the two. In some cases this was resolved through legislation. For example, in England, judicial hearings in the form of 'forest eyres' date back to 1166; these were hearings where forest officials presented evidence that offences had been committed against the king's deer and their habitat. The laws were formalized in a document called the 'Prima Assisa', subsequently replaced by the Assize of Woodstock in 1184. The 1217 Charter of the Forest in England is sometimes cited as an example of early legislation protecting the rights of forest-dependent people, but the Charter was really more about curtailing the rights of the King of England to declare royal hunting forests (usually at the expense of lands held by his nobles), and limiting the activities of the foresters (those entrusted with managing and protecting the game in forests, which involved the corporal punishment of anyone found poaching). The close association between the management of game and the management of forests continues in some parts of the world to this day, especially in central Europe.

As urban populations continued to expand, demand for forest products continued to grow, and in both Europe and Japan, systems for managing the supply of wood evolved. This appears to have been the origin of modern forms of science-based forest management, although priorities have since evolved substantially in some parts of the world. At the same time as forest management practices were evolving and being refined, the phase of colonial expansion occurred, as did the expansion westward of settler populations in North America. These were often based on the exploitation of forest resources (Williams, 1992). The establishment of colonial policies in the Americas, Africa and Australasia were normally accompanied by the nationalization of ancestral forest lands; in many cases, the forest remained nationalized after independence, creating the problems associated with property rights in forests that we still see today (Enuoh and Bisong, 2015).

The loss of forests was accompanied by a realization that forests provided more than timber. In particular, deforestation was associated with increased soil erosion and increased flooding. This was already evident 2,000 years ago in the Mediterranean basin and was repeated elsewhere, such as in the Kyoto area of Japan 1,000 years ago. It has also been evident in more recent periods of deforestation, and the value of maintaining the forest cover of catchments used to supply water to urban areas is now widely recognized.

Losses of forest cover were also associated with losses in biodiversity, and it is no coincidence that today many threatened species are forest dependent. For example, according to BirdLife International, 76% of threatened bird species are found in forests. The majority of deforestation is related to conversion of forest to agriculture or rangeland, and there is increasing recognition among a diverse array of stakeholders of the need to halt deforestation. This led to the New York Declaration on Forests in September 2014, where a range of governments, corporations, non-governmental organizations and charities pledged to cut natural forest loss by half by 2020, and to halt the loss of the world's natural forests by 2030. Those signing the declaration included a number of corporations with supply chains that have in the past or still today been associated with deforestation, including the Kellogg Company, Marks & Spencer, Danone, Unilever, McDonald's, Barclays, Nestlé, Cargill and Asia Pulp and Paper. The involvement of such companies is critical as they rely heavily on the four commodities most responsible for deforestation driven by agriculture: palm oil, soy, cattle, and timber and pulp. Some progress has already been made with palm oil and timber and pulp, but commitments related to soy and cattle have been less forthcoming (Supply Change/Forest Trends, 2015).

Forest managers are not in a strong position to halt deforestation, although if they work together with others in managing landscapes they can play an important role (Kettle and Poh, 2014). This will ensure that landscapes, particularly in tropical areas, continue to meet the needs of local people while also providing the goods and services that we have come to expect of them.

Forest managers do however have a major role to play in ensuring that forest degradation does not occur. This is a difficult topic, as any management activity within a natural forest could be considered as degradation. This is an extreme viewpoint, and belies the fact that the majority of the world's population is dependent on land that was once covered in forest or woodland, whether it be urban areas, agricultural land or rangeland. In the 21st century, it is apparent that to maintain the world's population (or support the inevitable further increase in population) and to support the increasing standard of living that much of the world's population seeks, a shift to an economy that is based much more on renewable resources is urgently required. There is global recognition of this, although the will to change is not quite as evident. Forests provide a range of goods and services, all of which are renewable if utilized carefully. Natural forests can also be utilized, but much greater care is needed than in many planted forests where priority is normally given to a particular management objective (e.g. environmental remediation or the production of fibre).

Good forest management is therefore an essential part not only of landscape management but of the development of a green economy. The various chapters of this book have emphasized the important aspects of forests that a forester today is required to maintain, including biodiversity, ecosystem productivity, forest health and vitality, soil and water, carbon and the complex suite of social, cultural and economic values associated with forests. While some argue that every stand of trees should be able to contribute to every forest value, a more realistic approach sees individual stands as a part of the landscape mosaic, and that it is the forests across the landscape that should be able to provide every value. Even this is questionable, given that planted forests are now being established at a landscape scale in some parts of the world.

The information presented in this book shows just how complex it is to manage all the values in a forest simultaneously. However, foresters are faced with an added challenge: climate change. There is an increasing number of reports of recent climate change influencing forest ecosystems. As might be expected, it is the more mobile species that are showing the quickest changes, and there are numerous reports of impacts on birds (BirdLife International and National Audubon Society, 2015).

While birds and animals can move, individual trees obviously cannot, and the observed responses that have been attributed to climate change include a range of physiological disorders and in the most severe cases, mortality. With trees, climate acts in conjunction with other stresses and, as described in Chapter 5, it is likely that the actual cause of death is an agent such as an insect or pathogen, with climate having stressed the tree, making it more susceptible to attack. Drought, flooding and wind can all cause mortality, but attributing individual weather events to climate change remains challenging. Adapting forest management to meet these challenges is a major task for today's forest managers (Peterson *et al.*, 2014).

There are also concerns that climate change will disrupt the ability of forests to provide ecological services (Chiabai, 2015). The maintenance of forest carbon is of great importance, and REDD+ should ensure that developing countries are compensated for maintaining their forests, rather than converting them to other forms of land

*Figure 16.2 Keel-billed toucan (*Ramphastos sulfuratus*), a species formerly of lowland rainforest in Costa Rica but now found up to elevations of 1,500 m, most likely because of climate change.*

use. Increasing numbers of non-timber forest products are being harvested. The açaí palm (*Euterpe oleracea*) is an example of a rainforest fruit that has increased rapidly in popularity, with the result that the palms are now cultivated specifically for the fruit, and some are concerned that a plantation model will be adopted with all the attendant risks. The interactions between forests and water are of particular concern, yet there are many uncertainties associated with this relationship (Malhi and Phillips, 2005), especially outside the temperate zone, where most research to date has been undertaken.

Forest managers must navigate their way through the complexities of the biophysical relationships within forest ecosystems. However, they must also deal with the complexity of forest interactions with people. The Declaration arising from the Seventh World Forestry Congress in 1972 stated that 'forestry is concerned not with trees, but with how trees can serve people'. As we have seen, trees can serve people in many ways, with the production of wood being only one. As Westoby (1987) has argued, managing that relationship is a critically important role for foresters.

Given the importance of forests to global climate and to the livelihoods of some 2 billion people, forest managers must always bear the public interest in mind. This requires that forest managers have strong codes of ethics, and many professional forestry associations require their members to follow such codes. Unfortunately, this does not prevent some forest managers from putting their own or their employers' interests ahead of the public interest. As discussed in Chapter 12, the governance of forests is a critical issue. Not only is illegal logging widespread, but government authorities in both the exporting and importing countries either pay little attention to the issue or collude in the trade. This has prompted some regions to develop legislation specifically to combat this problem – the 2008 amendment to the Lacey Act in the USA is an example, as is the European Union's Timber Regulation. However, some of the major consumers of illegally logged wood have yet to take action, despite the efforts of programs such as the European Union's Forest Law Enforcement, Governance and Trade (FLEGT) program.

While forest managers have traditionally worked in the forest, there are now all sorts of expectations for them to work in a wider range of environments. First, they need to adopt a landscape perspective, which means working with others in the landscape, including farmers, developers, planners and a range of other stakeholders. It will involve developing a better understanding of the principles and tools used in landscape planning, and integrating these with the tools used in forest planning. Second, they need to be much more engaged with their communities, ensuring that the goods and services provided by forests bring benefits to the community. There is much guidance that will help forest managers do this (e.g. Macqueen *et al.*, 2012). It will require the acquisition of sound communication and business skills, areas that foresters are not always particularly adept in.

Sayer and Maginnis (2005b) identify seven trends that they consider will play a major role in the future of sustainable forest management. Although already 10 years old, these trends are still just as relevant today as they were when they were first formulated.

Forest management objectives are broadening. While a number of different objectives have existed in the past, including timber production, hunting and protection from natural hazards, and multiple use forestry has attempted to combine these, more and more objectives are being formulated. These cross a number of scales, and local objectives may be inconsistent with global objectives. In particular, the scale of forest management is changing, covering longer timescales and larger areas, and involving multiple land uses (the landscape approach).

Society, in the form of regulators, certifiers and civil society, is increasingly specifying what it expects of forest management. Criteria and indicators represent a good example of this, especially in the extent to which they can be used to measure the quality of forest management or the health of forest ecosystems. This trend has been emphasized in previous chapters, as it increasingly defines and determines sustainable forest management. Different stakeholder groups increasingly have differing expectations, and the forest manager is left with the task of meeting these expectations.

While in the past there have been attempts to advocate particular forms of forest management, there is increasing recognition of the uniqueness of forests, especially if the ecological nature of forests is combined with their economic, cultural and social characteristics. This means that a single best type of management is unlikely to work, and even that there is no optimal form of tenure.

Decentralization and divestment of forests is accelerating. Countries are choosing to empower local communities, or have lost the capacity to manage centrally (or a combination of the two). This is not always successful, and can even lead to forest degradation if done badly. In many situations, there is insufficient local capacity to take on the management role, and considerable capacity building is still required.

Globalization is a pervasive trend and is having significant effects on forest management (Nikolakis and Innes, 2014). Considerable consolidation has occurred in the forest industry, and the biggest forestry corporations control increasing amounts of production forest. Companies such as Asia Pulp and Paper are increasingly important and expanding their reach, sometimes through subsidiary companies. Even in places such as the USA, consolidation continues to occur, as demonstrated by the 2015–2016 purchase (for $8.4 billion) of Plum Creek Timber Co. by Weyerhaeuser Corporation. This has increased Weyerhaeuser's US landholdings from 2.83 million ha to 5.26 million ha. Trade agreements, such as the 2016 Trans-Pacific Partnership, may have major effects on the economics of forestry in both partner and non-partner countries. International payments for ecosystem services are increasing, and regional agreements on forests may well be a prelude to global agreements. The importance of sustainable forest management is specifically mentioned in the global Sustainable Development Goals agreed in September 2015.

Climate change is again placing forests on the political agenda. Forests were mentioned by an unprecedented number of world leaders in their statements to the Conference of the Parties to the United Nations Framework Convention on Climate Change in Paris in December 2015. Forests are now seen as playing a critical role in addressing climate change, through their potential to sequester carbon and as a result of the carbon that is released when they are destroyed or degraded.

Finally, Sayer and Maginnis (2005b) identify changes in governance as an important trend. There is an increasing focus on problems such as illegal exploitation, land conversion and corruption, and this has resulted in significant changes to local management practices (through regulation) and to trade (through mechanisms such as the 2008 Lacey Act amendments in the USA). The capability to enforce regulations is increasing, but corruption remains a major issue in many countries, and is hindering the enforcement of laws passed to improve forest management.

These trends are all changing the nature of forest management. The forest manager of the future is going to need many more skills than are currently recognized, and will need to be able to adapt to new situations quickly. Fortunately, there is an increasing number of tools available, but the skill sets required to use these tools effectively are more difficult to acquire. Online courses will help, provided that these are updated regularly as new knowledge becomes available. This book, and the associated critical reading, should also contribute to the acquisition of the necessary skills by forest managers.

Further reading

Minang, P. A., van Noordwijk, M., Freeman, O. E., Mbow, C., de Leeuw, J., Catacutan, D. (eds.) 2015. *Climate-Smart Landscapes: Multifunctionality in Practice*. Nairobi: World Agroforestry Centre (ICRAF).

Ndubisi, F. 2002. *Ecological Planning: A Historical and Comparative Synthesis*. Baltimore: Johns Hopkins University Press.

Sayer, J. A., Maginnis, S. (eds.) 2005. *Forests in Landscapes: Ecosystem Approaches to Sustainability*. London: Earthscan.

Bibliography

Ackoff, R.L. 1974. *Redesigning the Future: A Systems Approach to Societal Problems*. New York: John Wiley and Sons.

Adamowicz, W., Beckley, T., MacDonald, D.H., Just, L., Luckert, M., Murray, E., Phillips, W. 1998. In search of forest resource values of indigenous peoples: Are nonmarket valuation techniques applicable? *Society and Natural Resources* 11(1): 51–66.

Adams, W.M. 2003. Nature and the colonial mind. In *Decolonizing Nature: Strategies for Conservation in a Post-Colonial Era*, W.M. Adams, M. Mulligan (eds.): 16–50. London: Earthscan Publications.

Aerts, R., Honnay, O. 2011. Forest restoration, biodiversity and ecosystem functioning. *BMC Ecology* 11: Article 29.

Agrawal, A., Cashore, B., Hardin, R., Shepherd, G., Benson, C., Miller, D. 2013. *Economic contributions of forests*. Background Paper No. 1. New York: United Nations Forum on Forests. Available at: http://www.un.org/esa/forests/pdf/session_documents/unff10/Eco ContrForests.pdf. Accessed on 30 December 2015.

Aguilar, L. 2016. Foreword. In *Gender and Forests: Climate Change, Tenure, Value Chains and Emerging Issues*, C.J.P. Colfer, B. Sijapati Basnett, M. Elias (eds.): xxv–xxix. London: Earthscan from Routledge.

Agustino, S., Mataya, B., Senelwa, K., Achigan-Dako, E.G. 2011. *Non-Wood Forest Products and Services for Socio-Economic Development: A Compendium for Technical and Professional Forestry Education*. Nairobi: African Forest Forum. Available at: http://www.afforum. org/sites/default/files/English/English_14.pdf. Accessed on 30 December 2015.

Aiama, D., Edwards, S., Bos, G., Ekstrom, J., Krueger, L., Quétier, F., Savy, C., Semroc, B., Sneary, M., Bennun, L. 2015. *No Net Loss and Net Positive Impact Approaches for Biodiversity: Exploring the Potential Application of These Approaches in the Commercial Agriculture and Forestry Sectors*. Gland, Switzerland: IUCN.

Alber, G. 2015. *Gender and Urban Climate Policy: Gender-Sensitive Policies Make a Difference*. Bonn: Deutsche Gesellschaft für Internationale Zusammenarbeit GmbH.

Albion, R.G. 1926/2000. *Forests and Sea Power: The Timber Problem of the Royal Navy, 1652–1862*. Annapolis, MD: Naval Institute Press.

Allendorf, F.W., Luikart, G., Aitken, S.N. 2011. *Conservation and the Genetics of Populations*. 2nd edition. Oxford: Wiley Blackwell.

Anonymous 1991. Discussion. In *Landscape Linkages and Biodiversity*, W.E. Hudson (ed.): 72–77. Washington, DC: Island Press.

Armitage, D., Berkes, F., Doubleday, N. 2007. *Adaptive Co-Management: Collaboration, Learning and Multi-Level Governance*. Vancouver: UBC Press.

Arno, S.F., Fiedler, C.E. 2005. *Mimicking Nature's Fire: Restoring Fire-Prone Forests in the West*. Washington, DC: Island Press.

Arnold, J., Townson, M., Liedholm, C., Mead, D. 1994. *Structure and growth of small enterprises in the forest sector in Southern and Eastern Africa*. Oxford Forestry Institute Occasional Paper 47. Oxford: University of Oxford. Available at: http://www.bodley.ox.ac.uk/users/millsr/isbes/ODLF/OP47.pdf. Accessed on 30 December 2015.

Arnstein, S. R. 1969. A ladder of citizen participation. *American Institute of Planners Journal* 35(4): 216–224.

Ashton, M. S., Mendelsohn, R., Singhakumara, B.M.P., Gunatilleke, C.V.S., Gunatilleke, I., Evans, A. 2001. A financial analysis of rain forest silviculture in Southwestern Sri Lanka. *Forest Ecology and Management* 154: 431–441.

Australian Forestry Standard 2013. *Australian Standard: 'Sustainable Forest Management'*. Available at: www.forestrystandard.org.au/resources/standards/AS4708-2013/AS4708-2013-Publish.pdf. Assessed 20 March 2014.

Azevedo-Ramos, C., Silva, J.N.M., Merry, F. 2015. The evolution of Brazilian forest concessions. *Elementa: Science of the Anthropocene*. DOI 10.12952/journal.elementa.000048.

Barnes, B. V., Zak, D. R., Denton, S. R., Spurr, S. H. 1998. *Forest Ecology*. 4th edition. New York: John Wiley.

Bass, S., Thornber, K., Markopoulos, M., Roberts, S., Grieg-Gran, M. 2001. *Certification's Impacts on Forests, Stakeholders and Supply Chains, Instruments for Sustainable Private Sector Forestry*. London: International Institute for Environment and Development.

Bauhus, J., Pokorny, B., Van der Meer, P.J., Kanowski, P.J., Kanninen, M. 2010. Ecosystems goods and services – the key for sustainable plantations. In *Ecosystem Goods and Services from Plantation Forests*, J. Bauhus, P. Van der Meer, M. Kanninen (eds.): 205–227. London: Earthscan.

BC Forest Practices Board 2015. *District manager's authority over forest operations*. Special Report FPB/SR/52. Victoria: Forest Practices Board. Available at: hpb.ca/sites/default/files/reports/SR52-Resource-District-Managers.pdf. Accessed on 28 December 2015.

Beer, C., Reichstein, M., Tomelleri, E., Ciais, P., Jung, M., Carvalhais, N., Rödenbeck, C., Arain, M. A., Baldocchi, D., Bonan, G. B., Bondeau, A., Cescatti, A., Lasslop, G., Lindroth, A., Lomas, M., Luyssaert, S., Margolis, H., Oleson, K. W., Roupsard, O., Veenendaal, E., Viovy, N., Williams, C., Woodward, F. I., Papale, D. 2010. Terrestrial gross carbon dioxide uptake: Global distribution and covariation with climate. *Science* 329: 834–838.

Beese, W. J., Dunsworth, B. J., Zielke, K., Bancroft, B. 2003. Maintaining attributes of old-growth forests in coastal B.C. through variable retention. *Forestry Chronicle* 79: 570–578.

Begon, M., Townsend, C. R., Harper, J. L. 2006. *Ecology: From Individuals to Ecosystems*. 4th edition. New York: John Wiley and Sons.

Beierle, T. C., Cayford, J. 2002. *Democracy in Practice: Public Participation in Environmental Decisions*. Washington, DC: Resources for the Future.

Belcher, B., Ruíz Pérez, M., Achdiawan, R. 2005. Global patterns and trends in the use and management of commercial NTFPs: Implications for livelihoods and conservation. *World Development* 33(9): 1435–1452.

Bell, D. T., Williams, J. E. 1997. Eucalypt ecophysiology. In *Eucalypt Ecology: Individuals to Ecosystems*, J. E. Williams, J.C.Z. Woinarski (eds.): 168–196. Cambridge: Cambridge University Press.

Bellassen, V., Viovy, N., Luyssaert, S., Le Maire, G., Schelhaas, M.-J., Ciais, P. 2011. Reconstruction and attribution of the carbon sink of European forests between 1950 and 2000. *Global Change Biology* 17: 3274–3292.

Belton, V., Stewart, T. J. 2002. *Multiple Criteria Decision Analysis: An Integrated Approach*. Dordrecht: Kluwer.

Bengston, D. N. 1994. Changing forest values and ecosystem management. *Society & Natural Resources* 7: 515–533.

Berger, A. L., Palik, B., D'Amato, A. W., Fraver, S., Bradford, J.B., Nislow, K., King, D., Brooks, R. T. 2013. Ecological impacts of energy-wood harvests: Lessons from whole-tree harvesting and natural disturbance. *Journal of Forestry* 111(2): 139–153.

Beschta, R. L., Bilby, R. E., Brown, G. W. 1987. Stream temperature and aquatic habitat: Fisheries and forestry interactions. In *Streamside Management: Forestry and Fishery Interactions*, E. O. Salo, T. W. Cundy, (eds.): 191–232. Seattle: University of Washington.

Bettinger, P., Boston, K., Siry, J. P., Grebner, D. L. 2009. *Forest Management and Planning*. Burlington, MA: Academic Press.

Bi, J., Blanco, J. A., Kimmins, J. P., Ding, Y., Seely, B., Welham, C. 2007. Yield decline in Chinese Fir plantations: A simulation investigation with implications for model complexity. *Canadian Journal of Forest Research* 37: 1615–1630.

Biermann, F., Chan, S., Mert, A., Pattberg, P. 2007. Multi-stakeholder partnerships for sustainable development: Does the promise hold? In *Partnerships, Governance and Sustainable Development: Reflections on Theory and Practice*, P. Glasbergen, F. Biermann, A. Mol (eds.): 239–260. Cheltenham, UK: Edward Elgar.

Biggs, T., Shaw, M. 2006. *African Small and Medium Enterprises, Networks, and Manufacturing Performance*. Washington, DC: World Bank.

Binkley, D., Fisher, R. F. 2013. *Ecology and Management of Forest Soils*. 4th edition. Chichester: Wiley-Blackwell.

BirdLife International and National Audubon Society 2015. *The Messengers: What Birds Tell Us about Threats from Climate Change and Solutions for Nature and People*. Cambridge, UK, and New York, USA: BirdLife International and National Audubon Society.

Bishop, P., Hines, A., Collins, T. 2007. The current state of scenario development: An overview of techniques. *Foresight* 9(1): 5–25.

Blackman, A. 2012. Does eco-certification boost regulatory compliance in developing countries? ISO 14001 in Mexico. *Journal of Regulatory Economics* 42(3): 242–263.

Blanco, J. A. 2012. Forests may need centuries to recover their original productivity after continuous intensive management: An example from Douglas-fir. *Science of the Total Environment* 437: 91–103.

Blanco, J. A., Imbert, J. B., Castillo, F. J. 2011. Thinning affects *Pinus sylvestris* needle decomposition rates and chemistry differently depending on site conditions. *Biogeochemistry* 106: 397–414.

Blanco, J. A., Zavala, M. A., Imbert, J. B., Castillo, F. J. 2005. Sustainability of forest management practices: Evaluation through a simulation model of nutrient cycling. *Forest Ecology and Management* 213: 209–228.

Blombäck, P., Poschen, P., Lövgren, M. 2003. *Employment trends and prospects in the European forest sector: A study prepared for the European Forest Sector Outlook Study (EFSOS)*. Report No. ECE/TIM/DP/29. New York, Geneva: United Nations. Available at: http://www.fao.org/3/a-ae888e.pdf. Accessed on 30 December 2015.

Bluffstone, R. A., Robinson, E. J. Z., Purdon, M. 2015. Introduction: Local forest reform. In *Forest Tenure Reform in Asia and Africa: Local Control for Improved Livelihoods, Forest Management, and Carbon Sequestration*, R. A. Bluffstone, E. J. Z. Robinson (eds.): 1–19. Washington, DC: RFF Press.

Bode, W. (ed.) 1997. *Naturnahe Waldwirtschaft: Prozeßschutz oder Biologische Nachhaltigkeit?* Holm: Deukalion.

Bohnet, I. C., Roebeling, P. C., Williams, K. J., Holzworth, D., van Grieken, M. E., Pert, P. L., Kroon, F. J., Westcott, D. A., Brodie, J. 2011. Landscapes Toolkit: An integrated modelling framework to assist stakeholders in exploring options for sustainable landscape development. *Landscape Ecology* 26: 1179–1198.

Boot, R. G. A., Gullison, R. E. 1995. Approaches to developing sustainable extraction systems for tropical forest products. *Ecological Applications* 5(4): 896–903.

Bosch, J. M., Hewlett, J. D. 1982. A review of catchment experiments to determine the effect of vegetation changes on water yield and evapotranspiration. *Journal of Hydrology* 55: 3–23.

Bowler, J. M., Johnston, H., Olley, J. M., Prescott, J. R., Roberts, R. G., Shawcross, W., Spoone, N. A. 2003. New ages for human occupation and climatic change at Lake Mungo, Australia. *Nature* 21: 837–840.

Bowman, D. M. J. S. 1993. Biodiversity: Much more than biological inventory. *Biodiversity Letters* 1: 163.

Boyland, M., Nelson, J., Bunnell, F. L., D'Eon, R. G. 2006. An application of fuzzy set theory for seral-class constraints in forest planning models. *Forest Ecology and Management* 223: 395–402.

Brand, D. G. 1997. Criteria and indicators for the conservation and sustainable management of forests: Progress to date and future directions. *Biomass and Bioenergy* 13(4/5): 247–253.

Bréda, N., Granier, A., Aussenac, G. 1995. Effects of thinning on soil and tree water relations, transpiration and growth in an oak forest (*Quercus petraea* (Matt.) Liebl.). *Tree Physiology* 15: 295–306.

Broekhoven, G., Savenije, H., von Scheliha, S. (eds.) 2012. *Moving Forward with Forest Governance*. Wageningen: Tropenbos International.

Brynte, B. 2002. C. L. Obbarius – ein deutscher Pionier in schwedischen Wäldern. *Forst und Holz* 57(12): 395–402.

Buchanan, R. 1992. Wicked problems in design thinking. *Design Issues* 8(2): 5–21.

Bull, G., Bazett, M., Schwab, O., Nilsson, S., White, A., Maginnis, S. 2006. Industrial forest plantation subsidies: Impacts and implications. *Forest Policy and Economics* 9(1): 13–31.

Bunnell, F. L. 1976. The myth of the omniscient forester. *Forestry Chronicle* 52: 150–152.

Bunnell, F. L. 1998a. Overcoming paralysis by complexity when establishing operational goals for biodiversity. *Journal of Sustainable Forestry* 7(3/4): 145–164.

Bunnell, F. L. 1998b. Managing forests to sustain biodiversity: Substituting accomplishment for motion. *Forestry Chronicle* 74: 822–827.

Bunnell, F. L. 1998c. Setting goals for biodiversity in managed forests. In *The Living Dance: Policy and Practices for Biodiversity in Managed Forests*, F. L. Bunnell, J. F. Johnson (eds.): 117–153. Vancouver: University of British Columbia Press.

Bunnell, F. L. 1999a. Let's kill a panchreston: Giving fragmentation meaning. In *Forest Fragmentation: Wildlife and Management Implications*, J. A. Rochelle, L. A. Lehmann, J. Wisniewski (eds.): vii–xiii. Leiden, Netherlands: Brill.

Bunnell, F. L. 1999b. What habitat is an Island? In *Forest Fragmentation: Wildlife and Management Implications*, J. A. Rochelle, L. A. Lehmann, J. Wisniewski (eds.): 1–31. Leiden, Netherlands: Brill.

Bunnell, F. L., Dunsworth, G. B. 2004. Making adaptive management for biodiversity work: The example of Weyerhaeuser in coastal British Columbia. *Forestry Chronicle* 80: 37–43.

Bunnell, F. L., Dunsworth, G. B. (eds.) 2009. *Forestry and Biodiversity: Learning How to Sustain Biodiversity in Managed Forests*. Vancouver: University of British Columbia Press.

Bunnell, F. L., Dunsworth, G. B., Huggard, D. J., Kremsater, L. L. 2009a. Introduction. In *Forestry and Biodiversity: Learning How to Sustain Biodiversity in Managed Forests*, F. L. Bunnell, G. B. Dunsworth (eds.): 5–16. Vancouver: University of British Columbia Press.

Bunnell, F. L., Houde, I. 2010. Down wood and biodiversity – implications to forest practices. *Environmental Reviews* 18: 397–421.

Bunnell, F. L., Huggard, D. J. 1999. Biodiversity across spatial and temporal scales: Problems and opportunities. *Forest Ecology and Management* 115(2/3): 113–126.

Bunnell, F. L., Huggard, D. J., Dunsworth, G. B. 2009b. Effectiveness monitoring: An introduction. In *Forestry and Biodiversity: Learning How to Sustain Biodiversity in Managed Forests*, F. L. Bunnell, G. B. Dunsworth (eds.): 75–82. Vancouver: University of British Columbia Press.

Bunnell, F. L., Huggard, D. J., Kremsater, L. L. 2009c. Summary: Progress and lessons learned. In *Forestry and Biodiversity: Learning How to Sustain Biodiversity in Managed Forests*, F. L. Bunnell, G. B. Dunsworth (eds.): 276–293. Vancouver: University of British Columbia Press.

Bunnell, F. L., Kremsater, L. L. 1994. Tactics for maintaining biodiversity in forested ecosystems. *XXI International Union of Game Biologists Congress* 1: 62–72.

Bunnell, F. L., Kremsater, L. L. 2012b. Actions to promote climate resilience in forests of British Columbia, *Journal of Ecosystems and Management* 31(2): 1–10.

Bunnell, F. L., Kremsater, L. L., Boyland, M. 1998. *An Ecological Rationale for Changing Forest Management on MacMillan Bloedel's Forest Tenure*. Publication R-22. Vancouver: Centre for Applied Conservation Biology, University of British Columbia.

Bunnell, F. L., Kremsater, L. L., Moy, A., Vernier, P. 2009d. *Coarse-Filter Assessment of the Contribution of Dying and Dead Wood to Sustaining Biodiversity on TFL 48*. Available at: http://www.for.gov.bc.ca/hfd/library/fia/2009/FSP_Y093014c.pdf. Accessed on 30 December 2015.

Bunnell, F. L., Kremsater, L. L., Moy, A., Vernier, P. 2009e. *Coarse-Filter Assessment of Contribution of Understory to Sustaining Biodiversity on TFL 48*. Available at: http://www.for.gov.bc.ca/hfd/library/FIA/2009/FSP_Y093014b.pdf. Accessed on 30 December 2015.

Bunnell, F. L., Kremsater, L. L., Moy, A., Vernier, P. 2010. *Biodiversity Sustainability Analysis of the Contribution of Hardwoods to Sustaining Biodiversity in the Fort St John TSA*. Available at: http://www.for.gov.bc.ca/hfd/library/FIA/2010/LBIP_8047009h.pdf. Accessed on 30 December 2015.

Bunnell, F. L., Squires, K. A., Houde, I. 2004. *Evaluating the effects of large-scale salvage logging for mountain pine beetle on terrestrial and aquatic vertebrates*. Mountain Pine Beetle Initiative Working Paper 2004–2. Victoria, Canada: Natural Resources Canada, Canadian Forest Service, Pacific Forestry Centre.

Burton, P. J., Macdonald, S. E. 2011. The restoration imperative: Challenges, objectives and approaches to restoring naturalness in forests. *Silva Fennica* 45: 843–863.

Bush, M. B. 1997. *Ecology of a Changing Planet*. Upper Saddle River, NJ: Prentice Hall.

Callicott, J. B. 2005. Non-anthropocentric value theory and environmental ethics. In *The Earthscan Reader in Environmental Values*, L. Kalof, T. Satterfield (eds.): 67–80. London: Earthscan.

Campoe, O. C., Stape, J. L., Albaugh, T. J., Allen, H. L., Fox, T. R., Rubilar, R., Binkley, D. 2013. Fertilization and irrigation effects on tree level aboveground net primary production, light interception and light use efficiency in a loblolly pine plantation. *Forest Ecology and Management* 288: 43–48.

Canby, K. 2006. Investing in natural tropical industries. *International Tropical Timber Organization (ITTO) Tropical Forest Update* 16(2): 4–8.

Cannell, M.G.R., Dewar, R.C., Pyatt, D.G. 1993. Conifer plantations on drained peatlands in Britain – a net gain or loss of carbon. *Forestry* 66: 353–369.

Cardinale, B. J., Duffy, J. E., Gonzalez, A., Hooper, D. U., Perrings, C., Venail, P., Narwani, A., Mace, G. M., Tilman, D., Wardle, D. A. 2012. Biodiversity loss and its impact on humanity. *Nature* 486: 59–67.

Carroll, M., Milakovsky, B., Finkral, A., Evans, A., Ashton, M. S. 2012. Managing carbon sequestration and storage in temperate and boreal forests. In *Managing Forest Carbon in a Changing Climate*, M. S. Ashton, M. L. Tyrrell, D. Spalding, B. Gentry (eds.): 205–226. Dordrecht: Springer.

Carrow, R. 1999. Canada's model forest program: Challenges for phase II. *Forestry Chronicle* 75(1): 73–79.

Castañeda, F. 2000. Criteria and indicators for sustainable forest management: International processes, current status and the way ahead. *Unasylva* 51(203): 34–40.

Castree, N. 2000. Marxism and the production of nature. *Capital and Class* 24(3): 5–36.

Caswell, S. 2014. The impacts of criteria and indicators. *ITTO Tropical Forest Update* 22(4): 3–10.

Chan, K., Satterfield, T., Goldstein, J. 2012. Rethinking ecosystem services to better address and navigate cultural values. *Ecological Economics* 74: 8–18.

Chandran, A., Innes, J. L. 2014. The state of the forest: Reporting and communicating the state of forests by Montreal Process countries. *International Forestry Review* 16(1): 103–111.

Chang, S. C., Yeh, C. F., Wu, M. J., Hsia, Y. J., Wu, J. T. 2006. Quantifying fog water deposition by in situ exposure experiments in a mountainous coniferous forest in Taiwan. *Forest Ecology and Management* 224(1/2): 11–18.

Chao, S. 2012. *Forest Peoples: Numbers Across the World*. Moreton-in-Marsh: Forest Peoples Programme.

Charnley, S. 2005. Industrial plantation forestry. *Journal of Sustainable Forestry* 21(4): 35–57.

Chartrand, P.L.A.H. 2010. The 'race' for recognition: Toward a policy of recognition of Aboriginal peoples in Canada. In *Aboriginal Title and Indigenous Peoples: Canada, Australia and New Zealand*, L. A. Knafla, H. J. Westra (eds.): 125–145. Vancouver: UBC Press.

Chazdon, R. L. 2014. *Second Growth: The Promise of Tropical Forest Regeneration in an Age of Deforestation*. Chicago: University of Chicago Press.

Chen, J., Innes, J. L., Tikina, A. 2010. Private cost-benefits of voluntary forest product certification. *International Forestry Review* 12(1): 1–12.

Cheng, B., Le Clue, S. 2010. *Forestry in Asia: Issues for Responsible Investors*. Singapore: Responsible Research.

Chiabai, A. (ed.) 2015. *Climate Change Impacts on Tropical Forests in Central America*. London: Earthscan.

Childs, S. W., Flint L. E. 1987. Effect of shadecards, shelterwoods, and clearcuts on temperature and moisture environments. *Forest Ecology and Management* 18: 205–217.

Choge, S. K. 2004. The wood carving industry in Kenya. In *Forest Products, Livelihood and Conservation: Case Studies of Non-Timber Forest Products*, T. Sunderland, O. Ndoye (eds.): 149–168. Bogor, Indonesia: CIFOR.

Chomitz, K. 2007. *At Loggerheads: Agricultural Expansion, Poverty Reduction, and Environment in Tropical Forests.* Washington, DC: World Bank.

Christy, L. C., Di Leva, C. E., Lindsay, J. M., Takoukam, P. T. 2007. *Forest Law and Sustainable Development Addressing Contemporary Challenges through Legal Reform: The International Bank for Reconstruction and Development.* Available at: www.worldbank.org.

Churchman, C. W. 1967. Wicked problems. *Management Science* 14(4): B-141–B-142.

Ciais, P., Rayner, P., Chevallier, F., Bousquet, P., Logan, M., Peylin, P., Ramonet, M. 2010. Atmospheric inversions for estimating CO_2 fluxes: methods and perspectives. *Climate Change* 103: 69–92(24).

Ciais, P., Sabine, C., Bala, G., Bopp, L., Brovkin, V., Canadell, J., Chhabra, A., DeFries, R., Galloway, J., Heimann, M., Jones, C., Le Quéré, C., Myneni, R. B., Piao, S., Thornton, P. 2013. Carbon and other biogeochemical cycles. In *Climate Change 2013: The Physical Science Basis: Contribution of Working Group I to the Fifth Assessment Report of the Intergovernmental Panel on Climate Change*, T. F. Stocker, D. Qin, G.-K. Plattner, M. Tignor, S. K. Allen, J. Boschung, A. Nauels, Y. Xia, V. Bex, P. M. Midgley (eds.): 465–570. Cambridge, NY: Cambridge University Press.

CITES 1973. *Convention on International Trade in Endangered Species of Wild Fauna and Flora.* Available at: https://www.cites.org/eng/disc/text.php. Accessed on 30 December 2015.

Clarke, P. J., Lawes, M. J., Midgley, J. J., Lamont, B. B., Ojeda, F., Burrows, G. E., Enright, N. J., Knoc, K. J. 2013. Resprouting as a key functional trait: How buds, protection and resources drive persistence after fire. *New Phytologist* 197: 19–35.

Cobo, M. 1981. *Study of the Problem of Discrimination Against Indigenous Populations.* New York: United Nations Sub-Commission on Prevention of Discrimination and Protection of Minorities.

Colchester, M. 2004. Conservation policy and indigenous peoples. *Environmental Science and Policy* 7(3): 145–153.

Coleman, D. C., Crossley, D. A. 1996. *Fundamentals of Soil Ecology.* San Diego: Academic Press.

Colfer, C.J.P., Dahal, G. R., Capistrano, D. (eds.) 2008. *Lessons from Forest Decentralization: Money, Justice and the Quest for Good Governance in Asia-Pacific.* London: Earthscan.

Colfer, C.J.P., Elias, M., Sijapati Basnett, B. 2016. A gender box analysis of forest management and conservation. In *Gender and Forests: Climate Change, Tenure, Value Chains and Emerging Issues*, C.J.P. Colfer, B. Sijapati Basnett, M. Elias (eds.): 3–16. London: Earthscan from Routledge.

Collaborative Partnership on Forests 2012a. *SFM and the Multiple Functions of Forests.* SFM Fact Sheet 1. Available at: http://www.cpfweb.org/32819-045ba23e53cbb67809cef3b724bef9cd0.pdf. Accessed on 30 December 2015.

Collaborative Partnership on Forests 2012b. *SFM, Food Security and Livelihoods.* SFM Fact Sheet 3. Available at: http://www.cpfweb.org/32824-0ddc38a4ca15d81db68e3c4301cbc2efd.pdf. Accessed on 30 December 2015.

Collaborative Partnership on Forests 2012c. *SFM and Indigenous Peoples.* SFM Fact Sheet 4. Available at: http://www.cpfweb.org/32825-0c3cd8211ec6ccb8cae04bf30e975a362.pdf. Accessed on 30 December 2015.

Colloff, M. J. 2014. *Flooded Forest and Desert Creek: Ecology and History of the River Red Gum.* Collingwood: CSIRO.

Commission on Sustainable Development 1997. *Report of the Ad Hoc Intergovernmental Panel on Forests on Its Fourth Session.* New York: United Nations Economic and Social Council Document E/CN.17/1997/12. Available at: http://daccess-dds-ny.un.org/doc/UNDOC/GEN/N97/069/11/PDF/N9706911.pdf?OpenElement.

Cook, B. I., Anchukaitis, K. J., Kaplan, J. O., Puma, M. J., Kelley, M., Gueyffier, D. 2012. Pre-Columbian deforestation as an amplifier of drought in Mesoamerica. *Geophysical Research Letters* 39: L16706.

Corace, R. G., III, Goebel, P. C., McCormick, D. L. 2010. Kirtland's warbler habitat management and multi-species bird conservation: Considerations for planning and management across Jack Pine (Pinus banksiana Lamb.) habitat types. *Natural Areas Journal* 30(2): 174–190.

Côté, P., Tittle, R., Messier, C., Kneeshaw, D. D., Fall, A., Fortin, M.-J. 2010. Comparing different forest zoning options for landscape-scale management of the boreal forest: Possible benefits of the TRIAD. *Forest Ecology and Management* 259: 418–427.

Cramb, R. A. 2007. *Land and Longhouse: Agrarian Transformation in the Uplands of Sarawak.* Copenhagen: Nordic Institute of Asian Studies.

Cronon, W. 1996. The trouble with wilderness: Or, getting back to the wrong nature. *Environmental History* 1(1): 7–28.

Dalgleish, H. J., Nelson, C. D., Scrivani, J. A., Jacobs, D. F. 2016. Consequences of shifts in abundance and distribution of American Chestnut for restoration of a foundation forest tree. *Forests* 7(1): Article 4.

Daniels, S. E., Walker, G. B. 2001. *Working through Environmental Conflict: The Collaborative Learning Approach.* Westport, CT: Praeger.

Dasmann, R. F. 1968. *A Different Kind of Country.* New York: Macmillan Company.

Davies, E., Mabel, M., Halverson, H. 2011. Draft Framework for Sharing Approaches for Better Multi-Stakeholder Participation Practices. UN-REDD PROGRAMME. http://www.un-redd.org

DeFries, R., Rosenzweig, C. 2010. Toward a whole-landscape approach for sustainable land use in the tropics. *Proceedings of the National Academy of Sciences of the USA* 107: 19627–19632.

DeGraff, J.V., Sidle, R. C., Ahmad, R., Scatena, F. N. 2012. Recognizing the importance of tropical forests in limiting rainfall-induced debris flows. *Environmental Earth Science* 67: 1225–1235.

Deinet, S., Ieronymidou, C., McRae, L., Burfield, I. J., Foppen, R. P., Collen, B., Böhm, M. 2013. *Wildlife comeback in Europe: The recovery of selected mammal and bird species.* Final Report to Rewilding Europe by ZSL, BirdLife International and the European Bird Census Council. London: ZSL.

Del Cid-Liccardi, C., Kramer, T., Ashton, M. S., Griscom, B. 2012. Managing carbon sequestration in tropical forests. In *Managing Forest Carbon in a Changing Climate*, M. S. Ashton, M. L. Tyrrell, D. Spalding, B. Gentry (eds.): 183–204. Dordrecht: Springer Science.

Delli Carpini, M. X., Cook, F. L., Jacobs, L. R. 2004. Public deliberation, discursive participation, and citizen engagement: A review of the empirical literature. *Annual Review of Political Science* 7: 315–344.

Demarest, A. 2004. *Ancient Maya: The Rise and Fall of a Rainforest Civilization.* Cambridge: Cambridge University Press.

Desprez-Loustau, M.-L., Marcais, B., Nageleisen, L.-M., Piou, D., Vannini, A. 2006. Interactive effects of drought and pathogens in forest trees. *Annals of Forest Science* 63(6): 597–612.

Diamond, J. 2005. *Collapse: How Societies Choose to Fail or Succeed.* New York: Viking Penguin.

Dickie, I. A., Martinez-Garcia, L. B., Koele, N., Grelet, G.-A., Tylianakis, J. M., Peltzer, D. A., Richardson, S. J. 2013. Mycorrhizas and mycorrhizal fungal communities throughout ecosystem development. *Plant and Soil* 367(1–2): 11–39.

Dickmann, D. I. 2006. Silviculture and biology of short-rotation woody crops in temperate regions: Then and now. *Biomass and Bioenergy* 30: 696–705.

Dietz, T., Fitzgerald, A., Shwom, R. 2005. Environmental values. *Annual Review of Environmental Resources* 30: 335–372.

Dolman, A. J., van der Werf, G. R., van der Molen, M. K., Ganssen, G., Erisman, J. W., Strengers, B. 2010. A carbon cycle science update since IPCC AR-4. *Ambio* 39: 402–412.

Donovan, J., Stoian, D., Macqueen, D., Grouwels, S. 2006. *The Business Side of Sustainable Forest Management: Small and Medium Forest Enterprise Development for Poverty Reduction.* London: Overseas Development Institute.

Drew, A. P., Zsuffa, L., Mitchell, C. P. 1987. Terminology relating to woody plant biomass and its production. *Biomass* 12: 79–82.

Dudley, N., Schlaepfer, R., Jackson, W., Jeanrenaud, J.-P., Stolton, S. 2006. *Forest Quality: Assessing Forests at a Landscape Scale.* London: Earthscan.

Durán-Medina, E., Mas, J., Velázquez, A. 2005. Land use/cover change in community-based forest management regions and protected areas in Mexico. In *The Community Forests of Mexico: Managing for Sustainable Landscapes*, D. Bray, L. Marino-Pérez, D. Barry (eds.): 215–238. Austin: University of Texas Press.

Edmonds, R. L. 1994. *Patterns of China's Lost Harmony: A Survey of the Country's Environmental Degradation and Protection.* London, Routledge.

Edmunds, R. L., Agee, J. K., Gara, R. I. 2000. *Forest Health and Protection*. Boston, MA: McGraw-Hill.

Ehrlich, P., Ehrlich, A. 1981. *Extinction: The Causes and Consequences of the Disappearance of Species*. New York: Random House.

El-Lakany, H., Jenkins, M., Richards, M. 2007. *Means of Implementation: Financing Sustainable Forest Management: Putting Proposals and Recommendations into Action*. Background Paper contribution by PROFOR (The World Bank) to UNFF-7. Available at: http://www.fao.org/forestry/14704-08294ea8ab45a4ca4f4a3cef22e7a8d71.pdf. Accessed on 30 December 2015.

Elliott, S., Blakesley, D., Hardwick, K. 2013. *Restoring Tropical Forests: A Practical Guide*. Kew, UK: Royal Botanic Gardens.

Elvin, M. 2004. *The Retreat of the Elephants: An Environmental History of China*. New Haven, CT: Yale University Press.

Enuoh, O. O. O., Bisong, F. E. 2015. Colonial forest policies and tropical deforestation: The case of Cross River State, Nigeria. *Open Journal of Forestry* 5: 66–79.

Fa, J. E., Juste, J., Pérez del Val, J., Castroviejo, J. 1995. Impact of market hunting on mammal species in Equatorial Guinea. *Conservation Biology* 9: 1107–1115.

Fahrig, L. 1997. Relative effects of habitat loss and fragmentation on population extinction. *Journal of Wildlife Management* 61: 603–610.

FAO 2003. Employment Trends and Prospects in the European Forest Sector. Geneva timber and forest discussion papers. ECE/TIM/DP/29. 45 p. Available at: http://www.fao.org/3/aae888e.pdf

FAO 2006. *Responsible management of planted forests: Voluntary guidelines*. Working Paper FP/37/E. Rome: Food and Agriculture Organization of the United Nations.

FAO 2007. *Tenure Security for Better Forestry: Understanding Forest Tenure in South and Southeast Asia*. Bangkok: Food and Agriculture Organization of the United Nations.

FAO 2010. *Global Forest Resources Assessment 2010*. Rome: Food and Agriculture Organization of the United Nations. Available at: www.fao.org.

FAO 2011. *Reforming forest tenure: Issues, principles and process*. FAO Forestry Paper 165. Rome: Food and Agriculture Organization of the United Nations.

FAO 2013. *Improving Governance of Forest Tenure: A Practical Guide*. Governance of Tenure Technical Guide 2. Rome: Food and Agriculture Organization of the United Nations.

FAO 2014. *State of the World's Forests: Enhancing the Socioeconomic Benefits from Forests*. Rome: Food and Agriculture Organization of the United Nations. Available at: http://www.fao.org/3/cf470fab-cc3c-4a50-b124–16a306ee11a6/i3710e.pdf. Accessed on 1 June 2014.

FAO 2015. *Global Forest Resources Assessment 2015: How Are the World's Forests Changing?* Rome: Food and Agriculture Organization of the United Nations.

FAOStat 2014. *FAOSTAT database*. Available at: http://faostat.fao.org/. Accessed on 1 June 2014.

FAOStat 2015. *FAOSTAT database*. Available at: http://faostat.fao.org/. Accessed on 24 December 2015.

Farley, K. A., Kelly, E. F., Hofstede, R.G.M. 2004. Soil organic carbon and water retention following conversion of grasslands to pine plantations in the Ecuadoran Andes. *Ecosystems* 7: 729–739.

Federer, C. A., Hornbeck, J. W., Tritton, L. M., Martin, C. W., Pierce, R. S. 1989. Long-term depletion of calcium and other nutrients in eastern US forests. *Environmental Management* 13: 593–601.

Feller, M. C. 2005. Forest harvesting and streamwater inorganic chemistry in western North America: A review. *Journal of the American Water Resources Association* 41: 785–811.

Fernow, B. E. 1907. *History of Forestry*. Toronto: Toronto University Press.

Finnigan, D., Gunton, T., Williams, P. W. 2003. Planning in the public interest: An evaluation of civil society participation in collaborative land use planning in British Columbia. *Environments* 31(3): 13–29.

Flagler, R. B. (ed.) 1998. *Recognition of Air Pollution Injury to Vegetation: A Pictorial Atlas*. 2nd edition. Pittsburgh: Air and Waste Management Association.

Foley, T. G., Richter, D., Galik, C. S. 2009. Extending rotation age for carbon sequestration: A cross-protocol comparison of North American forest offsets. *Forest Ecology and Management* 259: 201–209.

Ford, R.M., Williams, K.J.H., Bishop, I.D., Hickey, J.E. 2009. Effects of information on the social acceptability of alternatives to clearfelling in Australian wet eucalypt forests. *Environmental Management* 44: 1149–1162.

Forest Products Association of Canada (FPAC) 2014. *Forest Management Certification in Canada Over Time*. Available at: http://certificationcanada.org/en/statistics/canadian-statistics/. Accessed on 10 January 2016.

Forest Stewardship Council 2014. *Global FSC Certificates: Types and Distribution*. January 2014. Available at: https://ic.fsc.org/download.facts-and-figures-january-2014.a-2877.pdf. Accessed on 15 February 2014.

Forest Stewardship Council (FSC) Russia 2014. *Forest Stewardship Council Russia*. Available at: www.fsc.ru. Accessed on 20 March 2014.

Franklin, J.F., Spies, T.A., Van Pelt, R., Carey, A.B., Thornburgh, D.A., Berg, D.R., Lindenmayer, D.B., Harmon, M.E., Keeton, W.S., Shaw, D.C., Bible, K., Chen, J. 2002. Disturbances and structural development of natural forest ecosystems with silvicultural implications, using Douglas-fir forests as an example. *Forest Ecology and Management* 155: 399–423.

Frazer, J.G. 1958. *The Golden Bough*. 3rd edition, 1906–1915, Vols. 12. London: Macmillan Publishing Company.

Freemark, K.E., Merriam, G. 1986. Importance of area and habitat heterogeneity to bird assemblages in temperate forest fragments. *Biological Conservation* 3: 115–141.

Frey, B., Kremer, J., Rüdt, A., Sciaca, S., Matthies, D., Lüscher, P. 2009. Compaction of forest soils with heavy logging machinery affects soil bacterial community structure. *European Journal of Soil Biology* 5: 312–320.

Froehlich, H.A., McNabb, D.H. 1984. Minimizing soil compaction in Pacific Northwest forests. In *Forest Soils and Treatment Impacts*, E.L. Stone (ed.): 159–192. Knoxville, TN: Department of Forestry, Wildlife and Fisheries, University of Tennessee.

Fung, I.Y., Doney, S.C., Lindsay, K., John, J. 2005. Evolution of carbon sinks in a changing climate. *Proceedings of the National Academy of Sciences of the USA* 102: 11201–11206.

Gamfeldt, L., Snåll, T., Bagchi, R., Jonsson, M., Gustafsson, L., Kjellander, P., Ruiz-Jaen, M.C., Fröberg, M., Stendahl, J., Philipson, C.D. 2013. Higher levels of multiple ecosystem services are found in forests with more tree species. *Nature Communications* 4: Article 1340.

Gammage, B. 2011. *The Biggest Estate on Earth: How Aborigines Made Australia*. Sydney: Allen and Unwin.

Gerzon, M., Seely, B., McKinnon, A. 2011. The temporal development of old-growth structural attributes in second-growth stands: A chronosequence study in the Coastal Western Hemlock zone in British Columbia. *Canadian Journal of Forest Research* 41: 1534–1546.

Gilmour, D. 2016. *Forty years of community-based forestry: A review of its extent and effectiveness*. FAO Forestry Paper 176. Rome: Food and Agricultural Organization of the United Nations.

Gimmi, U., Poulter, B., Wolf, A., Portner, H., Weber, P., Bürgi, M. 2013. Soil carbon pools in Swiss forests show legacy effects from historic forest litter raking. *Landscape Ecology* 28: 835–846.

Glacken, C.F. 1967. *Traces on the Rhodian Shore: Nature and Culture in Western Thought from Ancient Times to the End of the Eighteenth Century*. Berkeley: University of California Press.

Glesinger, E. 1960. The role of forestry in world economic development. *Unasylva* 14(3): 99–103.

Global Footprint Network 2010. *The Ecological Wealth of Nations: Earth's Biocapacity as a New Framework for International Cooperation, GDP and Beyond: Measuring Progress in a Changing World*. Brussels: European Commission.

Gomi, T., Moore, R.D., Dhakal, A.S. 2006. Headwater stream temperature response to forest harvesting with different riparian treatments, coastal British Columbia, Canada. *Water Resources Research* 42: W08437.

Gössling, S. 1999. Ecotourism: A means to safeguard biodiversity and ecosystem functions? *Ecological Economics* 29: 303–320.

Gough, A.D., Innes, J.L., Allen, S.D. 2008. Development of common indicators of sustainable forest management. *Ecological Indicators* 8(4): 425–430.

Grabherr, G., Koch, G., Kirchmeir, H., Reiter, K. 1998. *Hemerobie. Österreichischer Waldökosysteme*. Veröffentlichungen des Österreichischen Maß-Programms, Vol. 17. Vienna: Österreichische Akademie der Wissenschaften.

Grainger, A. 2012. Forest sustainability indicator systems as procedural policy tools in global environmental governance. *Global Environmental Change* 22: 147–160.

Graybill, D. A., Idso, S. B. 1993. Detecting the aerial fertilization effect of atmospheric CO_2 enrichment in tree-ring chronologies. *Global Biogeochemical Cycles* 7: 81–95.

Grayson, D. K., Meltzer, D. J. 2002. Clovis hunting and large mammal extinction: A critical review of the evidence. *Journal of World Prehistory* 16: 313–359.

Greig, J. C. 1979. Principles of genetic conservation in relation to wildlife management in southern Africa. *South Africa Wildlife Research* 9: 57–78.

Griess, V. C., Knoke, T. 2013. Bioeconomic modelling of mixed Norway spruce – European beech stands: Economic consequences of considering ecological effects. *European Journal of Forest Research* 132(3): 511–522.

Grigal, D. F. 2000. Effects of extensive forest management on soil productivity. *Forest Ecology and Management* 138: 167–185.

Grove, R. H. 1995. *Green Imperialism: Colonial Expansion, Tropical Island Edens and the Origins of Environmentalism, 1600–1860.* Cambridge: Cambridge University Press.

Grove, R. H., Damodaran, V., Sangwan, S. (eds.) 1998. *Nature and the Orient: The Environmental History of South and Southeast Asia.* Oxford: Oxford University Press.

Guénette, J.-S., Villard, M.-A. 2005. Thresholds in forest bird response to habitat alteration as quantitative targets for conservation. *Conservation Biology* 19: 1168–1180.

Guha, R. 1993. The malign encounter: The Chipko Movement and competing visions of nature. In *Who Will Save the Forests? Knowledge, Power and Environmental Destruction*, T. Banuri, F. A. Marglin (eds.): 80–113. London: United Nations University/World Institute for Development Economics Research.

Guha, R. 2006. *How Much Should a Person Consume? Environmentalism in India and the United States.* Berkeley: University of California Press.

Gunn, J. M. 1995. *Restoration and Recovery of an Industrial Region: Progress in Restoring the Smelter-Damaged Landscape Near Sudbury, Canada.* New York: Springer Verlag.

Gustafson, E. J. 2013. When relationships estimated in the past cannot be used to predict the future: Using mechanistic models to predict landscape ecological dynamics in a changing world. *Landscape Ecology* 28: 1429–1437.

Hajehforooshnia, S., Soffianian, A., Mahiny, A. S., Fakheran, S. 2011. Multi objective land allocation (MOLA) for zoning Ghamishloo Wildlife Sanctuary in Iran. *Journal for Nature Conservation* 19: 254–262.

Hardin, G. 1968. The tragedy of the commons. *Science* 162(3859): 1243–1248.

Harmon, M. E., Franklin, J. F., Swanson, F. J., Sollins, P., Gregory, S. V., Lattin, J. D., Anderson, N. H., Cline, S. P., Aumen, N. G., Sedell, J. R., Lienkaemper, G. W., Cromack, K., Cummins, K. W. 1986. Ecology of coarse woody debris in temperate ecosystems. *Advances in Ecological Research* 15: 133–302.

Harr, R. D. 1982. Fog drip in the Bull Run Municipal Watershed, Oregon. *Water Resources Bulletin* 18(5): 785–789.

Harr, R. D., Harper, W. C., Krygier, J. T., Hsieh, F. S. 1975. Changes in storm hydrographs after road building and clear-cutting in the Oregon Coast Range. *Water Resources Research* 11: 436–444.

Harris, A. S. 1989. *Wind in the forests of Southeast Alaska and guides for reducing damage.* USDA Forest Service General Technical Report PNW-GTR-244. Portland, OR: USDA Forest Service.

Harris, D. D. 1977. *Hydrologic changes after logging in two small Oregon coastal watersheds.* Water Supply Paper 2037. Washington, DC: US Geological Survey.

Harshaw, H. W., Tindall, D. B. 2005. A network approach to understanding people's relationships to forests. *Journal of Leisure Research* 37(4): 426–449.

Harvey, D. 1996. *Justice, Nature, and the Geography of Difference.* Cambridge, MA: Blackwell Publishers.

Hassan, M. A., Hogan, D. L., Bird, S. A., May, C. L., Gomi, T., Campbell, D. 2005. Spatial and temporal dynamics of wood in headwater streams of the Pacific Northwest. *Journal of the American Water Resources Association* 41: 899–919.

Haynes, G. (ed.) 2009. *American Megafaunal Extinctions at the End of the Pleistocene.* Dordrecht: Springer.

Hazely, C. 2000. *Forest-Based and Related Industries of the European Union – Industrial Districts, Clusters and Agglomerations*. Helsinki: ETLA.

He, F., Vavrus, S. J., Kutzbach, J. E., Ruddiman, W. F., Kaplan, J. O., Krumhardt, K. M. 2014. Simulating global and local surface temperature changes due to Holocene anthropogenic land cover change. *Geophysical Research Letters* 41(2): 623–631.

Helgerson, O. T. 1990. Heat damage in tree seedlings and its prevention. *New Forests* 3: 333–358.

Hemingway, J. L. 1999. Leisure, social capital, and democratic citizenship. *Journal of Leisure Research* 31(2): 150–165.

Heske, F. 1938. *German Forestry*. New Haven: Yale University Press.

Hickey, G. M., Innes, J. L. 2008. Indicators for demonstrating sustainable forest management in British Columbia, Canada: An international review. *Ecological Indicators* 8(2): 131–140.

Hicks, B. J., Beschta, R. L., Harr, R. D. 1991. Long-term changes in streamflow following logging in western Oregon and associated fisheries implications. *Water Resources Bulletin* 27: 217–226.

Hjortso, C. N., Straede, S. 2001. Strategic multiple-use forest planning in Lithuania, applying multi-criteria decision-making and scenario analysis for decision support in an economy in transition. *Forest Policy and Economics* 3: 175–188.

Holl, K. D. 2013. Restoring tropical forest. *Nature Education Knowledge* 4(4): 4.

Holling, C. S. (ed.) 1978. *Adaptive Environmental Assessment and Management*. London: John Wiley and Sons.

Holling, C. S., Berkes, F., Folke, C. 1998. Science, sustainability and resource management. In *Linking Social and Ecological Systems: Management Practices and Social Mechanisms for Building Resilience*, F. Berkes, C. Folke, (eds.): 342–362. Cambridge: Cambridge University Press.

Holling, C. S., Gunderson, L. H. 2002. Resilience and adaptive cycles. In *Panarchy: Understanding Transformations in Human and Natural Systems*, L. Gunderson, C. S. Holling (eds.): 3–22. Washington, DC: Island Press.

Holmgren, S., Arora-Jonsson, S. 2016. The Forest Kingdom and values. Climate change and gender equality in a contested forest policy context. In *Gender and Forests: Climate Change, Tenure, Value Chains and Emerging Issues*, C.J.P. Colfer, B. Sijapati Basnett, M. Elias (eds.): 53–67. London: Earthscan from Routledge.

Horn, R., Vossbrink, J., Peth, S., Becker, S. 2007. Impact of modern forest vehicles on soil physical properties. *Forest Ecology and Management* 248: 56–63.

Hoskins, M. 2016. Gender and the roots of community forestry. In *Gender and Forests: Climate Change, Tenure, Value Chains and Emerging Issues*, C.J.P. Colfer, B. Sijapati Basnett, M. Elias (eds.): 17–32. London: Earthscan from Routledge.

Houghton, R. A., House, J. I., Pongratz, J., van der Werf, G. R., DeFries, R. S., Hansen, M. C., Le Quéré, C., Ramankutty, N. 2012. Carbon emissions from land use and land-cover change. *Biogeosciences* 9: 5125–5142.

Hudson, J., Agrawal, A., Miller, D. C. 2013. *Changing futures, choices, and contributions of forests*. Background paper prepared for the United Nations Forum on Forests UNFF -10. Available at: www.unff.org.

Huggard, D. J., Kremsater, L. L. 2009. Ecosystem representation: Sustaining poorly known species and functions. In *Forestry and Biodiversity: Learning How to Sustain Biodiversity in Managed Forests*, F. L. Bunnell, G. B. Dunsworth (eds.): 83–99. Vancouver: University of British Columbia Press.

Huggard, D. J., Kremsater, L. L., Bunnell, F. L. 2009. Designing a monitoring program. In *Forestry and Biodiversity: Learning How to Sustain Biodiversity in Managed Forests*, F. L. Bunnell, G. B. Dunsworth (eds.): 241–275. Vancouver: University of British Columbia Press.

Hunt, L., Haider, W. 2001. Fair and effective decision making in forest management planning. *Society & Natural Resources* 14(10): 873–887.

Hutchings, T. R., Moffat, A. J., French, C. J. 2002. Soil compaction under timber harvesting machinery: A preliminary report on the role of brash mats in its prevention. *Soil Use and Management* 18: 33–38.

Ice, G., Dent, L., Robben, J., Cafferata, P., Light, J., Sugden, B., Cundy, T. 2004. Programs assessing implementation and effectiveness of state forest practice rules and BMPs in the West. *Water, Air & Soil Pollution: Focus* 4(1): 143–169.

Ice, G. G., Schilling, E., Vowell, J. 2010. Trends for forestry Best Management Practices implementation. *Journal of Forestry* 108(6): 267–273.

IFMAT (Indian Forest Management Assessment Team) 2013. *An assessment of Indian Forests and Forest Management in the United States.* Available at: www.itcnet.org/file_download/4f8e541e-f355-4da6-92d3-131e0013828d. Accessed on 12 March 2014.

Inglehart, R. 1977. *The Silent Revolution: Changing Values and Political Styles among Western Publics.* Princeton, NJ: Princeton University Press.

Inglis, A. S., Lussignea, A. 1995. *Participation in Scotland: The Rural Development Forestry Programme.* PLA Notes. London: IIED. Available at: http://pubs.iied.org/pubs/pdfs/G01578.pdf. Accessed on 28 November 2012.

Innes, J. L. 2009. Is forestry a social science? *Journal of Tropical Forest Science* 21(2): v–vi.

Innes, J. L., Skelly, J. M., Schaub, M. 2001. *Ozone and Broadleaved Species: A Guide to the Identification of Ozone-Induced Foliar Injury.* Birmensdorf, Switzerland: WSL.

International Institute for Sustainable Development (IISD) 2014. Summary of the thirteenth session of the UN General Assembly Group on Sustainable Development Goals: 14–19 July 2014. *Earth Negotiations Bulletin* 32(13), published 22 July 2014. Available at: http://www.iisd.ca/vol32/enb3213e.html. Accessed on 24 December 2015.

ITC 2006. *International Trade Centre statistics.* Available at: http://langues.p-maps.org/pmaps/index.php?err=sess. Accessed on 26 October 2014.

ITTO 1992a. *ITTO Guidelines for the Sustainable Management of Natural Tropical Forests.* ITTO Policy Development Series 1. Yokohama: International Tropical Timber Organization. Available at: http://www.itto.int/policypapers_guidelines/.

ITTO 1992b. *Criteria for the Measurement of Sustainable Tropical Forest Management.* ITTO Policy Development Series No. 3. Yokohama: International Tropical Timber Organization. Available at: http://www.itto.int/policypapers_guidelines/.

ITTO 1998. *Criteria and Indicators for Sustainable Management of Natural Tropical Forests.* ITTO Policy Development Series No. 7. Yokohama: International Tropical Timber Organization. Available at: http://www.itto.int/policypapers_guidelines/.

ITTO 1999a. *Manual for the Application of Criteria and Indicators for Sustainable Management of Natural Tropical Forests. Part A: National Indicators.* ITTO Policy Development Series No. 9. Yokohama: International Tropical Timber Organization. Available at: http://www.itto.int/policypapers_guidelines/.

ITTO 1999b. *Manual for the Application of Criteria and Indicators for Sustainable Management of Natural Tropical Forests. Part B: Forest Management Unit Indicators.* ITTO Policy Development Series No. 10. Yokohama: International Tropical Timber Organization. Available at: http://www.itto.int/policypapers_guidelines/.

ITTO 2003. *ATO/ITTO Principles, Criteria and Indicators for the Sustainable Management of African Natural Tropical Forests.* ITTO Policy Development Series No 14. Yokohama: International Tropical Timber Organization. Available at: http://www.itto.int/policypapers_guidelines/.

ITTO 2005. *Revised ITTO Criteria and Indicators for the Sustainable Management of Tropical Forests Including Reporting Format.* ITTO Policy Development Series No. 15. Yokohama: International Tropical Timber Organization. Available at: http://www.itto.int/policypapers_guidelines/.

ITTO 2015. *Voluntary Guidelines for the Sustainable Management of Natural Tropical Forests.* ITTO Policy Development Series No. 20. Yokohama: International Tropical Timber Organization. Available at: http://www.itto.int/policypapers_guidelines/.

IUSS Working Group WRB 2006. *World reference base for soil resources 2006.* 2nd edition. World Soil Resources Reports No. 103. Rome: Food and Agriculture Organization of the United Nations. Available at: http://www.fao.org/ag/agl/agll/wrb/doc/wrb2006final.pdf.

Iwamoto, I. 2002. The development of Japanese forestry. In *Forestry and the Forest Industry in Japan*, Iwai, Y. (ed.): 3–9. Vancouver: UBC Press.

Jacobsen, T., Adams, R. M. 1958. Salt and silt in ancient Mesopotamian agriculture. *Science* 128: 1251–1258.

Jarvis, A., Mulligan, M. 2010. The climate of cloud forests. In *Tropical Montane Cloud Forests*, L. A. Bruijnzeel, F. N. Scatena, L. S. Hamilton (eds.): 39–56. Cambridge: Cambridge University Press.

Jeffries, S. B., Wentworth, T. R., Allen, H. L. 2010. Long-term effects of establishment practices on plant communities across successive rotations in a loblolly pine (*Pinus taeda*) plantation. *Forest Ecology and Management* 260: 1548–1556.

Jenny, H. 1941. *Factors of Soil Formation.* New York: McGraw-Hill.

Johann, E., Agnoletti, M., Bölöni, J., Yurdakol Erol, S., Holl, K., Kusmin, J., García Latorre, J., Molnár, Z., Rochel, X., Rotherham, I. D., Saratsi, E., Smith, M., Tarang, L., van Benthem, M., van Laar, J. 2012. Europe. In *Traditional Forest-Related Knowledge: Sustaining Communities, Ecosystems and Biocultural Diversity*, J. Parrotta, R. Trosper (eds.): 203–249. World Forests, Vol. 12. Dordrecht: Springer.

Johnson, D. W. 1992. Effects of forest management on soil carbon storage. *Water, Air and Soil Pollution* 64: 83–120.

Jones, J. A. 2000. Hydrologic processes and peak discharge response to forest removal, regrowth, and roads in 10 small experimental basins, western Cascades, Oregon. *Water Resources Research* 36(9): 2621–2642.

Kahle, P., Hildebrand, E., Baum, C., Boelcke, B. 2007. Long-term effects of short rotation forestry with willows and poplar on soil properties. *Archives of Agronomy and Soil Science* 53: 673–782.

Kangas, J. 1994. An approach to public participation in strategic forest management planning. *Forest Ecology and Management* 70: 75–88.

Kangas, J., Kangas, A. 2005. Multiple criteria decision support in forest management – the approach, methods applied, and experiences gained. *Forest Ecology and Management* 207: 133–143.

Kangas, A., Kangas, J., Kurttila, M. 2008. *Decision Support for Forest Management.* Berlin: Springer.

Kato, E., Kinoshita, T., Ito, A., Kawamiya, M., Yamagata, Y. 2013. Evaluation of spatially explicit emission scenario of land-use change and biomass burning using a process-based biogeochemical model. *Journal of Land Use Science* 8: 104–122.

Keeling, R. F., Piper, S. C., Heimann, M. 1996. Global and hemispheric CO_2 sinks deduced from changes in atmospheric O_2 concentrations. *Nature* 381: 218–221.

Kellomäki, S., Kilpeläinen, A., Alam, A. (eds.) 2013. *Forest Bioenergy Production: Management, Carbon Sequestration and Adaptation.* New York: Springer.

Kelly, L. A., Wentworth, T. A. 2009. Effects of mechanized pine straw raking on population densities of longleaf pine seedlings. *Forest Ecology and Management* 259: 1–7.

Kettle, C. J., Poh, L. P. (eds.) 2014. *Global Forest Fragmentation.* Wallingford: CABI Publishing.

Keyser, R. 2009. The transformation of traditional woodland management: Commercial sylviculture in Medieval Champagne. *French Historical Studies* 32: 353–384.

Kimmins, J. P. 2002. Future shock in forestry: Where have we come from; where are we going; is there a 'right way' to manage forests? Lessons from Thoreau, Leopold, Toffler, Botkin and Nature. *Forestry Chronicle* 78(2): 263–271.

Kimmins, J. P. 2004. *Forest Ecology: A Foundation for Sustainable Forest Management and Environmental Ethics in Forestry.* 3rd edition. Upper Saddle River, NJ: Prentice Hall.

Kimmins, J. P., Blanco, J. A., Seely, B., Welham, C., Scoullar, K. 2008. Complexity in modelling forest ecosystems: How much is enough? *Forest Ecology and Management* 256: 1646–1658.

Kimmins, J. P., Mailly, D., Seely, B. 1999. Modelling forest ecosystem net primary production: The hybrid simulation approach used in FORECAST. *Ecological Modelling* 122: 195–224.

Kindermann, G. E., McCallum, I., Fritz, S., Obersteiner, M. 2008. A global forest growing stock, biomass and carbon map based on FAO statistics. *Silva Fennica* 42: 387–396.

King, J. 1993. Learning to solve the right problems: The case of nuclear power in America. *Journal of Business Ethics* 13: 105–116.

Kirby, K. R., Potvin, C. 2007. Variations in carbon storage among tree species: Implications for the management of a small-scale carbon sink project. *Forest Ecology and Management* 246: 208–221.

Kirkby, C. A., Giudice, R., Day, B., Turner, K., Soares-Filho, B. S., Oliveira-Rodrigues, H., Yu, D. W. 2011. Closing the ecotourism-conservation loop in the Peruvian Amazon. *Environmental Conservation* 38(1): 6–17.

Kishor, N., Rosenbaum, K. 2012. *Assessing and Monitoring Forest Governance: A User's Guide to a Diagnostic Tool*. Washington, DC: Program on Forests.

Knapp, B.O., Wang, G.G., Hu, H., Walker, J.L., Tennant, C. 2011. Restoring longleaf pine (*Pinus palustris* Mill.) in loblolly pine (*Pinus taeda* L.) stands: Effects of restoration treatments on natural loblolly pine regeneration. *Forest Ecology and Management* 262(7): 1157–1167.

Knuchel, H. 1953. *Planning and Control in the Managed Forest*. Translated by M.L. Anderson. Edinburgh: Oliver and Boyd.

Kozak, R. 2007. *Small and Medium Forest Enterprises: Instruments of Change in the Developing World*. Washington, DC: Rights and Resources Initiative.

Kozak, R. 2009. Alternative business models for forest-dependent communities in Africa: A pragmatic consideration of small-scale enterprises and a path forward. *Madagascar Conservation and Development* 4(2): 76–81.

Kozlov, M., Zvereva, E., Zverev, V. 2009. *Impacts of Point Polluters on Terrestrial Biota*. Dordrecht: Springer.

Kracke, E.A., Jr. 1954. Sung society: Change within tradition. *Far Eastern Quarterly* 14: 479–488.

Kranabetter, J.M., De Montigny, L., Ross, G. 2013. Effectiveness of green-tree retention in the conservation of ectomycorrhizal fungi. *Fungal Ecology* 6(5): 430–438.

Kremsater, L.; Bunnell, F.L. 1999. Edge effects: theory, evidence and implications to management of western North American forests. In *Forest Wildlife and Fragmentation. Management Implications*, J.A. Rochelle, L.A. Lehmann, J. Wisniewski (eds.): 117–153. Leiden, Netherlands.

Kunstadter, P., Chapman, E.C. 2002. Problems of shifting cultivation and economic development in Northern Thailand. In *Foundations of Tropical Forest Biology: Classic Papers with Commentaries*, R.L. Chazdon, T.C. Whitmore (eds.): 779–800. Chicago: The University of Chicago Press.

Kunz, W., Rittel, H. 1970. *Issues as elements of information systems*. Technical Report S-78–2. Stutttgart: Institüt für Grundlagen der Planung I.A. Also available as: Working Paper 131, Institute of Urban and Regional Development, University of California at Berkeley, CA.

Kurz, W., Apps, M.J. 1999. A 70 year retrospective analysis of carbon fluxes in the Canadian forest sector. *Ecological Applications* 9: 526–547.

Kurz, W.A., Dymond, C.C., Stinson, G., Rampley, G.J., Neilson, E.T., Carroll, A.L., Ebata, T., Safranyik, L. 2008. Mountain pine beetle and forest carbon feedback to climate change. *Nature* 452(7190): 987–990.

Laclau, P. 2003. Biomass and carbon sequestration of ponderosa pine plantations and native cypress forests in northwest Patagonia. *Forest Ecology and Management* 180: 317–333.

Laiho, R., Prescott, C.E. 2004. Decay and nutrient dynamics of coarse woody debris in northern coniferous forests: A synthesis. *Canadian Journal of Forest Research* 34: 763–777.

Lamb, D. 2014. *Large-Scale Forest Restoration*. London: Earthscan.

Lamb, D. 2015. Restoration of forest ecosystems. In *Routledge Handbook of Forest Ecology*, K.S.-H. Peh, R.T. Corlett, Y. Bergeron (eds.): 397–410. London: Routledge.

Lamb, D., Stanturf, J., Madsen, P. 2012. What is forest restoration? In *Forest Landscape Restoration*, J. Stanturf, D. Lamb, P. Madsen (eds.): 3–23. Dordrecht: Springer.

Lambdon, P., Desprez-Loustau, M. 2010. Disentangling the role of environmental and human pressures on biological invasions across Europe. *Proceedings of the National Academy of Sciences of the United States of America* 107(27): 12157–12162.

Lambert, M., Turner, J. 2000. *Commercial Forest Plantations on Saline Lands*. Collingwood: CSIRO Publishing.

Lambin, E.F., Meyfroidt, P. 2010. Land use transitions: Socio-ecological feedback versus socio-economic change. *Land Use Policy* 27: 108–118.

Landsberg, J.J., Waring, R.H. 1997. A generalised model of forest productivity using simplified concepts of radiation-use efficiency, carbon balance and partitioning. *Forest Ecology and Management* 95: 209–228.

Larsen, I., MacDonald, L.H., Brown, E., Rough, D., Welsh, M.J., Pietraszek, J.H., Libohava, Z., Benavides-Solorio, J.D., Schaffrath, K. 2009. Causes of post-fire runoff and erosion: water repellency, cover, or soil sealing? *Soil Science Society of America Journal* 73: 1393–1407.

Larson, A. M., Barry, D., Dahal, G. R., Colfer, C. J.P. (eds.) 2010. *Forests for People: Community Rights and Forest Tenure Reform*. London, UK: Earthscan.

Lauber, T. B., Knuth, B. A. 1999. Measuring fairness in citizen participation: A case study of moose management. *Society & Natural Resources* 11(1): 19–37.

Laurance, W. F., Nascimento, H.E.M., Laurance, S. G., Andrade, A., Ewers, R. M., Harms, K. E., Luizao, R.C.C., Ribeiro, J. E. 2007. Habitat fragmentation, variable edge effects, and the landscape-divergence hypothesis. *PLoS ONE* 2(10): Article e1017.

Lawson, J., Cashore, B. 2003. Company choice on sustainable forestry forest certification: The case of JD Irving, Ltd. In *Forest Policy for Private Forestry: Global and Regional Challenge*, L. Teeter, B. Cashore, D. Zhang (eds.): 245–258. New York: CABI Publishing.

Leach, J. A., Moore, R. D. 2014. Winter stream temperature in the rain-on-snow zone: Influences of hillslope runoff and transient snow cover. *Hydrology and Earth Systems Science* 18: 819–838.

Levin, S. A. 1992. The problem of pattern and scale in ecology. *Ecology* 73: 1943–1967.

Levis, C., de Souza, P. F., Schietti, J., Emilio, T., Pinto, J.L.P.D., Clement, C. R., Costa, F.R.C. 2012. Historical human footprint on modern tree species composition in the Purus-Madeira Interfluve, Central Amazonia. *PLOS ONE* 7(11): Article e48559.

Lewis, K. J., Lindgren, B. S. 2002. Relationship between spruce beetle and tomentosus root disease: Two natural disturbance agents of spruce. *Canadian Journal of Forest Research* 32(1): 31–37.

Lewis, S. L., Lopez-Gonzalez, G., Sonké, B., Affum-Baffoe, K., Baker, T. R., Ojo, L. O., Phillips, O. L., Reitsma, J. M., White, L., Comiskey, J. A., Djuikouo, K., M.-N., Ewango, C.E.N., Feldpausch, T. R., Hamilton, A. C., Gloor, M., Hart, T., Hladik, A., Lloyd, J., Lovett, J.C., Makana, J.-R., Malhi, Y., Mbago, F. M., Ndangalasi, H. J., Peacock, J., Peh, K.S.-H., Sheil, D., Sunderland, T., Swaine, M. D., Taplin, J., Taylor, D., Thomas, S.C., Votere, R., Wöll, H. 2009. Increasing carbon storage in intact African tropical forests. *Nature* 457: 1003–1006.

Li, R., Buongiorno, J., Turner, J. A., Zhua, S., Prestemon, J. 2008. Long-term effects of eliminating illegal logging on the world forest industries, trade, and inventory. *Forest Policy and Economics* 10(7–8), 480–490.

Li, Y. Q., Ji, J. S., Chen, D. D. 1993. Growth responses of middle-aged Chinese fir plantation to fertilizer application. *Forest Research* 6: 390–396.

Lindenmayer, D. B., Burton, P. J., Franklin, J. F. 2008. *Salvage Logging and its Ecological Consequences*. Washington, DC: Island Press.

Lindenmayer, D. B., Franklin, J. F. 2002. *Conserving Forest Biodiversity: A Comprehensive Multiscaled Approach*. Washington, DC: Island Press.

Lindroth, A., Verwijst, T., Halldin, S. 1994. Water-use efficiency of willow: Variation with season, humidity and biomass allocation. *Journal of Hydrology* 156: 1–19.

Liu, L. 2007. *The Chinese Neolithic: Trajectories to Early States*. Cambridge: Cambridge University Press.

Lomský, B., Materna, J., Pfanz, H. 2002. *SO₂-Pollution and Forests Decline in the Ore Mountains*. Jiloviště-Strnady, Czech Republic: Forestry and Game Management Research Institute.

Loo, J. A. 2009. Ecological impacts of non-indigenous invasive fungi as forest pathogens. *Biological Invasions* 11(1): 81–96.

Lu, Y., Coops, N. C., Wang, T., Wang, G. 2015. A process-based approach to estimate Chinese Fir (*Cunninghamia lanceolata*) distribution and productivity in southern China under climate change. *Forests* 6: 360–379.

Lukac, M., Godbold, D. L. 2011. *Soil Ecology in Northern Forests: A Belowground View of a Changing World*. Cambridge: Cambridge University Press.

Lyons, C., Scott, J. M. 1994. Biodiversity defined. *Biological Diversity* (Newsletter of the Biological Diversity Working Group, The Wildlife Society) 1: 2–3.

MacLean, D. A., Seymour, R. S., Montigny, M. K., Messier, C. 2008. Allocation of conservation efforts over the landscape: The TRIAD approach. In *Setting Conservation Targets for Managed Forest Landscapes*, M.-A. Villard, B. G. Jonsson (eds.): 283–303. Cambridge: Cambridge University Press.

Macqueen, D. 2012. Enabling conditions for successful community forest enterprises. *Small-Scale Forestry* 12(1): 145–163.

Macqueen, D. (ed.), Baral, S., Chakrabarti, L., Dangal, S., du Plessis, P., Griffith, A., Grouwels, S., Gyawali, S., Heney, J., Hewitt, D., Kamara, Y., Katwal, P., Magotra, R., Pandey, S.S., Panta, N., Subedi, B., Vermeulen, S. 2012. *Supporting Small Forest Enterprises – A Facilitator's Toolkit: Pocket Guidance Not Rocket Science!* IIED Small and Medium Forest Enterprise Series No. 29. Edinburgh, UK: IIED.

Maini, J. 2004. *Future International Arrangement on Forests: Background Discussion Paper Prepared for the Country-Led Initiative in Support of the United Nations Forum on Forests on the Future of the International Arrangement on Forests*. New York: United Nations Forum on Forests.

Mäkelä, A., del Rio, M., Hynynen, J., Hawkins, M.J., Reyer, C., Soares, P., van Oijen, M., Tome, M. 2012. Using stand-scale forest models for estimating indicators of sustainable forest management. *Forest Ecology and Management* 285: 164–178.

Makeschin, F. 1994a. Soil ecological effects of energy forestry. *Biomass and Bioenergy* 6: 63–79.

Makeschin, F. 1994b. Soil ecological aspects of arable land afforestation with fast growing trees. In *Biomass for Energy and Environment, Agriculture and Industry*, D.O. Hall, G. Grassi, H. Scheer (eds.): 534–538. Bochum: Ponte Press.

Makeschin, F. 1999. Short rotation forestry in Central and Northern Europe ± introduction and conclusions. *Forest Ecology and Management* 121: 1–7.

Malhi, Y., Phillips, O. (eds.) 2005. *Tropical Forests and Global Atmospheric Change*. Oxford: Oxford University Press.

Malhi, Y., Roberts, J.T., Betts, R.A., Killeen, T.J., Wenhong, L., Nobre, C.A. 2008. Climate change, deforestation, and the fate of the Amazon. *Science* 319: 169–172.

Man, R., Colombo, S., Kayahara, G.J., Duckett, S., Velasquez, R., Dang, Q.-L. 2013. A case of extensive conifer needle browning in northwestern Ontario in 2012: Winter drying or freezing damage? *Forestry Chronicle* 89: 675–680.

Mander, M., Le Breton, G. 2006. Overview of the medicinal plants industry in Southern Africa. In *Commercialising Medicinal Plants: A Southern African Guide*, N. Diederichs (ed.): 3–8. Stellenbosch: Sun Press.

Manion, P.D. 2001. *Tree Disease Concepts*. 2nd edition. Englewood Cliffs, NJ: Prentice-Hall.

Manion, P.D. 2003. Evolution of concepts in forest pathology. *Phytopathology* 93(8): 1052–1055.

Manion, P.D., Griffin, D.H., Rubin, B.D. 2001. Ice damage impacts on the health of the northern New York State forest. *Forestry Chronicle* 77: 619–625.

Marcar, N., Crawford, D., Leppert, P., Jovanovic, T., Floyd, R., Farrow, R. 1995. *Trees for Saltland: A Guide to Selecting Native Species for Australia*. East Melbourne: CSIRO Press.

Marschner, P. 2012. *Marschner's Mineral Nutrition of Higher Plants*. 3rd edition. Amsterdam: Elsevier.

Martin, C. 2015. *On the Edge: The State and Fate of the World's Tropical Rainforests: A Report to the Club of Rome*. Berkeley and Vancouver: Greystone Books and the David Suzuki Foundation.

Martin, K., Aitken, K.E.H., Wiebe, K.L. 2004. Nest sites and nest webs for cavity-nesting communities in interior British Columbia, Canada: Nest characteristics and niche partitioning. *The Condor* 106(1): 5–19.

Martinez-Alier, J. 2002. *The Environmentalism of the Poor: A Study of Ecological Conflicts and Valuation*. Cheltenham: Edward Elgar.

Mascarenhas, M., Scarce, R. 2004. 'The intention was good': Legitimacy, consensus-based decision making, and the case of forest planning in British Columbia, Canada. *Society & Natural Resources* 17(1): 17–38.

Mather, A.S. 1990. *Global Forest Resources*. Portland: Timber Press.

Mather, A.S. 1992. The forest transition. *Area* 24: 367–379.

Mathey, A.H., Krcmar, E., Dragicevic, S., Vertinsky, I. 2008. An object-oriented cellular automata model for forest planning problems. *Ecological Modelling* 212: 359–371.

Mathey, A.H., Krcmar, E., Tait, D., Vertinsky, I., Innes, J.L. 2007. Forest planning using co-evolutionary cellular automata. *Forest Ecology and Management* 239: 45–56.

Matthews, J.D. 1989. *Silvicultural Systems*. Oxford: Clarendon Press.

Mayers, J. 2006. Small- and medium-sized forestry enterprises. *International Tropical Timber Organization (ITTO) Tropical Forest Update* 16(2): 10–11.

McCracken, A. R., Dawson, W. M. 2003. Rust disease (*Melampsora epitea*) of willow (*Salix* spp.) grown as short rotation coppice (SRC) in inter- and intra-species mixtures. *Annals of Applied Biology* 143: 381–393.

McDonald, G. T., Lane, M. B. 2005. Converging global indicators of sustainable forest management. *Forest Policy and Economics* 6(1): 63–70.

McFarlane, B. L., Boxall, P. C. 2000. Factors influencing forest values and attitudes of two stakeholder groups: The case of the Foothills Model Forest, Alberta, Canada. *Society & Natural Resources* 13(7): 649–661.

McKay, H. (ed.) 2011. *Short Rotation Forestry: Review of Growth and Environmental Impacts.* Forest Research Monograph 2. Farnham: Forest Research.

McShea, W. J., Underwood, H. B., Rappole, J. H. (eds.) 1997. *The Science of Overabundance: Deer Ecology and Population Management.* Washington, DC: Smithsonian Institution Press.

Meadows, D. H., Meadows, D. L., Randers, J. 1972. *Limits to Growth.* New York: Universe Books.

Meidinger, D., Pojar, J. 1991. *Ecosystems of British Columbia.* Special Report No. 6. Victoria: B.C. Ministry of Forests.

Meiggs, R. 1982. *Trees and Timber in the Ancient Mediterranean World.* Oxford: Oxford University Press.

Melo, F.P.L., Pinto, S.R.R., Brancalion, P.H.S., Castro, P. S., Rodrigues, R. R., Aronson, J., Tabarelli, M. 2013. Priority setting for scaling-up tropical forest restoration projects: Early lessons from the Atlantic Forest Restoration Pact. *Environmental Science & Policy* 33: 395–404.

Mendoza, G. A., Martins, H. 2006. Multi-criteria decision analysis in natural resource management: A critical review of methods and new modelling paradigms. *Forest Ecology and Management* 230: 1–22.

Mendoza, G. A., Prabhu, R. 2000. Multiple criteria decision making approaches to assessing forest sustainability using criteria and indicators: a case study. *Forest Ecology and Management* 131(1–3): 107–126.

Mendoza, G. A., Prabhu, R. 2003. Qualitative multicriteria approaches to assessing indicators of sustainable forest resource management. *Forest Ecology and Management* 174: 329–343.

Mendoza, G. A., Prabhu, R. 2006. Participatory modeling and analysis for sustainable forest management: Overview of soft system dynamics models and applications. *Forest Policy and Economics* 9: 179–196.

Merino, A., Balboa, M. A., Rodríguez Soalleiro, R., Alvarez Gonzalez, J. G. 2005. Nutrient exports under different harvesting regimes in fast-growing forest plantations in southern Europe. *Forest Ecology and Management* 207: 325–339.

Merriam, G. 1998. Biodiversity at the population level: A vital paradox. In *The Living Dance: Policy and Management for Biodiversity in Forested Systems*, F.L. Bunnell, J.F. Johnson (eds.): 45–65. Vancouver: University of British Columbia Press.

Meyerson, L., Mooney, H. 2007. Invasive alien species in an era of globalization. *Frontiers in Ecology and the Environment* 5(4): 199–208.

Michon, G., de Foresta, H. 1997. Agroforests: Pre-domestication of forest trees or true domestication of forest ecosystems? *Netherlands Journal of Agricultural Science* 45: 451–462.

Miller, A. 2012. *Ecotourism Development in Costa Rica: The Search for Oro Verde.* Plymouth: Lexington Books.

Miller, P. R., McBride, J. R. 1999. *Oxidant Air Pollution Impacts in the Montane Forests of Southern California: A Case Study of the San Bernardino Mountains.* New York: Springer.

Minang, P. A., van Noordwijk, M., Freeman, O. E., Mbow, C., de Leeuw, J., Catacutan, D. (eds.) 2015. *Climate-Smart Landscapes: Multifunctionality in Practice.* Nairobi: World Agroforestry Centre (ICRAF).

Minnesota Forest Resources Council 2005. *Sustaining Minnesota Forest Resources: Voluntary Site-Level Forest Management Guidelines for Landowners, Loggers and Resources Managers.* St. Paul, MN: Forest Resources Council.

Moeliono, M., Wollenberg, E., Limberg, G. (eds.) 2009. *The Decentralization of Forest Governance: Politics, Economics and the Fight for Control of Forests in Indonesian Borneo.* London: Earthscan.

Moffat, A., MacNeill, J. 1994. *Reclaiming Disturbed Land for Forestry.* Forestry Commission Bulletin 110. London: HMSO.

Molina, R., Marcot, B. G., Lesher, R. 2006. Protecting rare, old-growth, forest-associated species under the Survey and Manage Program guidelines of the Northwest Forest Plan. *Conservation Biology* 20: 306–318.

Montagnini, F., Finney, C. (eds.) 2011. *Restoring Degraded Landscapes with Native Species in Latin America*. Hauppauge, NY: Nova Science Publishers.

Moore, R. D., Richardson, J. S. 2012. Natural disturbance and forest management in riparian zones: Comparison of effects at reach, catchment and landscape scales. *Freshwater Science* 31: 239–247.

Morris, L. A., Miller, R. E. 1994. Evidence for long-term productivity change as provided by field trials. In *Impacts of Forest Harvesting on Long-Term Site Productivity*, W. J. Dyck, D. W. Cole, N. B. Comerford (eds.): 41–80. New York: Chapman and Hall.

Murphy, D. D. 1990. Conservation biology and scientific method. *Conservation Biology* 4: 203–204.

Nail, S. 2008. *Forest Policies and Social Change in England*. Berlin: Springer.

National Aboriginal Forestry Association 2007. *A Classification System for Forest Tenures on Crown Land*. Available at: http://www.fnforestrycouncil.ca/downloads/nafa_classification_of_forest_tenures.pdf.

National Forest Programme Facility 2012. *Final Report*. Rome: National Forest Programme Facility. Available at: www.nfpfacility.org.

Nepstad, D. C., Schwartzmann, S., Bamberger, B., Santilli, M., Ray, D. K., Schlesinger, P., Lefebvre, P. A., Alencar, A., Prinz, E., Fiske, G., Rolla, A. 2006. Inhibition of Amazon deforestation and fire by parks and Indigenous lands. *Conservation Biology* 20(1): 65–73.

Newman, R. H., Tate, K. R. 1984. The use of alkaline soil extracts for ^{13}C n.m.r. characterization of humic substances. *Journal of Soil Science* 34: 47–54.

Nikolakis, W., Innes, J. L. 2014. *Forests and Globalization: Challenges and Opportunities for Sustainable Development*. London: Earthscan.

Nilsson, S., Shvidenko, A., Jonas, M., McCallum, I., Thomson, A., Balzter, H. 2007. Uncertainties of a regional terrestrial biota full carbon account: A systems analysis. *Water, Air and Soil Pollution Focus* 7: 425–441.

Nitschke, C. R., Innes, J. L. 2008. Integrating climate change into forest management in South-Central British Columbia: An assessment of landscape vulnerability and development of a climate-smart framework. *Forest Ecology and Management* 256(3): 313–327.

Nolte, C., Agrawal, A., Silvius, K. M., Soares, B. S. 2013. Governance regime and location influence avoided deforestation success of protected areas in the Brazilian Amazon. *Proceedings of the National Academy of Sciences of the USA* 110(13): 4956–4961.

Noss, R. F., Cooperrider, A. Y. 1994. *Saving Nature's Legacy: Protecting and Restoring Biodiversity*. Washington, DC: Island Press.

Oliver, C. D., Larson, B. C. 1996. *Forest Stand Dynamics*. Updated edition. New York: John Wiley and Sons.

Olivier, J., Aardenne, J., Dentener, F., Ganzeveld, L., Peters, J. 2005. Recent trends in global greenhouse emissions: Regional trends 1970–2000 and spatial distribution of key sources in 2000. *Environmental Science* 2: 81–99.

Orange, C. 1987. *The Treaty of Waitangi*. Wellington, New Zealand: Allen and Unwin New Zealand.

Osmaston, F. C. 1968. *The Management of Forests*. London: Allen and Unwin.

Ostrom, E. 1990. *Governing the Commons: The Evolution of Institutions for Collective Action*. Cambridge: Cambridge University Press.

Ostrom, E. 1999. Coping with tragedies of the commons. *Annual Review of Political Science* 2: 493–535.

Otrosina, W. J., Cobb, F. W., Jr. 1989. Biology, ecology and epidemiology of Heterobasidion annosum. In *Proceedings of the Symposium on Research and Management of Annosus Root Disease (Heterobasidion annosum) in Western North America*, W. J. Otrosina, R. F. Scharpf, (eds.): 26–33. USDA Forest Service General Technical Report PSW-GTR-116. Berkeley, CA: USDA Forest Service, Pacific Southwest Forest and Range Experiment Station.

Outerbridge, R., Trofymo, J. A. 2004. Diversity of ectomycorrhizae on experimentally planted Douglas-fir seedlings in variable retention forestry sites on southern Vancouver Island. *Canadian Journal of Botany* 82: 1671–1681.

Ovaska, K., Sopuck, L. 2008. *Land Snails and Slugs as Ecological Indicators of Logging Practices: Recommendations for Adaptive Management*. Victoria: Forest Science Program of the BC Ministry of Forests Report. Available at: http://www.for.gov.bc.ca/hfd/library/FIA/2008/ FSP_Y083030a.pdf. Accessed on 30 December 2015.

Overdevest, C. 2000. Participatory democracy, representative democracy, and the nature of diffuse and concentrated interests: A case study of public involvement on a national forest district. *Society & Natural Resources* 13(7): 685–696.

Pacala, S., Canham, C., Saponara, J., Silander, J., Kobe, R., Ribbens, E. 1996. Forest models defined by field measurements: II. Estimation, error analysis and dynamics. *Ecological Monographs* 66: 1–44.

Paquette, A., Messier, C. 2010. The role of plantations in managing the world's forests in the Anthropocene. *Forest Ecology and Management* 8: 27–34.

Parrotta, J. A., Turnbull, J. W., Jones, N. 1997. Catalyzing native forest regeneration on degraded tropical lands. *Forest Ecology and Management* 99: 1–7.

Paumgarten, F., Shackleton, C. 2011. The role of non-timber forest products in household coping strategies in South Africa: The influence of household wealth and gender. *Population and Environment* 33: 108–131.

Payne, R. J., Graham, R. 1993. Visitor planning and management in parks and protected areas. In *Parks and Protected Areas in Canada: Planning and Management*, P. Dearden, R. Rollins (eds.): 185–210. Toronto: Oxford University Press.

Pearce, D. W., Putz, F., Vanclay, J. K. 2003. Sustainable forestry in the tropics: Panacea or folly? *Forest Ecology and Management* 172(2–3): 229–247.

Perlin, J. 2005. *A Forest Journey: The Story of Wood and Civilization*. Woodstock, VT: Countryman Press.

Perry, D. A. 1994. *Forest Ecosystems*. Baltimore: Johns Hopkins University Press.

Perttu, K. L. 1999. Environmental and hygienic aspects of willow coppice in Sweden. *Biomass & Bioenergy* 16: 291–297.

Peters, G. P., Andrew, R. M., Boden, T., Canadell, J. G., Ciais, P., Le Quéré, C., Marland, G., Raupach, M. R., Wilson, C. 2013. The challenge to keep global warming below 2°C. *Nature Climate Change* 3: 4–6.

Peterson, D. L., Vose, J. M., Patel-Weynand, T. (eds.) 2014. *Climate Change and United States Forests*. Dordrecht: Springer.

Peth, S., Horn, R. 2006. Consequences of grazing on soil physical and mechanical properties in forest and tundra environments. In *Reindeer Management in Northernmost Europe: Linking Practical and Scientific Knowledge in Socio-Ecological Systems*, B. C. Forbes, M. Bölter, L. Müller-Wille, J. Hukkinen, F. Müller, N. Gunslay, Y. Konstantinov (eds.): 217–244. Ecological Studies, Vol. 184. Berlin, Heidelberg: Springer Verlag.

Pickett, S. T., White, P. S. (eds.) 1985. *The Ecology of Natural Disturbance and Patch Dynamics*. Orlando: Academic Press.

Pike, R. G., Redding, T. E., Moore, R. D., Winkler, R. D., Bladon, K. D. (eds.) 2010. *Compendium of Forest Hydrology and Geomorphology in British Columbia*. Land Management Handbook 66. Victoria and Kamloops: B.C. Ministry of Forests and Range, Forest Science Program and FORREX Forum for Research and Extension in Natural Resources.

Pinard, M. A., Putz, F. E. 1996. Retaining forest biomass by reducing logging damage. *Biotropica* 29: 278–295.

Pinto, S. R., Melo, F., Tabarelli, M., Padovesi, A., Mesquita, C. A., de Mattos Scaramuzza, C. A., Castro, P., Carracosca, H., Calmon, M., Rodrigues, R., César, R. C., Brancalion, P.H.S. 2014. Governing and delivering a biome-wide restoration initiative: The case of Atlantic Forest Restoration Pact in Brazil. *Forests* 5(9): 2212–2229.

Pisek, A., Larcher, W., Moser, W., Pack, I. 1969. Kardinale Temperaturbereiche der Photosynthese and Grenztemperaturen des Lebens der Blatter verschiedener Spermatophyten. III Temperaturabhangigkeit und optimaler Temperaturbereich der Netto-Photosynthese. *Flora* B158: 608–630.

Pitman, R. M. 2006. Wood ash use in forestry – a review of the environmental impacts. *Forestry* 79(5): 563–588.

Pokorny, B. 2013. *Smallholders, Forest Management and Rural Development in the Amazon*. London: Earthscan.

Poncelet, E.C. 2001. Personal transformation in multi-stakeholder environmental partnerships. *Policy Sciences* 34: 273–30.

Pongratz, J., Reick, C.H., Raddatz, T., Claussen, M. 2009. Effects of anthropogenic land cover change on the carbon cycle of the last millennium. *Global Biogeochemical Cycles* 23: Article Gb4001.

Poore, D. 1989. The sustainable management of natural forests: The issues. In *No Timber Without Trees: Sustainability in the Tropical Forest*, D. Poore, P. Burgess, J. Palmer, S. Peitbergen, T. Synott (eds.): 1–27. London: Earthscan Publications.

Poore, D. 2003. *Changing Landscapes: The Development of the International Tropical Timber Organization and Its Influence on Tropical Forest Management*. London: Earthscan Publications.

Poudyal, M., Adhikari, B., Lovett, J. 2015. Community-based leasehold forestry in Nepal: A genuine tenure reform in progress? In *Forest Tenure Reform in Asia and Africa: Local Control for Improved Livelihoods, Forest Management, and Carbon Sequestration*, R.A. Bluffstone, E.J.Z. Robinson (eds.): 196–211. Washington, DC: RFF Press.

Poulter, B., Aragão, L., Heyder, U., Gumpenberger, M., Heinke, J., Langerwisch, F., Rammig, A., Thonicke, K., Cramer, W. 2010. Net biome production of the Amazon Basin in the 21st century. *Global Change Biology* 16: 2062–2075.

Prescott, C.E. 2010. Litter decomposition: What controls it and how can we alter it to sequester more carbon in forest soils? *Biogeochemistry* 101: 133–149.

Pretzsch, H. 2010. *Forest Dynamics, Growth and Yield*. Berlin: Springer Verlag.

Pretzsch, H., Schütze, G., Uhl, E. 2012. Resistance of European tree species to drought stress in mixed *versus* pure forests: Evidence of stress release by inter-specific facilitation. *Plant Biology* 15: 483–495.

Price, A.G., Carlyle-Moses, D.E. 2003. Measurement and modelling of growing-season canopy water fluxes in a mature mixed deciduous forest stand, southern Ontario, Canada. *Agricultural and Forest Meteorology* 119: 69–85.

Prideaux, B. (ed.) 2014. *Rainforest Tourism, Conservation and Management: Challenges for Sustainable Development*. London: Earthscan.

Priess, J., Then, C., Folster, H. 1999. Litter and fine-root production in three types of tropical premontane rain forest in SE Venezuela. *Plant Ecology* 143: 171–187.

Pritchett, W.I. 1979. *Properties and Management of Forest Soils*. New York: John Wiley.

Program on Forests (PROFOR) 2012. *Certification, verification and governance in forestry in Southeast Asia*. Working Paper. Washington, DC: Program on Forests. Available at: www.woldbank.org.

Programme for the Endorsement of Forest Certification (PEFC) 2014a. *Find Certified*. Available at: www.pefc.org/find-certified/certified-certificates. Accessed on 12 January 2016.

Programme for the Endorsement of Forest Certification (PEFC) 2014b. *PEFC Global Statistics: SFM & COC Certification*. Available at: www.pefc.org/images/documents/PEFC_Global_Certificates_-_January_2014.pdf. Accessed on 1 March 2014.

Puettmann, K.J., Coates, K.D., Messier, C. 2009. *A Critique of Silviculture: Managing for Complexity*. Washington, DC: Island Press.

Pyne, S.J. 1995. *World Fire: The Culture of Fire on Earth*. New York: Henry Holt.

Pyörälä, P., Kellomäki, S., Peltola, H. 2012. Effects of management on biomass production in Norway spruce stands and carbon balance of bioenergy use. *Forest Ecology and Management* 275: 87–97.

Quine, C.P. 1995. Assessing the risk of wind damage to forests: Practice and pitfalls. In *Wind and trees*, M.P. Coutts, J. Grace (eds.): 379–403. Cambridge: Cambridge University Press.

Rab, M.A. 2004. Recovery of soil physical properties from compaction and soil profile disturbance caused by logging of native forest in Victorian Central Highlands, Australia. *Forest Ecology and Management* 181: 329–340.

Rackham, O. 2006. *Woodlands*. London: HarperCollins.

Raffa, K.F., Aukema, B., Bentz, B.J., Carroll, A., Erbilgin, N., Herms, D.A., Hicke, J.A., Hofstetter, R.W., Katovich, S., Lindgren, B.S., Logan, J., Mattson, W., Munson, A.S., Robison, D.J., Six, D.L., Tobin, P.C., Townsend, P.A., Wallin, K.F. 2009. A literal use of 'forest health' safeguards against misuse and misapplication. *Journal of Forestry* 107(5): 276–277.

Rajala, R. A. 2006. *Up-Coast: Forests and Industry on British Columbia's North Coast, 1870–2005*. Victoria: Royal British Columbia Museum.

Ramankutty, N., Gibbs, H. K., Acard, F., DeFries, R., Foley, J. A., Houghton, R. A. 2007. Challenges to estimating carbon emissions from tropical deforestation. *Global Change Biology* 13: 51–66.

Rametsteiner, E., Mayer, P. 2004. Sustainable forest management and Pan-European forest policy. *Ecological Bulletins* 51: 51–57.

Ramsar Convention 1971. *Convention on Wetlands of International Importance Especially as Waterfowl Habitat*. Available at: http://www.ramsar.org/sites/default/files/documents/library/original_1971_convention_e.pdf. Accessed on 30 December 2015.

Rao, M. R., Palada, M. C., Becker, B. N. 2004. Medicinal and aromatic plants in agroforestry systems. *Agroforestry Systems* 61: 107–122.

Rayner, J., Buck, A., Katila, P. 2010. *Embracing Complexity: Meeting the Challenges of International Forest Governance*. Vienna: International Union of Forest Research Organizations.

Reichle, D. E. 1981. *Dynamic Properties of Forest Ecosystems*. Cambridge: Cambridge University Press.

Reinecke, J., Klemm, G., Heinken, T. 2014. Vegetation change and homogenization of species composition in temperate nutrient deficient Scots pine forests after 45 yr. *Journal of Vegetation Science* 25(1): 113–121.

Reyes, R., Nelson, H. 2014. A tale of two forests: Why forests and forest conflicts are both growing in Chile. *International Forestry Review* 16(4): 379–388.

Reynolds, K. M. 2005. Integrated decision support for sustainable forest management in the United States: Fact or fiction? *Computers and Electronics in Agriculture* 49: 6–23.

Richards, B. N. 1987. *The Microbiology of Terrestrial Ecosystems*. New York: John Wiley.

Richardson, J. S., Zhang, Y., Marczak, L. B. 2010. Resource subsidies across the land-freshwater interface and responses in recipient communities. *River Research and Applications* 26: 55–66.

Rittel, H. 1972. On the planning crisis: Systems analysis of the 'first and second generations'. *Bedrifts Okonomen* 8: 390–396.

Ritter, D. F., Kochel, R. C., Miller, J. R. 2011. *Process Geomorphology*. 5th edition. Long Grove, IL: Waveland Press.

Rockstrom, J., Steffen, W., Noone, K., Persson, A., Chapin, F. S., III, Lambin, E. F., Lenton, T. M., Scheffer, M., Folke, C., Schellnhuber, H. J., Nykvist, B., de Wit, C. A., Hughes, T., van der Leeuw, S., Rodhe, H., Sorlin, S., Snyder, P. K., Costanza, R., Svedin, U., Falkenmark, M., Karlberg, L., Corell, R. W., Fabry, V. J., Hansen, J., Walker, B., Liverman, D., Richardson, K., Crutzen, P., Foley, J. A. 2009. A safe operating space for humanity. *Nature* 461(7263): 472–475.

Rose, D. B. 1996. *Nourishing Terrains: Australian Aboriginal Views of Landscape and Wilderness*. Canberra: Australian Heritage Commission. Available at: http://www.environment.gov.au/heritage/ahc/publications/commission/books/pubs/nourishing-terrains.pdf.

Roukens, O., Worku, T., Amare, L. 2005. *Gums Naturally! Export Potential of Ethiopian Gums*. Addis Ababa: Export Promotion Department. Available at: http://www.eap.gov.et/sites/default/files/Export%2520potential%2520of%2520Ethiopian%2520Gums.pdf. Accessed on 26 October 2014.

Rousselet, J., Imbert, C.-E., Dekri, A., Garcia, J., Goussard, F., Vincent, B., Denux, O., Robinet, C., Dorkeld, F., Roques, A., Rossi, J.-P. 2013. Assessing species distribution using Google Street View: A pilot study with the pine processionary moth. *PLoS ONE* 8(10): Article e74918.

Rozenzweig, M. L., Abramsky, Z. 1993. How are diversity and productivity related? In *Species Diversity in Ecological Communities: Historical and Geographical Perspectives*, E. Ricklefs, D. Schluter (eds.): 52–65. Chicago: University of Chicago Press.

RRI 2016. *Closing the Gap: Strategies and Scale Needed to Secure Rights and Save Forests*. Washington, DC: Rights and Resources Initiative.

Rule, S., Brook, B. W., Haberle, S. G., Turney, C. S., Kershaw, A. P., Johnson, C. N. 2012. The aftermath of megafaunal extinction: Ecosystem transformation in Pleistocene Australia. *Science* 335(6075): 1483–1486.

Runyan, C. W., D'Odorico, P., Lawrence, D. 2012. Physical and biological feedbacks of deforestation. *Review of Geophysics* 50: Article RG4006.

Russell, P. H. 2005. *Recognizing Aboriginal Title: The Mabo Case and Indigenous Resistance to English-Settler Colonialism*. Sydney: UNSW Press.

Russell-Smith, J., Whitehead, P., Cooke, P. 2009. *Culture, Ecology and Economy of Fire Management in North Australian Savannas: Rekindling the Wurrk Tradition*. Collingwood, Victoria: CSIRO Publishing.

Saaty, T. L. 1977. A scaling method for priorities in hierarchical structures. *Journal of Mathematical Psychology* 15: 234–281.

Saenger, P. 2003. *Mangrove Ecology, Silviculture and Conservation*. Berlin: Springer.

Salwasser, H. 1988. Managing ecosystems for viable populations of vertebrates: A focus for biodiversity. In *Ecosystem Management for Parks and Wilderness*, J. K. Agee, D. R. Johnson (eds.): 87–104. Seattle: University of Washington Press.

Sands, R. 2013. A history of human interaction with forests. In *Forestry in a Global Context*, R. Sands (ed.): 1–36. 2nd edition. Wallingford: CAB International.

Sasikumar, K., Vijayalakshmi, C., Parthiban, K. T. 2004. Allelopathic effects of four Eucalyptus species on cowpea (*Vigna unguiculata*). *Journal of Tropical Forest Science* 16(4): 419–428.

Satterfield, T. 2002. *Anatomy of a Conflict: Identity, Knowledge, and Emotion in Old-Growth Forests*. Vancouver: UBC Press.

Sayer, J. A., Maginnis, S. (eds.) 2005a. *Forests in Landscapes: Ecosystem Approaches to Sustainability*. London: Earthscan Publications.

Sayer, J. A., Maginnis, S. 2005b. New challenges for forest management. In *Forests in Landscapes: Ecosystem Approaches to Sustainability*, 1–16. London: Earthscan Publications.

Schama, S. 1996. *Landscape and Memory*. Vintage Books Edition. New York: Vintage Books.

Schlager, E., Ostrom, E. 1992. Property-rights regimes and natural resources: A conceptual analysis. *Land Economics* 68(3): 249–262.

Schmithüsen, F., Hirsch, F. 2010. *Private forest ownership in Europe*. Geneva Timber and Forest Study paper 26. Report ECE/TIM/SP/26. Geneva: United Nations.

Schreckenberg, K. 2004. The contribution of shea butter (*Vitellaria paradoxa* C. F. Gaertner) to local livelihoods in Benin. In *Forest Products, Livelihoods and Conservation: Case Studies of Non-Timber Forest Product Systems: Volume 2 — Africa*, T. Sunderland, O. Ndoye (eds.): 91–114. Bogor: CIFOR.

Schuck, A., Päivinen, R., Hytönen, T., Pajari, B. 2002. *Compilation of forestry terms and definitions*. Internal Report No. 6. Joensuu: European Forest Institute.

Schulte, A., Ruhiyat, D. (eds.) 1998. *Soils of Tropical Forest Ecosystems: Characteristics, Ecology and Management*. Berlin, Heidelberg: Springer.

Schütz, J.-P. 1997. *Sylviculture 2. La gestion des forêts irrégulières et mélangées*. Lausanne: Presses Polytechniques et Universitaires Romandes.

Scowcroft, P. G., Yeh, J. T. 2013. Passive restoration augments active restoration in deforested landscapes: The role of root suckering adjacent to planted stands of *Acacia koa*. *Forest Ecology and Management* 305: 138–145.

Scurrah-Ehrhart, C., Blomley, T. 2006. *Amani Butterfly Forest-Based Enterprise, Tanga, Tanzania*. Yokohama and Washington, DC: International Tropical Timber Organization, Forest Trends, and Rights and Resources Initiative. Available at: http://www.rightsandresources.org/documents/files/doc_3212.pdf. Accessed on 30 December 2015.

Secretariat of the Convention on Biological Diversity 2010. *Global Biodiversity Outlook 3*. Montreal: Convention on Biological Diversity.

Sedjo, R. 1999. The potential of high-yield plantation forestry for meeting timber needs. *New Forests* 17: 339–359.

Seely, B., Nelson, J., Wells, R., Peter, B., Meitner, M., Anderson, A., Harshaw, H., Sheppard, S., Bunnell, F.L., Kimmins, J. P., Harrison, D. 2004. The application of a hierarchical, decision-support system to evaluate multi-objective forest management strategies: A case study in northeastern British Columbia, Canada. *Forest Ecology and Management* 199: 283–305.

Seidl, R., Lexer, M. J. 2013. Forest management under climatic and social uncertainty: Trade-offs between reducing climate change impacts and fostering adaptive capacity. *Journal of Environmental Management* 114: 461–469.

Selander, R. K. 1976. Genetic variation in natural populations. In *Molecular Evolution*, F. J. Ayala (ed.): 21–45. Sunderland: Sinauer Associates.

Sevigne, E., Gasol, C. M., Brun, F., Rovira, L., Pagés, J. M., Camps, F., Rieradevall, J., Gabarrell, X. 2011. Water and energy consumption of *Populus* spp. bioenergy systems: A case study in Southern Europe. *Renewable and Sustainable Energy Reviews* 15: 1133–1140.

Seymour, R. S., Hunter, M. L., Jr. 1992. *New Forestry in Eastern Spruce-Fir Forests: Principles and Applications to Maine*. Maine Miscellaneous Publication 716. Orono: Maine Agricultural Experiment Station.

Shackleton, C., Shackleton, S., Shanley, P. 2011. *Non-Timber Forest Products in the Global Context: Tropical Forestry*. Berlin: Springer Verlag.

Sheng, W. T. 1992. Soil degradation and its control techniques for timber plantations in China. In *Research on Site Degradation of Timber Plantation* (in Chinese), Sheng, W. T., (ed.), 1–7. Beijing: Science and Technology Press of China.

Siddiqui, K. M. 1994. Tree planting for sustainable use of soil and water with special reference to the problem of salinity. *Pakistan Journal of Forestry* 43: 54–64.

Sidle, R. C. 1992. A theoretical model of the effects of timber harvesting on slope stability, *Water Resources Research* 28(7): 1897–1910.

Sidle, R. C., Furuichi, T., Kono, Y. 2011. Unprecedented rates of landslide and surface erosion along a newly constructed road in Yunnan, China. *Natural Hazards* 57: 313–326.

Sidle, R. C., Noguchi, S., Tsuboyama, Y., Laursen, K. 2001. A conceptual model of preferential flow systems in forested hillslopes: Evidence of self-organization. *Hydrological Processes* 15: 1675–1692.

Sidle, R. C., Ochiai, H. 2006. *Landslides: Processes, Prediction, and Land Use*. Water Resources Monograph No. 18. Washington, DC: American Geophysical Union.

Sidle, R. C., Pearce, A. J., O'Loughlin, C. L. 1985. *Hillslope Stability and Land Use*. Water Resources Monograph 11. Washington, DC: American Geophysical Union.

Sidle, R. C., Sasaki, S., Otsuki, M., Noguchi, S., Abdul Rahim, N. 2004. Sediment pathways in a tropical forest: effects of logging roads and skid trails. *Hydrological Processes* 18: 703–720.

Sidle, R. C., Ziegler, A. D., Negishi, J. N., Abdul Rahim, N., Siew, R., Turkelboom, F. 2006. Erosion processes in steep terrain – truths, myths, and uncertainties related to forest management in Southeast Asia. *Forest Ecology and Management* 224(1–2): 199–225.

Sikor, T., Stahl, J. (eds.) 2011. *Forests and People: Property, Governance and Human Rights*. London: Earthscan.

Sist, P., Fimbel, R., Sheil, D., Nasi, R., Chevallier, M. H. 2003. Towards sustainable management of mixed dipterocarp forests of South-East Asia: Moving beyond minimum diameter cutting limits. *Environmental Conservation* 30: 364–374.

Skelly, J. M., Innes, J. L. 1994. Waldsterben in the forests of central Europe and eastern North America: Fantasy or reality? *Plant Disease* 78: 1021–1032.

Smethurst, P. J., Nambiar, E.K.S. 1990. Effects of contrasting silvicultural practices on nitrogen supply to young radiata pine. In *Impact of Intensive Harvesting on Forest Site Productivity*, W. J. Dyck, C. A. Mees (eds.): 85–96. IEA/BE T6/A6 Report Number 2, FRI Bulletin Number 159. Rotorua: Forest Research Institute.

Smith, D. M. 1962. *The Practice of Silviculture*. New York: Wiley and Sons.

Smith, N. 2008. *Uneven Development: Nature, Capital, and the Production of Space*. Atlanta, GA: University of Georgia Press.

Smith, P. D., McDonough, M. H. 2001. Beyond public participation: Fairness in natural resource decision making. *Society & Natural Resources* 14(3): 239–249.

Smith, R. D., Maltby, E. 2003. *Using the Ecosystem Approach to Implement the Convention on Biological Diversity*. Gland, Switzerland: IUCN.

Soil Science Society of America 1973. *Glossary of Soil Science Terms*. Madison, WS: Soil Science Society of America.

Soulé, M. E. 1985. What is conservation biology? *BioScience* 35: 727–734.

Soulé, M. E., Wilcox, B. A. 1980. *Conservation Biology: An Evolutionary-Ecological Perspective*. Sunderland: Sinauer Associates.

South Asia Enterprise Development Facility (SEDF) & Inter-cooperation (IC) 2003. Medicinal Plants Marketing in Bangladesh. A Market Study Report. SEDF-Inter-cooperation, Dhaka.

Spalding, D., Kendirli, E., Oliver, C. D. 2012. The role of forests in global carbon budgeting. In *Managing Forest Carbon in a Changing Climate*, M. S. Ashton, M. L. Tyrrell, D. Spalding, B. Gentry (eds.): 165–179. Dordrecht: Springer.

Stankey, G. H., Bormann, B. T., Ryan, C., Shindler, B., Sturtevant, V., Clark, R. N., Philpot, C. 2003. Adaptive management and the Northwest Forest Plan: Rhetoric and reality. *Journal of Forestry* 101: 40–46.

Stanturf, J., Lamb, D., Madsen, P. (eds.) 2012. *Forest Landscape Restoration: Integrating Natural and Social Sciences*. Dordrecht: Springer.

Stanturf, J. A., Palik, B. J., Dumroese, R. K. 2014. Contemporary forest restoration: A review emphasizing function. *Forest Ecology and Management* 331: 292–323.

Stephens, B. B., Gurney, K. R., Tans, P. P., Sweeney, C., Peters, W., Bruhwiler, L., Ciais, P., Ramonet, M., Bousquet, P., Nakazawa, T., Aoki, S., Machida, T., Inoue, G., Vinnichenko, N., Lloyd, J., Jordan, A., Heimann, M., Shibistova, O., Langenfelds, R. L., Steele, L. P., Francey, R. J., Denning, A. S. 2007. Weak northern and strong tropical land carbon uptake from vertical profiles of atmospheric CO_2. *Science* 316: 1732–1735.

Stirn, L. Z. 2006. Integrating the fuzzy analytic hierarchy process with dynamic programming approach for determining the optimal forest management decisions. *Ecological Modelling* 194: 296–305.

Stocker, B. D., Strassmann, K., Joos, F. 2011. Sensitivity of Holocene atmospheric CO_2 and the modern carbon budget to early human land use: Analyses with a process-based model. *Biogeosciences* 8: 69–88.

Strahler, A. H., Strahler, A. N. 1992. *Modern Physical Geography*. 4th edition. New York: John Wiley.

Stupak, I., Asikainen, A., Röser, D., Pasanen, K. 2008. Review of recommendations for forest energy harvesting and wood ash recycling. In *Managing Forest Ecosystems: Sustainable Use of Forest Biomass for Energy: A Synthesis with Focus on the Baltic and Nordic Region*, D. Röser, A. Asikainen, K. Raulund-Rasmussen, I. Stupak (eds.): 155–196. Dordrecht: Springer.

Sturtevant, B. R., Fall, A., Kneeshaw, D. D., Simon, N.P.P., Papaik, M. J., Berninger, K., Doyon, F., Morgan, D. G., Messier, C. 2007. A toolkit modeling approach for sustainable forest management planning: Achieving balance between science and local needs. *Ecology and Society* 12(2): Article 7.

Styles, C. V. 1995. The elephant and the worm. *BBC Wildlife* 13(3): 22–24.

Sugden, B. D., Ethridge, R., Mathieus, G., Heffernan, P.E.W., Frank, G., Sanders, G. 2012. Montana's forestry best management practices program: 20 years of continuous improvement. *Journal of Forestry* 110(6): 328–336.

Sunderlin, W. D., Angelsen, A., Belcher, B., Burgers, P., Nasi, R., Santoso, L., Wunder, S. 2005. Livelihoods, forests and conservation in developing countries: An overview. *World Development* 33(9): 1383–1402.

Supply Change/Forest Trends 2015. *Firm Commitments: Tracking Company Endorsers of the New York Declaration on Forests*. Washington, DC: Forest Trends.

Sustainable Development Solutions Network 2012. *A Framework for Sustainable Development*. New York: United Nations, Sustainable Development Solutions Network. Available at: http://unsdsn.org/wp-content/uploads/2014/02/121220-Draft-Framework-of-Sustainable-Development1.pdf. Accessed on 30 December 2015.

Sveiby, K.-E., Skuthorpe, T. 2006. *Treading Lightly: The Hidden Wisdom of the World's Oldest People*. Crows Nest, NSW: Allen and Unwin.

TEEB 2010. *The Economics of Ecosystems and Biodiversity: Mainstreaming the Economics of Nature: A Synthesis of the Approach, Conclusions and Recommendations of TEEB*. Wageningen: Institute for Environmental Studies. Available at: http://www.teebweb.org/our-publications/teeb-study-reports/synthesis-report/.

Ten Brink, P., Mazza, L., Badura, T., Kettunen, M., Withana, S. 2012. *Nature and Its Role in the Transition to a Green Economy*. Wageningen: Institute for Environmental Studies. Available at: http://www.teebweb.org/wp-content/uploads/2013/04/Nature-Green-Economy-Full-Report.pdf. Accessed on 26 October 2014.

Terradas, J. 2001. *Ecología de la Vegetación*. Barcelona: Ediciones Omega.

Thirgood, J. V. 1981. *Man and the Mediterranean Forest: A History of Resource Depletion*. London: Academic Press.

Tian, D., Xiang, W., Chen, X., Yan, W., Fang, X., Kang, W., Dan, X., Peng, C., Peng, Y. 2011. A long-term evaluation of biomass production in first and second rotations of Chinese fir plantations at the same site. *Forestry* 84: 411–418.

Tiessen, H., Cuevas, E., Chacon, P. 1994. The role of soil organic-matter in sustaining soil fertility. *Nature* 371: 783–785.

Tikina, A. V. 2006. *Assessing the Effectiveness of Forest Certification in the US Pacific Northwest and British Columbia, Canada*. PhD thesis, University of British Columbia, Canada.

Tikina, A. V., Innes, J. L. 2008. A framework for assessing the effectiveness of forest certification. *Canadian Journal of Forest Research* 38(6): 1357–1365.

Tikina, A. V., Innes, J. L., Trosper, R. L., Larson, B. C. 2010. Aboriginal people and forest certification: A review of the Canadian situation. *Ecology and Society* 15(3): Article 33.

Tindall, D. B. 2003. Social values and the contingent nature of public opinion and attitudes about forests. *Forestry Chronicle* 79: 692–705.

Tomaselli, M., Timko, J., Kozak, R. 2012. The role of government in the development of small and medium forest enterprises: Case studies from the Gambia. *Small-Scale Forestry* 11(2): 237–253.

Totman, C. D. 1989. *The Green Archipelago: Forestry in Pre-Industrial Japan*. Athens: Ohio University Press.

Trombulak, S. C., Frissell, C. A. 2000. Review of ecological effects of roads on terrestrial and aquatic communities. *Conservation Biology* 14: 18–30.

Trosper, R. L., Parrotta, J. A. 2012. Introduction: The growing importance of traditional forest-related knowledge. In *Traditional Forest-Related Knowledge: Sustaining Communities, Ecosystems and Biocultural Diversity*, J. Parrotta, R. Trosper (eds.): 1–36. World Forests, Vol. 12. Dordrecht: Springer.

Trubilowicz, J., Moore, R. D., Bottle, J. M. 2013. Prediction of streamflow regime using ecological classification zones. *Hydrological Processes* 27: 1935–1944.

Turner, N., Spalding, P. R. 2013. 'We might go back to this'; Drawing on the past to meet the future in northwestern North American indigenous communities. *Ecology and Society* 18(4): Article 29.

Turney, C.S.M., Flannery, T. F., Roberts, R. G., Reid, C., Fifield, L. K., Higham, T.F.G., Jacobs, Z., Kemp, N., Colhoun, E. A., Kalyn, R. M., Ogle, N. 2008. Late-surviving megafauna in Tasmania, Australia, implicate human involvement in their extinction. *Proceedings of the National Academy of Sciences of the USA* 105(34): 12150–12153.

Ulrich, A., Hill, F. J. 1967. Principles and practices of plant analysis. In *Soil Testing and Plant Analysis. Part II: Plant Analysis*, Soil Science Society of America (ed.): Special Publications Series No. 2. Madison, WI: Soil Science Society of America.

UN 1992a. *Report of the United Nations Conference on Environment and Development: Annex III: Non-Legally Binding Authoritative Statement on Principles for a Global Consensus on the Management, Conservation and Sustainable Development of All Types of Forests*. General Assembly A/CONF.151/26, Vol. 3. New York: United Nations.

UN 1992b. *Convention on Biological Diversity*. New York: United Nations. Available at: https://www.cbd.int/convention/text/.

UN 1992c. *The UN Framework Convention on Climate Change*. New York: United Nations. Available at: http://unfccc.int/resource/docs/convkp/conveng.pdf.

UN 1992d. *Rio Declaration on Environment and Development*. New York: United Nations. Available at: http://www.unep.org/Documents.Multilingual/Default.asp?documentid=78&articleid=1163.

UN 1996. *The UN Convention to Combat Desertification*. New York: United Nations. Available at: http://www.unccd.int/convention/text/convention.php.

UN 2000a. *The UN Declaration on the Rights of Indigenous Peoples*. AWG-LCA 2010, 24. New York: United Nations. Available at: http://www.hrc.co.nz/hrc_new/hrc/cms/files/documents/30-Jan-2008_10-39-35_UN_Declaration_on_Rights_of_Indigenous_People_english.pdf.

UN 2000b. *Millennium Development Goals*. New York: United Nations. Available at: http://www.un.org/millenniumgoals/.

UN 2007. *Non-Legally Binding Instrument on All Types of Forests*. New York: United Nations. Available at: http://www.fordaq.com/www/news/2007/UN_Instrument%20on%20all%20types%20of%20forests.pdf.

UN 2008. Non-legally binding instrument on all types of forests. UN General Assembly Sixty-Second Session Second Committee Agenda Item 54. A/RES/62/98. 31 January 2008. Available at: https://www.un.org/en/ga/second/62/draftproposals.shtml.

UNEP 2011. Oil palm plantations: Threats and opportunities for tropical ecosystems. *UNEP Global Environment Alert Service*, December 2011. Available at: http://www.unep.org/pdf/Dec_11_Palm_Plantations.pdf. Accessed on 30 December 2015.

UNEP 2012. Green Economy Sectoral Study: BioTrade – Harnessing the potential for transitioning to a green economy – The Case of Medicinal and Aromatic Plants in Nepal. http://www.unep.org/greeneconomy/Portals/88/documents/research_products/Bio%20Trade%20Medicinal%20and%20Aromatic%20Plants%20in%20Nepal.pdf.

UNEP 2013. *Green Economy and Trade – Trends, Challenges and Opportunities*. Available at: http://www.unep.org/greeneconomy/GreenEconomyandTrade. Accessed on 30 December 2015.

UNESCO 1972. *Convention Concerning the Protection of the World Cultural and Natural Heritage*. New York: United Nations. Available at: http://whc.unesco.org/en/conventiontext/. Accessed on 30 December 2015.

UNFCCC 2015. Conference of the Parties. *Twenty-First Session Paris*, 30 November to 11 December 2015. Adoption of the Paris Agreement. Available at: http://unfccc.int/resource/docs/2015/cop21/eng/l09r01.pdf. Accessed on 15 January 2015.

UN Forum on Forests 2006, Report of the Sixth Session Report no. E/CN.18/2006/18. New York: United Nations.

UN Forum on Forests Tenth Session 2013. *Forests and economic development: Conclusions and recommendations for addressing key challenges of forests and economic development: Report of the secretary general*. Report no. E/CN.18/2013/5. New York: United Nations.

UN General Assembly 2000. *United Nations Millennium Declaration*. Resolution Adopted by the General Assembly, 18 September 2000. A/RES/55/2. New York: United Nations. Available at: http://www.refworld.org/docid/3b00f4ea3.html. Accessed on 30 December 2015.

United Nations General Assembly 2015. Resolution adopted by the General Assembly, 25 September 2015. 70/1. *Transforming our world: The 2030 Agenda for Sustainable Development*. Available at: http://www.un.org/ga/search/view_doc.asp?symbol=A/RES/70/1&Lang=E. Accessed on 15 January 2016.

UN-REDD Programme 2011. Draft framework for sharing approaches for better multi-stakeholder participation practices: Florence Davies with input from Marian Mabel and Elspeth Halverson. Forest Carbon Partnership Facility and UN-REDD Programme. Available at: http://forestcarbonasia.org/other-publications/draft-framework-for-sharing-approaches-for-better-multi-stakeholder-participation-practices/. Accessed on 15 January 2016.

UNSD 1992. *Agenda 21*. New York: United Nations. Available at: http://sustainabledevelopment.un.org/content/documents/Agenda21.pdf. Accessed on 30 December 2015.

Upton, C., Bass, S. 1996. *The Forest Certification Handbook*. Delray Beach: St. Lucie Press.

Vacik, H., Lexer, M. J. 2001. Application of a spatial decision support system in managing the protection forests of Vienna for sustained yield of water resources. *Forest Ecology and Management* 143: 65–76.

van der Werf, G. R., Randerson, J. T., Giglio, L., Collatz, G. J., Mu, M., Kasibhatla, P. S., Morton, D. C., DeFries, R. S., Jin, Y., van Leeuwen, T. T. 2010. Global fire emissions and the contribution of deforestation, savanna, forest, agricultural, and peat fires (1997–2009). *Atmospheric Chemistry and Physics* 10: 11707–11735.

van Huis, A., van Itterbeeck, J., Klunder, H., Mertens, E., Muir, G., Vantomme, P. 2013. *Edible Insects: Future Prospects for Food and Feed Security*. Rome: Food and Agriculture Organization of the United Nations. Available at: http://www.fao.org/docrep/018/i3253e/i3253e.pdf. Accessed on 28 October 2014.

van Minnen, J. G., Klein Goldewijk, K., Stehfest, E., Eickhout, B., van Drecht, G., Leemans, R. 2009. The importance of three centuries of land-use change for the global and regional terrestrial carbon cycle. *Climate Change* 97: 123–144.

Varhola, A., Coops, N. C., Weiler, M., Moore, R. D. 2010. Forest canopy effects on snow accumulation and ablation: An integrative review of empirical results. *Journal of Hydrology* 392: 219–233.

Vega, D. J., Dopazo, R., Ortiz, L. 2010. *Manual de cultivos Energéticos*. Vigo: Universidad de Vigo.

Venne, S. H. 1998. *Our Elders Understand Our Rights: Evolving International Law Regarding Indigenous Rights*. Penticton, British Columbia: Theytus Books.

Vertessy, R., Watson, F.G.R., O'Sullivan, S. K. 2001. Factors determining relations between stand age and catchment water balance in mountain ash forests. *Forest Ecology and Management* 143: 13–26.

Verwijst, T., Lundkvist, A., Edeldeldt, S., Alertsson, J. 2013. Development of sustainable willow short rotation forestry in northern Europe. In *Biomass Now – Sustainable Growth and Use*, M.D. Matovic (ed.): 479–502. Ryjeka: InTech.

Vidal, N., Bull, G., Kozak, R. 2010. Diffusion of corporate responsibility practices to companies: The experience of the forest sector. *Journal of Business Ethics* 94(4): 553–567.

Vita-Finzi, C. 1969. *The Mediterranean Valleys: Geological Changes in Historical Times*. Cambridge: Cambridge University Press.

Vogt, K., Larson, B., Gordon, J., Vogt, D., Fanzeres, A. (eds.) 2000. *Forest Certification: Roots, Issues, Challenges and Benefits*. Boca Raton: CRC Press.

Waeber, P.O., Nitschke, C.R., Le Ferrec, A., Harshaw, H., Innes, J.L. 2013. Evaluating alternative forest management strategies for the Champagne and Aishihik Traditional Territory, southwest Yukon. *Journal of Environmental Management* 120: 148–156.

Walker, B.H. 1992. Biological diversity and ecological redundancy. *Conservation Biology* 6: 18–23.

Wallace, L.L. (ed.) 2004. *After the Fires: The Ecology of Change in Yellowstone National Park*. Newhaven: Yale University Press.

Walters, C.J. 1986. *Adaptive Management of Renewable Resources*. New York: Macmillan.

Walters, C.J. 1997. Challenges in adaptive management of riparian and coastal ecosystems. *Conservation Ecology* 1(2): Article 1.

Walter, S., Cole, D., Kathe, W., Lovett, P., Paz Soldán, M. 2003. Impact of certification on the sustainable use of NTFP: Lessons-learnt from three case studies. Paper submitted for presentation at the International Conference on Rural Livelihoods, Forests and Biodiversity. 19–23 May 2003, Bonn, Germany. Available at: http://foris.fao.org/static/pdf/NWFP/CIFOR_pres.pdf. Accessed on 30 December 2015.

Wang, T., Campbell, E.M., O'Neill, G.A., Aitken, S.N. 2012. Projecting future distributions of ecosystem climate niches: Uncertainties and management applications. *Forest Ecology and Management* 279: 128–140.

Wang, T., O'Neill, G.A., Aitken, S.N. 2010. Integrating environmental and genetic effects to predict responses of tree populations to climate. *Ecological Applications* 20: 153–163.

Waring, R.H., Running, S.W. 2007. *Forest Ecosystems: Analysis at Multiple Scales*. 3rd edition. Burlington, MA, San Diego, CA and London, UK: Elsevier Academic Press.

Watkins, C. 2014. *Trees, Woods and Forests: A Social and Cultural History*. London: Reaktion Books.

Watts, J.D., Colfer, C.J.P. 2011. The governance of tropical forested landscapes. In *Collaborative Governance of Tropical Landscapes*, C.J.P. Colfer, J.-L. Pfund (eds.): 35–54. London: Earthscan.

Weaver, J.C. 2006. *The Great Land Rush and the Making of the Modern World, 1650–1900*. Montreal: McGill-Queens University Press.

Wei, X., Blanco, J.A. 2014. Significant increase in ecosystem C can be achieved with sustainable forest management in subtropical plantation forests. *PLOS ONE* 9(2): Article e89688.

Wei, X., Blanco, J.A., Jiang, H., Kimmins, J.P. 2012. Effects of nitrogen deposition on carbon sequestration in Chinese fir forests. *Science of the Total Environment* 416: 351–361.

Welham, C., Blanco, J.A., Seely, B., Bampfylde, C. 2012. Oil sands reclamation and the projected development of wildlife habitat attributes. In *Reclamation and Restoration of Boreal Ecosystems: Attaining Sustainable Development*, D.H. Vitt, J.S. Bhatti (eds.): 336–356. Cambridge: Cambridge University Press.

Welham, C., Seely, B., Kimmins, J.P. 2002. The utility of the two-pass harvesting system: An analysis using the ecosystem simulation model FORECAST. *Canadian Journal of Forest Research* 32: 1071–1079.

Wells, R.W., Bunnell, F.L., Haag, D., Sutherland, G. 2003. Evaluating ecological representation within different planning objectives for the central coast of British Columbia. *Canadian Journal of Forest Research* 33: 141–150.

West, N.E. 1994. Biodiversity and land use. In *Sustainable Ecological Systems: Implementing an Ecological Approach to Land Management*, W.W. Covington, L.F. DeBano (tech. co-ords.): 21–26. USDA Forest Service, General Technical Report, GTR-RM-247. Fort Collins, CO: USDA Forest Service, Rocky Mountains Research Station.

Westoby, J. 1987. *The Purpose of Forests: Follies of Development*. Oxford: Basil Blackwell.

Weston, D. 1992. The biodiversity crisis: A challenge for biology. *Oikos* 63: 29–38.

White, A., Martin, A., 2002. *Who Owns the World's Forests?* Washington, DC: Forest Trends, Centre for Environmental Law.

Wicker, G. 2002. Motivation for private landowners. In *Southern Forest Resource Assessment*, D.N. Wear, J.G. Greis (eds.): 225–238. General Technical Report SRS-53. Asheville, NC: USDA Forest Service, Southern Research Station.

Wiersum, K.F. 1995. 200 years of sustainability in forestry: Lessons from history. *Environmental Management* 19(3): 321–329.

Wilkie, M.H. 2003. *Sustainable forest management and the ecosystem approach: Two concepts, one goal*. Forest Management Working Paper FM 25. Rome: Food and Agricultural Organization of the United Nations.

Willems-Braun, B. 1997. Buried epistemologies: The politics of nature in (post) colonial British Columbia. *Annals of the Association of American Geographers* 87(1): 3–31.

Williams, M. 1992. *Americans and Their Forests*. Cambridge: Cambridge University Press.

Williams, M. 2003. *Deforesting the Earth: From Prehistory to Global Crisis*. Chicago: University of Chicago Press.

Williams, M.I., Dumroese, R.K. 2014. Planning the future's forests with assisted migration. In *Forest Conservation and Management in the Anthropocene*, V.A. Sample, R.P. Bixler (eds.): 133–144. USDA Forest Service, Rocky Mountain Research Station, Proceedings RMRS-P-71. Fort Collins: USDA Forest Service.

Williams, P.W., Penrose, R.W., Hawkes, S. 1998. Shared decision-making in tourism land use planning. *Annals of Tourism Research* 25(4): 860–889.

Williams, R.A., Jr. 1990. *The American Indians in Western Legal Thought: The Discourses of Conquest*. Oxford: Oxford University Press.

Wilson, E.O. 1988. The current state of biological diversity. In *Biodiversity*, E.O. Wilson, F.M. Peter (eds.): 3–18. Washington, DC: National Academy Press.

Wittemyer, G., Northrup, J.M., Blanc, J., Douglas-Hamilton, I., Omondi, P., Burnham, K.P. 2014. Illegal killing for ivory drives global decline in African elephants. *Proceedings of the National Academy of Science of the USA* 111(36): 13117–13121.

Wittmann, F., Schongart, J., Montero, J.C., Motzer, T., Junk, W.J., Piedade, M.T.F., Queiroz, H.L., Worbes, M. 2006. Tree species composition and diversity gradients in white-water forests across the Amazon Basin. *Journal of Biogeography* 33: 1334–1347.

Wondolleck, J.M., Yaffee, S.L. 2000. *Making Collaboration Work: Lessons from Innovation in Natural Resource Management*. Washington, DC: Island Press.

Woods, A. 2011. Is the health of British Columbia's forests being influenced by climate change? If so, was this predictable? *Canadian Journal of Plant Pathology* 33(2): 117–126.

World Bank 2008. *Forests Sourcebook: Practical Guidance for Sustaining Forests in Development Cooperation*. Washington, DC: The World Bank.

World Bank 2009a. *Rethinking Forest Partnerships and Benefit Sharing Insights on Factors and Context that Make Collaborative Arrangements Work for Communities and Owners*. Washington, DC: The World Bank.

World Bank 2009b. *Framework for Governance Reforms: Overall Effects of Poor Forest Governance*. Report no. 49572-glb. Washington, DC: The World Bank, Agriculture and Rural Development Department.

World Commission on Forests and Sustainable Development 1999. *Our Forests – Our Future: Summary Report*. Winnipeg: World Commission on Forests and Sustainable Development. Available at: www.iisd.org/pdf/wcfsdsummary.pdf.

Wunder, S. 2008. Payments for environmental services and the poor: Concepts and preliminary evidence. *Environment and Development Economics* 13(3): 279–297.

Xin, Z.-H., Juang, H., Jie, C.-Y., Wei, Z.-H., Blanco, J. 2011. Simulated nitrogen dynamics for a *Cunninghamia lanceolata* plantation with selected rotation ages. *Journal of Zhejiang A & F University* 28: 855–862.

Yang, X., Richardson, T.K., Jain, A.K. 2010. Contributions of secondary forest and nitrogen dynamics to terrestrial carbon uptake. *Biogeosciences* 7: 3041–3050.

Young, C.R. 1979. *The Royal Forests of Medieval England*. Philadelphia: University of Pennsylvania Press.

Zhu, Q., Sarkis, J., Lai, K. 2012. Green supply chain management innovation diffusion and its relationship to organizational improvement: An ecological modernization perspective. *Journal of Engineering and Technology Management* 29(1): 168–185.

Index